Entomology Monographs

Insects are the most diverse group of organisms, and many biological advances have been derived from this group. Although entomology is considered to be a classic science, recent developments in molecular methods, application of modern theoretical concepts and collaboration with related sciences have opened new directions in entomology. Japanese scientists play a significant role in these fields, and this book series will focus on such developments. The book series Entomology Monographs publishes refereed volumes on all aspects of entomology, including ecology, ethology, physiology, taxonomy, systematics, morphology, evolutionary developmental biology, genetics, biochemistry, and molecular biology in insects and related arthropods. Authors are not restricted to Japanese entomologists, and other international experts will also be considered on the basis of their recent contribution to these fields.

More information about this series at http://www.springer.com/series/15687

Yukio Ishikawa
Editor

Insect Sex Pheromone Research and Beyond

From Molecules to Robots

Springer

Editor
Yukio Ishikawa
Laboratory of Applied Entomology,
Graduate School of Agricultural and Life Sciences
The University of Tokyo
Tokyo, Japan

Laboratory of Applied Entomology,
Faculty of Agriculture
Setsunan University
Osaka, Japan

ISSN 2522-526X ISSN 2522-5278 (electronic)
Entomology Monographs
ISBN 978-981-15-3081-4 ISBN 978-981-15-3082-1 (eBook)
https://doi.org/10.1007/978-981-15-3082-1

This Springer imprint is published by the registered company Springer Nature Singapore Pte Ltd.
The registered company address is: 152 Beach Road, #21-01/04 Gateway East, Singapore 189721, Singapore

Aims and Scope of This Book

Background

Female moths were suggested more than 100 years ago to release a kind of scent, which humans cannot detect, to attract conspecific males from a long distance (Fabre 1913); however, chemical identification of the "scent" remained impracticable for a long time because of the extremely small quantity of the compound produced by the females. The first identification of the attractive compound, bombykol, from the silk moth *Bombyx mori* had to wait for the breakthrough made by Butenandt et al. (1959). In view of potent attractancy, high species specificity, relatively simple structure, and nonhazardous nature of the new class of biologically active compounds, "sex pheromones" (Karlson and Lüscher 1959), researchers soon considered the use of synthetic sex pheromones as a measure to control agricultural pest moths. Accordingly, efforts have been made to identify sex pheromones of many pest moth species. Currently, sex pheromones of approximately 700 moth species have been identified (Ando and Yamamoto 2019), and their use is an indispensable measure in integrated pest management of moths, as exemplified by their large-scale use in the control of codling moth *Cydia pomonella* and gypsy moth *Lymantria dispar* (Witzgall et al. 2010). The increase in knowledge on the sex pheromones of moth species aroused interest in the sex pheromones of non-lepidopteran insects such as flies, cockroaches, longhorn beetles, seed beetles, aphids, and mealybugs.

In addition to their application in pest control, sex pheromones attracted the interest of researchers in many fields of basic biology such as reproductive isolation, speciation, signal transduction, biosynthesis, hormonal control of production, and neural mechanisms that enable efficient finding of female moths from a long distance. *B. mori* has also been frequently used in these lines of research because of their ease of culture, availability of many strains and mutants, genome data, and molecular biological techniques that cannot be applied to other moth species. It is of note that several epoch-making findings, such as the identification of pheromone biosynthesis-activating neuropeptide (PBAN)(Kitamura et al. 1989; Nagasawa et al. 1988), fatty-acyl-CoA reductase, a key enzyme involved in sex pheromone biosynthesis (Moto et al. 2003), and sex pheromone receptor (Sakurai et al. 2004), were all from *B. mori*.

Aims and Scope

This book aims to give an overview of the recent progress in insect sex pheromone research, which spans from their identification, biosynthesis, and reception to the control of odor-source searching behavior and from molecules to robots. To achieve this aim, this book summarizes the progress of studies conducted using *B. mori* and a few groups of moths on the one hand and reviews sex pheromones of some non-lepidopteran insect groups of agricultural importance on the other. It should be noted that although application of sex pheromones in the control of agricultural pests is an important outcome of insect sex pheromone research, I did not intend to make this book comprehensive in terms of this aspect because excellent reviews and books are amply available (e.g., Baker et al. 2016; Kydonieus 2019; Mitchell 2012; Saha and Chandran 2017; Suckling 2015; Tabata 2018; Witzgall et al. 2010).

Organization of This Book

This book consists of four parts: "Chemistry of Sex Pheromones" (Part I), "Biosynthesis of Sex Pheromones" (Part II), "Reception of Sex Pheromones" (Part III), and "Mechanisms Controlling Behavior and Its Application to Robotics" (Part IV).

Part I overviews mainly the chemical aspects of lepidopteran and non-lepidopteran sex pheromones. In Chap. 1, classification of lepidopteran sex pheromones (Type 0, Type I, Type II, and Type III), which has become both possible and necessary due to the increase in knowledge of their chemical structures (Ando and Yamamoto 2019; Allison and Cardé 2016), is discussed with reference to their possible evolutionary history. Chapter 2 deals with the evolution of sex pheromone communication systems in hawk moths, a large group of moths with both nocturnal and diurnal species, varying life cycles, and unique flight behavior (hovering at flowers). In the following Chaps. 3, 4, and 5, sex pheromones of mealybugs, seed beetles, and longhorn beetles, non-lepidopteran insects for which knowledge of their sex pheromones has rapidly expanded, are described in detail. The readers can learn the similarities and dissimilarities of chemicals used as sex pheromones in these groups of insects and those used in lepidopteran species.

Part II summarizes our current knowledge on hormonal control of pheromone production and enzymes involved in the biosynthesis of sex pheromone components in moths. In Chap. 6, the progress in understanding the control of pheromone biosynthesis in *B. mori* is described in detail from the identification of PBAN to the elucidation of the signal transduction cascade that leads to the production of sex pheromone. Chapter 7 describes our current knowledge of sex pheromone biosynthetic pathways in a group of closely related moths, *Ostrinia* spp., which underlie the production of species-specific sex pheromones in terms of the combination of components and/or their blend ratios. In the last chapter of this part (Chap. 8), recent identification of an epoxidase, which functions in the last step of so-called Type II sex pheromone biosynthesis, is reported.

Part III deals with the molecular mechanisms of sex pheromone reception in moths, which underlie the ultrasensitive and highly specific detection of conspecific sex pheromone molecules (Chap. 9), evolutionary history of lepidopteran genes associated with sex pheromone recognition (Chap. 10), and the application of sex pheromone detection systems in sensing technologies (Chap. 11).

Part IV comprises three chapters, which deal with mechanisms controlling behavior and its application to robotics. The neuronal mechanisms that generate locomotor command for pheromone-source localization, which underlie sophisticated olfactory navigation tactics in *B. mori*, are discussed in Chap. 12. In Chap. 13, progress in comparative and evolutionary studies on sex pheromone preference coding in male moths is overviewed. The last chapter of this book (Chap. 14) reviews the searching strategy of *B. mori* in terms of the integration of visual and olfactory information and also introduces studies using a silk moth-driven hybrid robot, which revealed how silk moths adaptively behave under challenging circumstances.

The chapters of this book are arranged as indicated above with a clear intention; however, the readers may start reading from any chapter depending on their interest because each chapter is independent and provides the necessary background to understand the content. In order for the readers to feel the exciting and rapid progress in insect pheromone research, I invited authors who are on the cutting edge of their respective research fields. I hope this book gives the readers an overview of the forefront of insect sex pheromone research.

Tokyo, Japan
November 2019

Yukio Ishikawa

References

Allison JD, Cardé RT (2016) Pheromone communication in moths: evolution, behavior, and application. University of California Press, Oakland

Ando T, Yamamoto M (2019) Internet database: https://lepipheromone.sakura.ne.jp/pdb_top.html. Cited 15 Nov. 2019

Baker TC, Zhu JJ, Millar JG (2016) Delivering on the promise of pheromones. Journal of chemical ecology 42(7):553–556

Butenandt A, Beckmann R, Stamm D, Hecker E (1959) Über den Sexuallockstoff des Seidenspinners. *Bombyx mori*: Reindarstellung und Konstitution. Z Naturforsch 14:283–284

Fabre JH (1913) The great peacock moth. In: Teale EW (ed) The Insect World of J. Henri Fabre, 1964. Dodd, Mead & Co., New York

Karlson P, Lüscher M (1959) Pheromones: a new term for a class of biologically active substances. Nature. 183(4653):55–56

Kitamura A, Nagasawa H, Kataoka H, Inoue T, Matsumoto S, Ando T, Suzuki A (1989) Amino acid sequence of pheromone-biosynthesis-activating neuropeptide (PBAN) of the silkworm, *Bombyxmori*. Biochemical and biophysical research communications 163(1):520–526

Kydonieus AF (2019) Insect suppression with controlled release pheromone systems (Vol. 1). CRC Press, Boca Raton

Mitchell E (ed) (2012) Management of insect pests with semiochemicals: concepts and practice. Springer Science & Business Media, Boston

Moto K, Yoshiga T, Yamamoto M, Takahashi S, Okano K, Ando T, Nakata T, Matsumoto S (2003) Pheromone gland-specific fatty-acyl reductase of the silkmoth, *Bombyx mori*. Proceedings of the National Academy of Sciences of the United States of America. 100, 9156–9161. https://doi.org/10.1073/pnas.1531993100

Nagasawa H, Kitamura A, Inoue T et al (1988) Isolation of pheromone biosynthesis activating neuropeptide of the silkworm, *Bombyx mori*. Agricultural and Biological Chemistry 52:2985–2987

Saha T, Chandran N (2017) Chemical ecology and pest management: A review. International Journal of Chemical Studies 5(6):618–621

Sakurai T, Nakagawa T, Mitsuno H, Mori H, Endo Y, Tanoue S, Yasukochi Y, Touhara K, Nishioka T (2004) Identification and functional characterization of a sex pheromone receptor in the silkmoth *Bombyx mori*. Proceedings of the National Academy of Sciences of the United States of America 101:16653–16658

Suckling DM (2015) Sex pheromones and semiochemicals offer an elegant future for pest management and biosecurity. Acta Horticulturae (1105):275–281

Tabata J (2018) Mating disruption: concepts and keys for effective application. In: Chemical Ecology of Insects. CRC Press, Boca Raton, pp 197–233

Witzgall P, Kirsch P, Cork A (2010) Sex pheromones and their impact on pest management. Journal of Chemical Ecology 36:80–100. https://doi.org/10.1007/s10886-009-9737-y

Contents

Part I
Chemistry of Sex Pheromones

Chapter 1
Chemical Divergences in the Sex Pheromone Communication Systems in Moths

Hideshi Naka and Takeshi Fujii

Abstract Moth female sex pheromones have been classified into type I, type II, and miscellaneous, the former two of which are defined by the presence and absence of a terminal functional group, respectively. The classification of moth female sex pheromones is subject to refinement with the increase in our knowledge on the chemical structures of the pheromone molecules and biosynthetic pathways that underlie the formation of chemical structure. Recently, the addition of type 0 and type III to the classification has been proposed. Type-0 pheromones are short-chain secondary alcohols and their corresponding methyl ketones, which have been found from some primitive moths. Type-III pheromones are those containing one or more methyl branches in their carbon chain. In this chapter, we review the chemical structure and classification of moth female sex pheromones, and discuss scenarios for the evolution of different types of sex pheromones.

Keywords Moths · Female sex pheromones · Chemical structure · Evolution

1.1 Introduction

The order Lepidoptera is the second largest insect group comprising approximately 160,000 described species, most of which are moths (Kristensen et al. 2007; Komai 2011). Although moths are highly diverged in morphology and biology, most species use female-produced sex pheromones for long-distance communication between the sexes. As visual cues are practically useless for nocturnal moths, sex pheromone–mediated olfactory cues play an important role in the search and

H. Naka (✉)
Laboratory of Applied Entomology, Faculty of Agriculture, Tottori University, Tottori, Japan
e-mail: chun@tottori-u.ac.jp

T. Fujii
Laboratory of Applied Entomology, Graduate School of Agricultural and Life Sciences, The University of Tokyo, Tokyo, Japan

Y. Ishikawa (ed.), *Insect Sex Pheromone Research and Beyond*, Entomology Monographs, https://doi.org/10.1007/978-981-15-3082-1_1

recognition of conspecific mates. Studies on sex pheromones of moths began approximately 100 years ago. The French entomologist Jean-Henri Fabre discovered that females of the two species of peacock moths, *Saturnia pyri* and *Saturnia pavonia* (Saturniidae), release a kind of scent, which humans cannot perceive, to attract conspecific males (Fabre 1913). Approximately 50 years after Fabre's discovery, the German biochemist Adolf Butenandt and his colleagues succeeded in isolating and identifying a compound named 'bombykol' from ≈500,000 females of the silkmoth *Bombyx mori* (Bombycidae) (Butenandt et al. 1959, 1961a, b). Bombykol is released from a gland located at the abdominal tip of *B. mori* females (Percy-Cunningham and MacDonald 1987), and exhibits strong attractiveness for conspecific males at a minute amount. Karlson and Lüscher (1959), colleagues of Butenandt, proposed the new term 'pheromone' for the chemicals that mediate communication between individuals of the same species. In the early days of pheromone research, a marked amount of female moths were needed to identify the sex pheromone. Currently, the development of analytical techniques, particularly the spread of the use of gas chromatography-mass spectrometry (GC/MS) equipped with capillary columns, has enabled us to identify the sex pheromone from only several female moths. To date, sex pheromones have been identified from approximately 700 moth species, and two databases on moth sex pheromones (https://lepi-pheromone.sakura.ne.jp/pdb_top.html, http://www.pherobase.com/) are open to the public (Ando and Yamamoto 2019; El-Sayed 2019).

1.2 Chemical Structures and Classification of Moth Sex Pheromones

Ando et al. (2004) classified moth female sex pheromones into 'type I', 'type II', and miscellaneous, the former two of which are defined by the presence and absence of a terminal functional group, respectively. To date, type-I and -II pheromones have been identified from approximately 600 and 100 moth species, respectively. The former group of moths includes species belonging to the families Bombycidae, Noctuidae, Sesiidae, Tortricidae, and several others, whereas the latter group includes species belonging to Erebidae, Geometridae, and Tischeriidae (Ando and Yamamoto 2019). Type-I compounds are characterized by a C_{10}–C_{18} straight chain with 0–3 double bonds (Fig. 1.1a; Ando et al. 2004). Typical type-I compounds have a functional group, such as an acetyl, hydroxyl, or aldehyde group, at the terminal position. Examples of type-I pheromones are (3*E*,13*Z*)-3,13-octadecadienyl acetate (E3,Z13-18:OAc) identified from several Sesiidae species (Tumlinson et al. 1974; Mozūraitis and Karalius 2007; Naka et al. 2008), (10*E*,12*Z*)-10,12-hexadecadien-1-ol (E10,Z12-16:OH, bombykol) from the silkmoth *B. mori* (Butenandt et al. 1959), and (*Z*)-11-hexadecanal (Z11-16:Ald) from several moth species (e.g., Roelofs et al. 1974; Uehara et al. 2015). In contrast, type-II compounds are characterized by polyunsaturated (2–5 double bonds) hydrocarbons with a C_{17}–C_{25} straight chain and their corresponding epoxy

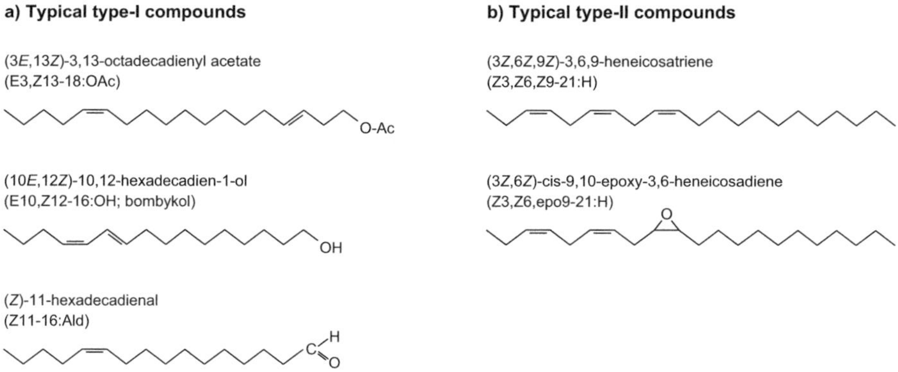

Fig. 1.1 Chemical structures of typical type-I and type-II sex pheromones in moths

derivatives (Fig. 1.1b; Ando et al. 2004). Examples are (3Z,6Z,9Z)-3,6,9-henicosatriene (Z3,Z6,Z9-21:H) and cis-9,10-epoxy-(3Z,6Z)-3,6-henicosadiene (Z3,Z6,epo9-21:H) from several Erebidae and Geometridae moth species (Ando and Yamamoto 2019).

The chemical classification of pheromones by Ando et al. (2004) has increased the interest of biochemists in the difference in biosynthetic pathways of type-I and -II pheromones (Fig. 1.2). Most type-I pheromones are biosynthesized from a common saturated fatty-acyl CoA via enzymatic steps such as desaturation, reduction, and acetylation. In contrast, most type-II pheromones are biosynthesized from dietary linoleic or linolenic acid via carbon chain elongation, decarboxylation, and subsequent optional epoxidation (Millar 2000, Rule and Roelofs 1989; Wei et al. 2003). Thus, the double bonds characteristic to type-II compounds originate from dietary essential fatty acids. Although the biosynthesis of type-I pheromones is completed in the pheromone gland, type-II pheromones are biosynthesized in the oenocytes from dietary essential fatty acids, and the products are transported via the hemolymph to the pheromone gland before the optional epoxidation and release from the gland (Chino 1985; Schal and Sevala 1998; Foster 2016).

As described above, the classification of 'typical' type-I and type-II pheromones is generally well supported by the difference in the biosynthetic pathways (Jurenka 2004; Foster 2016); however there are some compounds that have characteristics of both type-I and type-II. One example is (9Z,12Z,15Z)-9,12,15-octadecatrienal (Z9,Z12,Z15-18:Ald) from the fall webworm *Hyphantria cunea* (Erebidae: Arctiinae). Although this compound is classified as type I on the basis of the presence of a terminal functional group, this compound also has double bonds characteristic of type II. The position of double bonds and carbon chain length strongly indicate that this compound is derived from α-linolenic acid as are other type-II pheromones; however, this compound may not be produced in the oenocytes or transported to pheromone gland as typical type-II pheromones are (Kiyota et al. 2011). In this chapter, we consider Z9,Z12,Z15-18:Ald and other aldehydes derived from dietary essential fatty acids as type I following the classification by Ando et al. (2004).

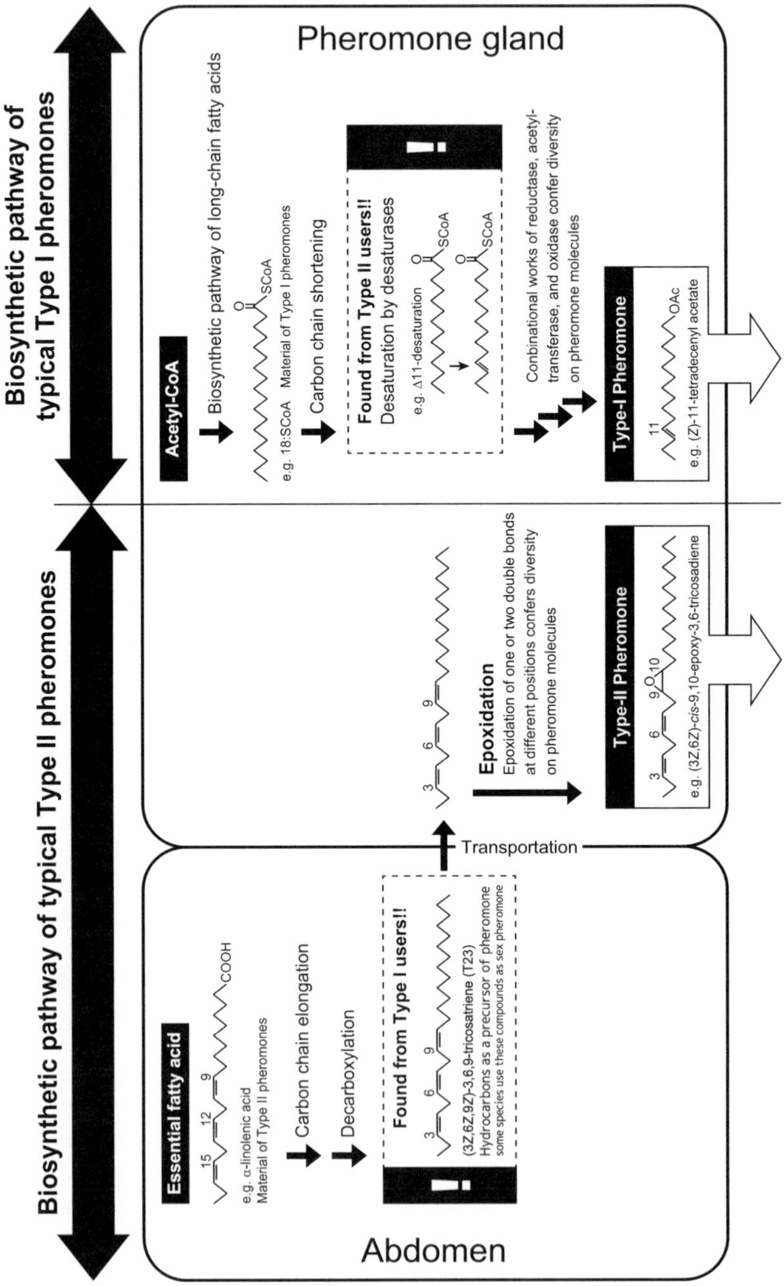

Fig. 1.2 Typical biosynthetic pathways of type-I and type-II sex pheromones

The classification of moth female sex pheromones is subject to refinement with the increase in our knowledge on the chemical structures of the pheromone molecules and biosynthetic pathways that underlie the formation of chemical structure. Löfstedt et al. (2016) proposed the incorporation of two new types, 'type 0' and 'type III', into the classification of moth female sex pheromones. Type-0 pheromones are short-chain secondary alcohols and their corresponding methyl ketones, which have been found from some primitive moths (Tóth et al. 1995; Zhu et al. 1995; Kozlov et al. 1996). The addition of 'type 0' is supported by the clear difference in the length of the carbon chain skeleton, which suggests different biosynthetic pathways from those of type I and type II. The difference in the location of the pheromone gland (sternal pheromone gland: Melnitsky and Deev 2009) also suggests the different evolutionary origin of type-0 pheromones. We will discuss this in detail below. Type-III pheromones are those containing one or more methyl branches in their carbon chain (Löfstedt et al. 2016). Methyl-branched pheromones have been found from some Erebidae, Geometridae, and Lyonetiidae moth species. In arctiids *Virbia* (formerly *Holomelina*) spp., which use 2-methyl-heptadecane (type-III) as a pheromone component, Charlton and Roelofs (1991) reported that the methyl branch in this compound is derived from the amino acid leucine. Later, Schal and Sevala (1998) revealed that this pheromone is biosynthesized in the oenocytes and transported to the pheromone gland, as in the case of type-II pheromones. However, current information is insufficient to determine whether all pheromones with methyl branch(es) represent a clear group of compounds that share the same evolutionary origin. We therefore refrain from discussing the evolution of "type-III" pheromones in this chapter.

1.3 Enzymes and Source Materials Commonly Found in Type-I and Type-II Users

We often find enzymes specifically used for the biosynthesis of type-I pheromones in type-II users, and vice versa. For example, most moth females that use type-I pheromones have Δ11-desaturase (see Fig. 1.2), a key enzyme for the generation of type-I pheromones, which inserts a double bond into pheromone precursors. However, transcripts of a Δ11-desaturase gene, *Asdesat1*, were found from the pheromone gland of the giant looper *Ascotis selenaria cretacea* (Geometridae) (Fujii et al. 2013). This species uses a typical type-II pheromone, 3,4-epoxy-(*Z*,*Z*)-6,9-nonadecadiene, and a desaturation step is not involved in their pheromone biosynthetic pathway. It is of interest to clarify why *A. selenaria* females possess Δ11-desaturase and for what purpose this enzyme is used in *A. selenaria* females. On the other hand, an alkene (3*Z*,6*Z*,9*Z*)-3,6,9-tricosatriene (hereafter 'T23'), which has been reported as a female sex pheromone or its precursor from several moth species using type-II pheromones, is found in the hemolymph of several type-I users (Fujii et al. 2015). These findings suggest that most moth species have the latent ability to biosynthesize both types of pheromones. If this suggestion holds true, many more moth species may have utilized a mixture of type I and type II as their sex pheromones. We next discuss the use of a mixture of type-I and type-II compounds in moths.

1.4 Hybrid Pheromone System (Hybrid Type)

The sex pheromone of *H. cunea* is a mixture of type-I and type-II compounds (Hill et al. 1982). Ando et al. (2004) referred to such pheromones composed of different types as the "hybrid pheromone system" (hereafter, hybrid type). The hybrid-type pheromones have been found from several Arctiinae moths, including *H. cunea*, several Pyraloidea moths, and a tortricid, the spruce budworm *Choristoneura fumiferana* (Ando and Yamamoto 2019). Although Erebidae is a large family, most of the members use type-II pheromone; the hybrid-type pheromone has been found only from the species belonging to the subfamily Arctiinae. Furthermore, although female arctiid moths use many type-II compounds, only a single type-I compound, Z9,Z12,Z15-18:Ald, is used as the pheromone component. This suggests that hybrid-type users in arctiid lack regular biosynthetic pathways for type-I pheromone, and they acquired Z9,Z12,Z15-18:Ald as a pheromone component after the split of the arctiid lineage from the other subfamilies in Erebidae. Moths of the superfamily Pyraloidea use T23 and related triene hydrocarbons (Crambidae) or pentaene hydrocarbons (Pyralidae) as type-II components in combination with numerous type-I compounds (Table 1.1). Accordingly, the "hybrid pheromone system" of arctiids and that of pyraloids are different. The hybrid type in the superfamily Pyraloidea will be discussed below.

1.5 The Moth Lineages and Chemical Structures of Pheromones

Mapping of the pheromone types (Type 0, type I, type II, Hybrid type, and others) onto a phylogenetic tree of all lepidopteran families proposed by Regier et al. (2013) highlighted the close relationship between the chemical structures of moth sex pheromones and the phylogeny of lepidopteran families (Fig. 1.3). Chemical divergence in the use of sex pheromones may have occurred at least twice during the evolution of moths. The first was the acquisition of type-I pheromones. Some primitive moth species use type-0 pheromones (Löfstedt et al. 2016), which are short-chain secondary alcohols and their corresponding methyl ketones. The sex pheromones in primitive moths were discovered in two independent studies by Tóth et al. (1995) and Zhu et al. (1995). Tóth et al. (1995) reported that *Stigmella mallela* in the family Nepticulidae uses a mixture of (2*S*,6*Z*)-nona-6,8-dien-2-ol and (2*S*,6*E*)-nona-6,8-

Table 1.1 List of species that use hybrid-type pheromones

Classification	Scientific name	Pheromone components	Reference
Typical type-II compounds + aldehydes delivered from dietary linoleic or linolenic acid			
Erebidae: Catocalinae	*Achaea janata*	Z3,Z6,Z9-21:H + Z6,Z9-21:H + 21:H + Z9,Z12-18:Ald	Persoons et al. (1993) and Krishnakumari et al. (1998)
Erebidae: Arctiinae	*Amsacta albistriga*	Z3,Z6,Z9-21:H + Z9,Z12,Z15-18:Ald + Z9,Z12-18:Ald + 18:Ald	Persoons et al. (1993)
Erebidae: Arctiinae	*Estigmene acrea*	Z3,Z6,epo9-21:H + Z9,Z12,Z15-18:Ald + Z9,Z12-18:Ald	Hill et al. (1981)
Erebidae: Arctiinae	*Hyphantria cunea*	Z3,Z6,epo9-21:H, (9S,10R)-isomer + Z9,Z12,Z15-18:Ald + Z9,Z12-18:Ald	Hill et al. (1982) and Einhorn et al. (1982)
Erebidae: Arctiinae	*Spilarctia* (formally *Diacrisia*) *obliqua*	Z3,Z6,Z9-21:H + Z3,Z6,epo9-21:H + 1,Z3,Z6,epo9-21:H + Z9,Z12,Z15-18:Ald + Z9,Z12-18:Ald	Persoons et al. (1993)
Typical type-I compounds + T23			
Crambidae: Odontiinae	*Deanolis sublimbalis*	Z11-16:Ald + Z3,Z6,Z9-23:H	Gibb et al. (2007)
Crambidae: Spilomelinae	*Conogethes pluto*	E10-16:Ald + E10,E12-16:Ald + Z3,Z6,Z9-23:H	El-Sayed et al. (2013)
Crambidae: Spilomelinae	*Leucinodes orbonalis*	E11-16:OAc + Z3,Z6,Z9-23:H	Vang et al. (2018)
Crambidae: Spilomelinae	*Neoleucinodes elegantalis*	E11-16:OH + Z3,Z6,Z9-23:H	Cabrera et al. (2001)
Crambidae: Spilomelinae	*Omphisa anastomosalis*	E10,E14-16:Ald + Z3,Z6,Z9-23:H	Yan et al. (2014)
Crambidae: Spilomelinae	*Rehimena surusalis*	E10,Z12-16:OAc + E10,Z12-16:Ald + Z3,Z6,Z9-23:H	Honda et al. (2015)
Typical type-I compounds + type-II pentaenes			
Pyralidae: Epipaschiinae	*Orthaga achatina*	Z11-16:OAc + Z3,Z6,Z9,Z12,Z15-23:H	Yan et al. (2018)
Pyralidae: Phycitinae	*Amyelois transitella*	Z11,Z13-16:Ald + Z11,Z13-16:OH + Z11,E13-16:OH + Z3,Z6,Z9,Z12,Z15-23:H	Leal et al. (2005) and Kuenen et al. (2010)
Pyralidae: Phycitinae	*Dioryctria abietella*	Z9,E11-14:OAc + Z3,Z6,Z9,Z12,Z15-25:H	Löfstedt et al. (2012)
Pyralidae: Phycitinae	*Dioryctria abietivorella*	Z9,E11-14:OAc + Z3,Z6,Z9,Z12,Z15-25:H	Millar et al. (2005)
Pyralidae: Phycitinae	*Dioryctria amatella*	Z11-16:OAc + Z3,Z6,Z9,Z12,Z15-25:H	Miller et al. (2010)
Pyralidae: Phycitinae	*Dioryctria mendacella*	Z9,E11-14:OAc + Z3,Z6,Z9,Z12,Z15-25:H	Hall et al. (2017)
Type-II compounds enhance attractiveness within close range			
Crambidae: Spilomelinae	*Conogethes punctiferalis*	E10-16:Ald + Z10-16:Ald + Z3,Z6,Z9-23:H + Z9-27:H	Xiao et al. (2012)
Tortricidae: Tortricinae	*Choristoneura fumiferana*	E11-14:Ald + Z11-14:Ald + Z11-16:Ald + Z3,Z6,Z9-23:H + Z5-23:H	Silk et al. (2017)

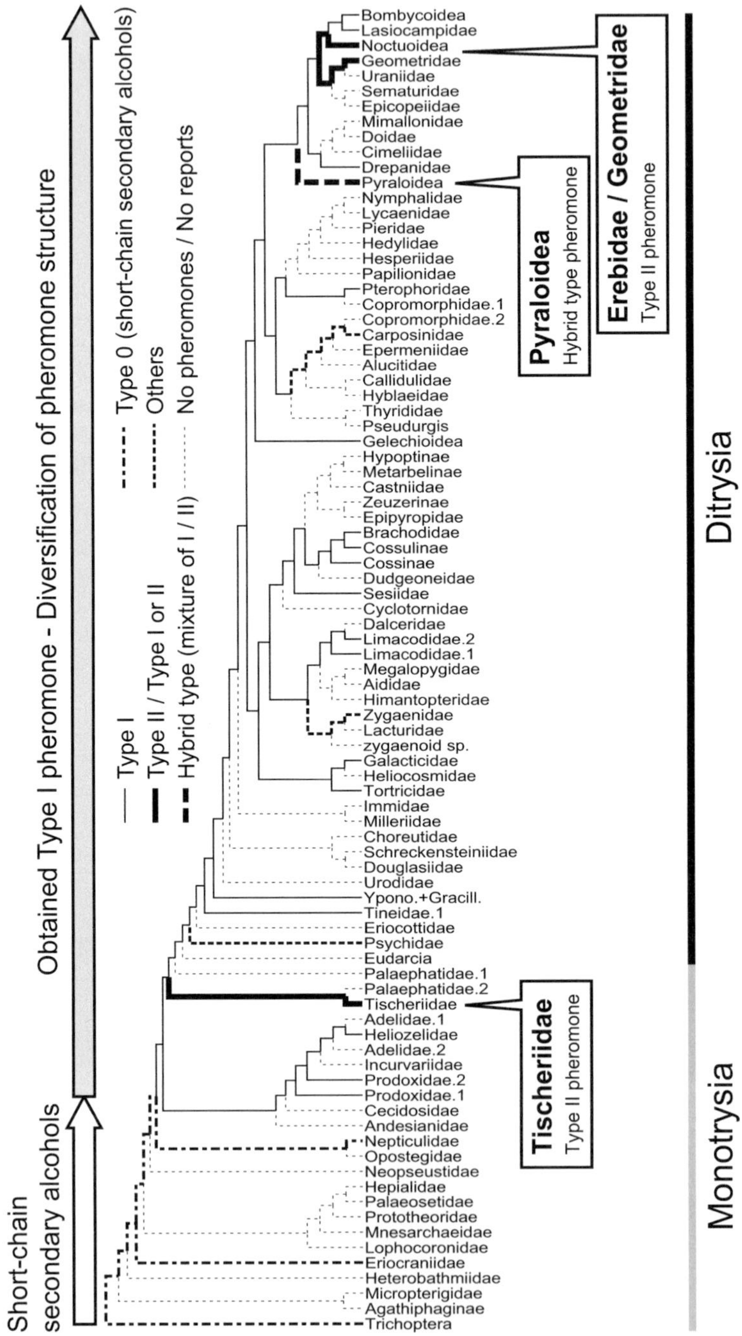

Fig. 1.3 Relationship between the saltational changes in the chemical structures of moth sex pheromones and the phylogeny of lepidopteran families

dien-2-ol as the sex pheromone. A mixture of these two compounds also attracted other Nepticulidae moths, *Stigmella crataegella* and *Trifurcula melanoptera*, in the field tests. Similar compounds have been reported as sex pheromone components of three species in the family Eriocraniidae, *Eriocrania cicatricella* (Zhu et al. 1995), *Eriocrania semipurpurella*, and *Eriocrania sangii* (Kozlov et al. 1996). *E. cicatricella* uses a mixture of (2*R*,4Z)-hept-4-en-2-ol and (*R*)-2-heptanol, and the other two species use mixtures of (2*S*,6*Z*)-non-6-en-2-ol and (2*S*,6*E*)-non-6-en-2-ol with different blend ratios. The chemical structure of these pheromone compounds is very similar to that of the sex pheromones of the species in Trichoptera, the sister group of Lepidoptera (e.g., Löfstedt et al. 2008). Moreover, females of these primitive moths release sex pheromones from the 4th and 5th abdominal sternites (Kozlov et al. 1996). The position of pheromone gland is the same as that of the sternal pheromone gland in Trichoptera (Melnitsky and Deev 2009), notably different from the typical pheromone gland of moths. These reports are consistent with the hypothesis that sex pheromones of primitive moths share the same origin as that of trichopteran species.

Moths may have acquired type-I pheromones and the pheromone gland in the abdominal tip earlier than the appearance of Ditrysia in the moth lineage; typical type-I sex pheromones are identified from the moths belonging to two monotrysian families, Prodoxidae and Heliozelidae, in the superfamily Adeloidea. Löfstedt et al. (2004) identified a mixture of (9*Z*,11*Z*)-9,11-tetradecadien-1-ol and its corresponding aldehyde and acetate as the sex pheromone of the currant shoot borer *Lampronia capitella* (Prodoxidae). Wang et al. (2015) also identified a mixture of (*Z*)-5-tetradienal and (*Z*)-7-tetradienal as the sex pheromone of a heliozeliid leaf miner, *Holocasista capensis* (Heliozelidae). Moreover, Löfstedt et al. (2004) reported that female *L. capitella* displayed a calling behavior similar to that reported for many ditrysian moths, and demonstrated that the sex pheromone was released from the abdominal tip. Short-chain secondary alcohols have not been reported as sex pheromones in moth groups more advanced than Adeloidea, and the majority of moths in advanced groups use type-I pheromones (Fig. 1.3). Type-I pheromones may have become predominant because the use of type-I pheromones has some advantages over the use of type-0 pheromones. The change in the pheromone communication system to the use of type-I pheromones in the ancestor of Adeloidea may have enabled the subsequent diversification of ditrysian moths. The shift in the use of type-0 to type-I pheromones may have been saltational; however, it should have been achieved via several important evolutionary steps; e.g., evolution of the pheromone gland on abdominal tip, use of saturated fatty acyl-CoA as the source material for pheromone biosynthesis, and acquisition of enzymes involved in biosynthesis such as desaturases (Δ11-desaturase in particular), reductases, and enzymes for limited β-oxidation. None of these are indispensable for the production of typical type-I pheromones. In the phylogenetic tree of moths, the lengths of the branches of Adeloidea and related moth taxa are long (Fig. 1.3; Regier et al. 2013), indicating that most primitive moths that may have had only a part of these characteristics became extinct or have not yet been discovered. It is unclear which characteristic evolved first, but specific expression of Δ11-desaturase (Liu et al. 2004) and epoxidase (Rong et al. 2014) in the pheromone gland suggests that evolution of the pheromone gland on the abdominal tip occurred first and the enzymes specifically involved in the biosynthesis of sex pheromones were acquired later.

The second chemical divergence in the moth sex pheromone was the switch to the use of type-II pheromones. Although typical type-II pheromones have been found from species in the families Erebidae and Geometridae, type-II users are also found in families other than these two (Fig. 1.3). A monotrysian moth, *Tischeria ekebladella* (Tischeriidae), uses a mixture of two type-II compounds, T23 and (3Z,6Z,9Z,19Z)-3,6,9,19-tricosatetraene (Z3,Z6,Z9,Z19-23:H), for their sex pheromone (Molnár et al. 2012). Although T23 is a typical type-II compound, the tetraene component Z3,Z6,Z9,Z19-23:H has been found only from this species. The characteristic chemical structure with a (19Z)-double bond suggests that *T. ekebladella* acquired the type-II pheromone independently from the other type-II users, i.e., species in Erebidae and Geometridae. Some species in Pyraloidea also utilize type-II compounds as pheromone components. Before discussing the evolutionary scenario for the evolution of type-II pheromones in moths, the discovery of the use of type-II compounds in Pyraloidea needs to be explained because it is different from the cases of species in Erebidae and Geometridae. Pyraloidea are a large superfamily composed of two families, Pyralidae and Crambidae, and in total, comprises more than 15,576 described species worldwide (Nuss et al. 2018). All pyraloids were once thought to be type-I users, and until recently, female sex pheromones of many (mainly agricultural pests) pyraloid species have been analyzed on this premise. Although sex pheromones of a large number of pyraloid species have been reported to date (Ando and Yamamoto 2019), the attractiveness of synthetic pheromones for some species is inferior to virgin females and cannot be used for practical use for monitoring pest populations in the field. Accordingly, although sex pheromones have been implemented for population monitoring or mating disruption of species in moth families, such as Tortricidae, Sesiidae, and Noctuidae, the use of pheromone lures for pyraloids, especially species in the subfamilies Pyraustinae and Spilomelinae (both in Crambidae), failed. Cabrera et al. (2001) identified a mixture of (*E*)-11-hexadecen-1-ol (E11-16:OH) and T23 as the sex pheromone of the tomato fruit borer *Neoleucinodes elegantalis* (Spilomelinae); the addition of T23 to E11-16:OH, the main component, increased the male capture approximately threefold. After this finding, pyraloids, Pyraustinae and Spilomelinae species in particular, which use hybrid-type pheromones, have been subsequently found (Table 1.1; e.g., Löfstedt et al. 2012; El-Sayed et al. 2013; Yan et al. 2014; Honda et al. 2015). We presume that the type-II components were overlooked in previous studies on the sex pheromone of Pyraustinae and Spilomelinae pest species, and it hampered the practical use of sex pheromones in the control of these pests. Based on this hypothesis, we started comprehensive identification of the sex pheromone of Spilomelinae species, and discovered hybrid-type pheromones from several species (Naka et al. unpublished). Recently, Silk et al. (2017) reported that the spruce budworm *C. fumiferana* (Tortricidae: Tortricoidea) uses two type-II compounds, T23 and (3Z,6Z,9Z)-3,6,9-pentacosatriene, in addition to two type-I pheromone components, in close-range attraction of males. This finding suggests that moth species in a wide range of taxa use type-II compounds for mate recognition.

1.6 The Origin of Type-II Pheromones

Tracking of the use of pheromone types in the lepidopteran phylogenetic tree (Fig. 1.3) suggests that the type-II pheromones have evolved at least three times (in Tischeriidae, Pyraloidea, and Noctuoidea + Geometroidea clade) independently. This suggests that acquisition of type-II compounds as sex pheromone components may not have been difficult for moths. What evolutionary processes have moths gone through to acquire the type-II pheromones? We think the origin of type-II pheromone may be cuticular hydrocarbons (CHCs), which are widely used as contact sex pheromones by many holometabolous insects. CHCs have been used for recognition of conspecific species and/or mating partners from the distant past in the holometabolous insect lineage. For example, ants have developed a sophisticated chemical communication system using CHCs, which has enabled them to recognize nestmates (Ozaki et al. 2005). The order Hymenoptera (ants, wasps, and bees) is the sister lineage to all other holometabolous insects (Ishiwata et al. 2011). Furthermore, the chemical structure of CHCs used as contact pheromones in some longicorn beetles is very similar to that of lepidopteran type-II compounds (Fig. 1.4; Yasui et al. 2003). Ants perceive these CHCs by the antennae, whereas longicorn beetles do so by the forelegs. This suggests that lepidopteran species, which have a common ancestor with ants and longicorn beetles, perceive some CHCs as contact pheromones via the antennae or forelegs. As mentioned above, T23 was found from the hemolymph of some type-I users (Fujii et al. 2015). It is possible that T23 plays a role as a contact pheromone, causing abdominal contact and the following copulation in longicorn beetles. Indeed, Grant et al. (1987) reported that C_{21}-C_{29} alkanes,

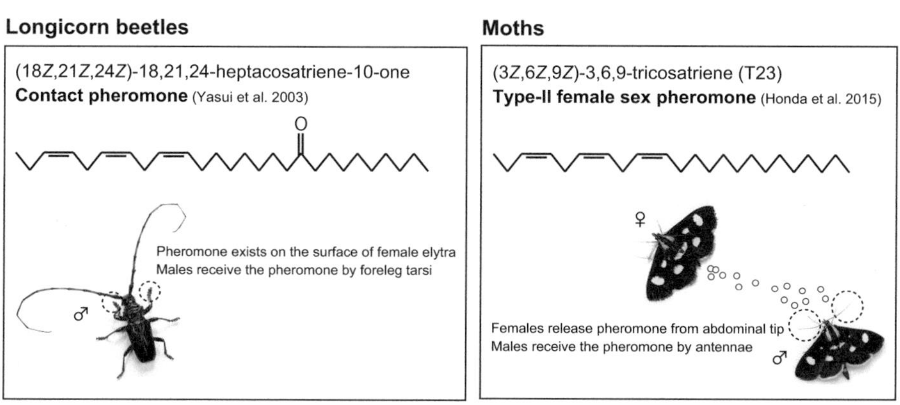

Fig. 1.4 Similarity between the chemical structures of moth sex pheromones and contact sex pheromones in longhorn beetles

compounds similar to T23, stimulated male whitemarked tussock moth *Orgyia leucostigma* (Erebidae: Lymantriinae) to attempt copulation. Moreover, T23 and its related compounds enhanced the attractiveness of female sex pheromones in the yellow peach moth *Conogethes punctiferalis* (Crambidae) (Xiao et al. 2011, 2012) and the spruce budworm *C. fumiferana* (Silk et al. 2017) in close range. These studies demonstrated that type-I users can utilize CHCs to recognize mating partners. Exploitation of compounds that had been used as contact pheromones or short-range attractants as long-range attractant pheromones may have been easier for moths than inventing entirely new compounds for a new purpose. If males can perceive hydrocarbons in the air by the antennae, hydrocarbons may have been exploited as long-range attraction pheromones without much difficulty.

In this chapter, we reviewed the chemical structure and classification of moth female sex pheromones, and discussed scenarios for the evolution of type-I and type-II pheromones. During the evolution of moths, saltational shift from the use of type 0 to the use of type I has occurred once, whereas the use of type II may have evolved several times independently in the moth lineage. As shown in Fig. 1.3, however, the chemical structures of pheromones in many moth families/superfamilies are still unknown. To date, for example, no information is available on the female sex pheromones of species in a large superfamily Drepanoidea. Likewise, the sex pheromones of many primitive moth families also remain unknown. Information on the sex pheromones of species in these taxa may help us to reconstruct a reliable evolutionary history of the sex pheromone communication systems in moths.

Acknowledgments We thank Emeritus Prof. Yukio Ishikawa of the University of Tokyo for providing us the opportunity to write this chapter. We also thank Drs. Mei Yamaguchi and Masayuki Goshima for supplying relevant information. We gratefully acknowledge the work of all collaborators, Drs. Tetsu Ando, Hiroshi Honda, Takeshi Sakurai, Shigehiro Namiki, Rong Yu, Yoichi Seki, and Mr. Takuya Nirazawa.

References

Ando T, Yamamoto M (2019) Internet database: https://lepipheromone.sakura.ne.jp/pdb_top.html. Cited 15 Nov 2019

Ando T, Inomata S, Yamamoto M (2004) Lepidopteran sex pheromones. Top Curr Chem 239:51–96

Butenandt A, Beckmann R, Stamm D, Hecker E (1959) Über den Sexuallockstoff des Seidenspinners. *Bombyx mori*: Reindarstellung und Konstitution. Z Naturforsch 14:283–284

Butenandt A, Beckmann R, Hecker E (1961a) Über den Sexuallockstoff des Seidenspinners, I. Der biologische Test und die Isolierung des reinen Sexuallockstoffes Bombykol. Hoppe-Seyler Z Physiolog Chemie 324:71

Butenandt A, Beckmann R, Hecker E (1961b) Über den Sexuallockstoff des Seidenspinners. II Konstitution und Konfiguration des Bombykols. Hoppe-Seyler Z Physiolog Chemie 324:84–87

Cabrera A, Eiras AE, Gries G, Gries R, Urdaneta N, Mirás B, Badji C, Jafee K (2001) Sex pheromone of tomato fruit borer, Neoleucinodes elegantalis. J Chem Ecol 27:2097–2107

Charlton RE, Roelofs WL (1991) Biosynthesis of a volatile, methyl-branched hydrocarbon sex pheromone from leucine by arctiid moths (*Holomelina* spp.). Arch Insect Biochem 18:81–97

Chino H (1985) Lipid transport: biochemistry of hemolymph lipophorin. In: Kerkut GA, Gilbert LI (eds) Comprehensive insect physiology, biochemistry and pharmacology, vol 10. Pergamon Press, New York, pp 115–135

Einhorn J, Lallemand J-Y, Zagatti P, Gallois M, Virelizier H, Riom J, Menassieu P (1982) Isolation and identification of the sex pheromone blend of Hyphantria cunea (Drury) (Lepidoptera, Arctiidae). C R Acad Sci Ser 2(294):41–44

El-Sayed AM (2019) Pherobase: http://www.pherobase.com/. Cited 15 Nov 2019

El-Sayed AM, Gibb AR, Mitchell VJ, Manning LAM, Revell J, Thistleton B, Suckling DM (2013) Identification of the sex pheromone of *Conogethes pluto*: a pest of Alpinia. Chemoecology 23:93–101

Fabre JH (1913) The great peacock moth. In: Teale EW (ed) The Insect World of J. Henri Fabre, 1964. Dodd, Mead & Co., New York

Foster SP (2016) Toward a quantitative paradigm for sex pheromone production in moths. In: Allison FD, Carde RT (eds) Pheromone communication in moths. University of California Press, Berkeley, pp 113–126

Fujii T, Suzuki MG, Katsuma S, Ito K, Rong Y, Matsumoto S, Ando T, Ishikawa Y (2013) Discovery of a disused desaturase gene from the pheromone gland of the moth *Ascotis selenaria*, which secretes an epoxyalkenyl sex pheromone. Biochem Biophys Res Commun 441:849–855

Fujii T, Yamamoto M, Nakano R, Nirazawa T, Rong Y, Dong S-L, Ishikawa Y (2015) Alkenyl sex pheromone analogs in the hemolymph of an arctiid *Eilema japonica* and several non-arctiid moths. J Insect Physiol 82:109–113

Gibb AR, Pinese B, Tenakanai D, Kawi AP, Bunn B, Ramankutty P, Suckling DM (2007) (Z)-11-Hexadecenal and (3Z,6Z,9Z)-tricosatriene: sex pheromone components of the red banded mango caterpillar Deanolis sublimbalis. J Chem Ecol 33:579–589

Grant GG, Frech D, MacDonald L, Slessor KN, King GGS (1987) Copulation releaser pheromone in body scales of female whitemarked tussock moth, *Orgyia leucostigma* (Lepidoptera: Lymantriidae): identification and behavioral role. J Chem Ecol 13:345–356

Hall DR, Farman D, Dominguez JC, Pajares JA (2017) Female sex pheromone of the cone moth, Dioryctria mendacella: investigation of synergism between Type I and Type II pheromone components. J Chem Ecol 43:433–442

Hill AS, Roelofs WL (1981) Sex pheromone of the saltmarsh caterpillar moth, Estigmene acrea. J Chem Ecol 7:655–668

Hill AS, Kovalev BG, Nikolaeva LN, Roelofs RL (1982) Sex pheromone of the fall webworm moth, *Hyphantria cunea*. J Chem Ecol 8:383–396

Honda H, Yamasaki R, Sumiuchi Y, Uehara T, Matsuyama S, Ando T, Naka H (2015) Hybrid sex pheromones of the hibiscus flower-bud borer, *Rehimena surusalis*. J Chem Ecol 41:1043–1049

Ishiwata K, Sasaki G, Ogawa J, Miyata T, Su Z-H (2011) Phylogenetic relationships among insect orders based on three nuclear protein-coding gene sequences. Mol Phylogenet Evol 58:169–180

Jurenka R (2004) Insect pheromone biosynthesis. Top Curr Chem 239:97–132

Karlson P, Lüscher M (1959) 'Pheromones': a new term for a class of biologically active substances. Nature 183:55–56

Kiyota R, Arakawa M, Yamakawa R, Yasmin A, Ando T (2011) Biosynthetic pathways of the sex pheromone components and substrate selectivity of the oxidation enzymes working in pheromone glands of the fall webworm, *Hyphantria cunea*. Insect Biochem Mol Biol 41:362–369

Komai F (2011) Phylogeny above superfamilies and higher classification: history. In: Komai F, Yoshiyasu Y, Nasu Y, Saito T (eds) A guide to the Lepidoptera of Japan. Tokai University Press, Hiratsuka, pp 59–64. (in Japanese)

Kozlov MV, Zhu J, Philipp P, Francke W, Zvereva EL, Hannson BS, Löfstedt C (1996) Pheromone specificity in *Eriocrania semipurpurella* (Stephens) and *E. sangii* (Wood) (Lepidoptera: Eriocraniidae) based on chirality of semiochemicals. J Chem Ecol 22:431–454

Krishnakumari B, Prasuna AL, Jyothi KN, Valli MY, Sighamony S, Prasad AR, Yadav JS (1998) Behavioural and electrophysiological responses of Achaea janata Linn. (Lepidoptera: Noctuidae) males to synthetic female-produced sex pheromone components. J Entomol Res 22:197–202

Kristensen NP, Scoble MJ, Karsholt O (2007) Lepidoptera phylogeny and systematics: the state of inventorying moth and butterfly diversity. Zootaxa 1668:699–747

Kuenen LPS, McElfresh JS, Millar JG (2010) Identification of critical secondary components of the sex pheromone of the navel orangeworm (Lepidoptera: Pyralidae). J Econ Entomol 103:314–330

Leal WS, Parra-Pedrazzoli AL, Kaissling K-E, Morgan TI, Zalom FG, Pesak DJ, Dundulis EA, Burks CS, Higbee BS (2005) Unusual pheromone chemistry in the navel orangeworm: novel sex attractants and a behavioral antagonist. Naturwissenschaften 92:139–146

Liu W, Rooney AP, Xue B, Roelofs WL (2004) Desaturases from the spotted fireworm moth (*Choristoneura parallela*) shed light on the evolutionary origins of novel moth sex pheromone desaturases. Gene 342:303–311

Löfstedt C, Zhu J, Kozlov MV, Buda V, Jirle EV, Hellqvist S, Löfqvist J, Plass E, Franke S, Francke W (2004) Identification of the sex pheromone of the currant shoot borer *Lampronia capitella*. J Chem Ecol 30:643–658

Löfstedt C, Bergmann J, Francke W, Jirle E, Hannson BS, Ivanov VD (2008) Identification of a sex pheromone produced by sternal glands in females of the caddisfly *Molanna angustata* Curtis. J Chem Ecol 34:220–228

Löfstedt C, Svensson GP, Jirle EV, Rosenberg O, Roques A, Millar JG (2012) (3Z,6Z,9Z,12Z,15Z)-pentacosapentaene and (9Z,11*E*)-tetradecadienyl acetate: sex pheromone of the spruce coneworm *Dioryctria abietella* (Lepidoptera: Pyralidae). J Appl Entomol 136:70–78

Löfstedt C, Wahlberg N, Millar JG (2016) Evolutionary pattern of pheromone diversity in Lepidoptera. In: Allison FD, Carde RT (eds) Pheromone communication in moths. University of California Press, Berkeley, pp 43–78

Melnitsky SI, Deev RV (2009) The fine structure of sternal pheromone glands in the two caddisfly species from the Rhyacophilidae and Limnephilidae families (Insecta: Trichoptera). Russ Entomol 18:107–116

Millar JG (2000) Polyene hydrocarbons and epoxides: a second major class of lepidopteran sex attractant pheromones. Annu Rev Entomol 45:575–604

Millar JG, Grant GG, McElfresh JS, Strong W, Rudolph C, Stein JD, Moreira JA (2005) (3Z,6Z,9Z,12Z,15Z)-Pentacosapentaene, a key pheromone component of the fir coneworm moth, Dioryctria abietivorella. J Chem Ecol 31:1229–1234

Miller DR, Millar JG, Mangini A, Crowe CM, Grant GG (2010) (3Z,6Z,9Z,12Z,15Z)-Pentacosapentaene and (Z)-11-hexadecenyl acetate: sex attractant blend for Dioryctria amatella (Lepidoptera: Pyralidae). J Econ Entomol 103:1216–1221

Molnár BP, Tröger A, Toshova TB, Subchev M, van Nieukerken EJ, (Sjaak) Koster JC, Szöcs G, Tóth M, Francke W (2012) Identification of the female-produced sex pheromone of Tischeria ekebladella, an oak leafmining moth. J Chem Ecol 38:1298–1305

Mozūraitis R, Karalius V (2007) Identification of minor sex pheromone components of the poplar clearwing moth *Paranthrene tabaniformis* (Lepidoptera, Sesiidae). Zeit Naturforsch C 62:138–142

Naka H, Horie Y, Mochizuki F, Vang LV, Yamamoto M, Saito T, Watrai T, Tsuchida K, Arita Y, Ando T (2008) Identification of the sex pheromone secreted by *Synanthedon hector* (Lepidoptera: Sesiidae). Appl Entomol Zool 43:467–474

Nuss M, Landry B, Mally R, Vegliante F, Trankner A, Bauer F, Hayden J, Segerer A, Schouten R, Li H, Trofimova T, Solis MA, De Prins J, Speidel W (2018) Global Information System on Pyraloidea: www.pyraloidea.org. Cited 2 Nov 2018

Ozaki M, Wada-Katsumata A, Fujikawa K, Iwasaki M, Yokohari F, Satoji Y, Nisimura T, Yamaoka R (2005) Ant nestmate and non-nestmate discrimination by a chemosensory sensillum. Science 309:311–314

Percy-Cunningham JE, MacDonald JA (1987) Biology and ultrastructure of sex pheromone-producing glands. In: Prestwich GD, Blomquist GJ (eds) Pheromone biochemistry. Academic, Orlando, FL, pp 27–75

Persoons CJ, Vos JD, Yadav JS, Prasad AR, Sighomony S, Jyothi KN, Prasuna AL (1993) Indo-Dutch cooperation on pheromones of Indian agricultural pest insects: sex pheromone components of Diacrisia obliqua (Arctiidae), Achaea janata (Noctuidae) and Amsacta albistriga (Arctiidae). Bulletin Oilb Srop 16:136–140

Regier JC, Mitter C, Zwick A, Bazinet AL, Cummings MP, Kawahara AY, Sohn J-C, Zwickl DJ, Cho S, Davis DR, Baixeras J, Brown J, Parr C, Weller S, Lees DC, Mitter K (2013) A large-scale, higher-level, molecular phylogenetic study of the insect order Lepidoptera (moths and butterflies). PLoS One 8:e58568

Roelofs WL, Hill AS, Cardé RT, Baker TC (1974) Two sex pheromone components of tobacco budworm moth, Heliothis virescens. Life Sci 14:1555–1562

Rong Y, Fujii T, Katsuma S, Yamamoto M, Ando T, Ishikawa Y (2014) CYP341B14: a cytochrome P450 involved in the specific epoxidation of pheromone precursors in the fall webworm *Hyphantria cunea*. Insect Biochem Mol Biol 54:122–128

Rule GS, Roelofs RL (1989) Biosynthesis of sex pheromone components from linolenic acid in arctiid moths. Arch Insect Biochem 12:89–97

Schal C, Sevala V (1998) Novel and highly specific transport of a volatile sex pheromone by hemolymph lipophorin in moths. Naturwissenschaften 85:339–342

Silk PJ, Eveleigh E, Roscoe L, Burgess K, Weatherby S, Leclair G, Mayo P, Brophy M (2017) Unsaturated cuticular hydrocarbons enhance responses to sex pheromone in spruce budwork, *Choristoneura fumiferana*. J Chem Ecol 43:753–762

Tóth M, Szöcs G, van Nieukerken EJ, Philipp P, Schmidt F, Francke W (1995) Novel type of sex pheromone structure identified from *Stigmella malella* (Stainton) (Lepidoptera: Nepticulidae). J Chem Ecol 21:13–27

Tumlinson JH, Yonce CE, Doolittle RE, Heath RR, Gentry CR, Mitchell ER (1974) Sex pheromones and reproductive isolation of the lesser peachtree borer and the peachtree borer. Science 185:614–616

Uehara T, Naka H, Matsuyama S, Ando T, Honda H (2015) Identification of the pheromone of the diurnal halk moth, *Hemaris affinis*. J Chem Ecol 41:9–14

Vang LV, Yan Q, Nghia NYN, Khanh CNQ, Ando T (2018) Unsaturated cuticular hydrocarbon components of the sex pheromone of eggplant fruit borer, Leucinodes orbonalis Guenee (Lepidoptera: Crambidae). J Chem Ecol 44:631–636

Wang H-L, Geertsema H, van Nieukerken EJ, Löfstedt C (2015) Identification of the female-produced sex pheromone of the leafminer *Holocacista capensis* infesting grapevine in South Africa. J Chem Ecol 41:724–731

Wei W, Miyamoto T, Endo M, Murakawa T, Pu GQ, Ando T (2003) Polyunsaturated hydrocarbons in the hemolymph: biosynthetic precursors of epoxy pheromones of geometrid and arctiid moths. Insect Biochem Mol Biol 33:397–405

Xiao W, Honda H, Matsuyama S (2011) Monoenyl hydrocarbons in female body wax of the yellow peach moth as synergists of aldehyde pheromone components. Appl Entomol Zool 46:239–246

Xiao W, Matsuyama S, Ando T, Millar JG, Honda H (2012) Unsaturated cuticular hydrocarbons synergize responses to sex attractant pheromone in the yellow peach moth, *Conogethes punctiferalis*. J Chem Ecol 38:1143–1150

Yan Q, Vang LV, Nguyen C, Khanh Q, Naka H, Ando T (2014) Reexamination of the female sex pheromone of the sweet potato vine borer moth: identification and field evaluation of a tricosatriene. J Chem Ecol 40:590–598

Yan Q, Li H-D, Chen Y, Ye Z-F, You X-Y, Zhou J, Mu L-F, Liu S-J, Kong X-B, Khuhro SA, Dong S-L (2018) Identification and field evaluation of the sex pheromone of Orthaga achatina (Lepidoptera: Pyralidae). J Chem Ecol 44:886–893

Yasui H, Akino T, Yasuda T, Fukaya M, Ono H, Wakamura S (2003) Ketone components in the contact sex pheromone of the white-spotted longicorn beetle, *Anoplophora malasiaca*, and pheromonal activity of synthetic ketones. Entomol Exp Appl 107:167–176

Zhu J, Kozlov MV, Philipp P, Francke W, Löfstedt C (1995) Identification of a novel moth sex pheromone in *Eriocrania cicatricella* (Zett.) (Lepidoptera: Eriocraniidae) and its phylogenetic implications. J Chem Ecol 21:29–43

Chapter 2
Sex Pheromone Communication System in Hawk Moths

Takuya Uehara and Hiroshi Honda

Abstract Sphingidae (hawk moths) is one of the largest families in Bombycoidea and includes approximately 1450 species worldwide. Hawk moths have varying life cycles, interesting flight behavior (hovering at flowers), include both nocturnal and diurnal species, and have a large body size and long proboscis, making them valuable subjects for studies across a broad range of biological fields. Hawk moths also provide useful subjects for studies on sex pheromone communication systems; however, they remain to be further explored. We have determined the chemical structures of sphingid sex pheromone components, and evaluated their biological activity in both the field and the laboratory. The compound (10*E*,12*Z*)-hexadecadienal, known as bombykal, and its analogues are sex pheromone components in many hawk moth species. In this chapter, we discuss the sex pheromone communication system of hawk moths with reference to the similarity and dissimilarity of pheromone components.

Keywords Hawk moths · Sex pheromone · Biosynthesis · Calling time

2.1 Introduction

Hawk moths (Sphingidae) are also known as hummingbird moths or hornworm moths because of their characteristic hovering flight and feeding with a long proboscis, and the horn-shaped ornament in the larval stage (Pittaway 1993). Approximately 1450 species of hawk moths have been described worldwide (Nieukerken et al. 2011). In Japan, there are approximately 80 hawk moth species representing three

T. Uehara (✉)
Institute of Agrobiological Sciences, NARO, Tsukuba, Ibaraki, Japan
e-mail: tue@affrc.go.jp

H. Honda
Faculty of Life Sciences, Department of Chemistry for Life Science and Agriculture, Tokyo University of Agriculture, Tokyo, Japan

Y. Ishikawa (ed.), *Insect Sex Pheromone Research and Beyond*, Entomology Monographs, https://doi.org/10.1007/978-981-15-3082-1_2

subfamilies, Sphinginae, Smerinthinae, and Macroglossinae. They include both nocturnal and diurnal species, most of which are common and easily accessible.

Hawk moths have been the subject of many types of scientific research. For example, their relatively large body size is suitable for surgical manipulation and has greatly contributed to our understanding of insect physiology such as molting and metamorphosis (Truman and Riddiford 2002). Their long proboscis, which is suited for feeding on nectar, and their flight control during foraging have been studied with regard to pollination ecology and aerodynamics (e.g. Wang et al. 2008; Johnson et al. 2017). However, studies on the sex pheromone communication system in hawk moths are limited.

An important advantage of using hawk moths in the study of sex pheromones is that they belong to the superfamily Bombycoidea along with the silkmoth, *Bombyx mori*, whose pheromone system has been extensively studied. In addition, the chemical structures of sex pheromones in hawk moths and silkmoths share a strong similarity. The pheromone component of *B. mori* is (10*E*,12*Z*)-hexadecadien-1-ol, known as bombykol, whereas many hawk moths use an aldehyde derivative of bombykol, known as bombykal.

In this chapter, we review our current understanding of the sex pheromone system of hawk moths, and recent progress in the reception and biosynthesis of hawk moth pheromones.

2.2 Diversity in Sex Pheromone Components

Lepidopteran sex pheromones are categorized into two main types, Type I and Type II (Millar 2000; Ando et al. 2004; Löfstedt et al. 2016). Type-I pheromones are straight long-chain unsaturated aliphatic derivatives (C12–C18), such as alcohol, acetate, and aldehyde, whereas Type-II pheromones are long-chain hydrocarbons (C17–C23). Although many lepidopteran species use a single type of sex pheromone, recent studies have reported that some crambid moths use both types together (e.g. Millar et al. 2005; Honda et al. 2015; Hall et al. 2017; Yan et al. 2018a, b).

The sex pheromone components identified from hawk moths thus far are all Type-I pheromones (Table 2.1). Starratt et al. (1979) first identified (10*E*,12*Z*)-10,12-hexadecadienal (E10,Z12-16:Ald, bombykal) from *Manduca sexta* (Linnaeus), and Tumlinson et al. (1989) found (10*E*,12*E*,14*Z*)-10,12,14-hexadecatrienal (E10,E12,Z14-16:Ald) to be a minor component. These two components are essential for male attraction in this species (Tumlinson et al. 1994). Bombykal and its geometric isomers have been detected from most hawk moth species investigated thus far (Table 2.1). In addition to these components, C16 monoenyl aldehydes, such as (*E*)- and (*Z*)-11-hexadecenals (E11- and Z11-16:Ald), have been detected in many species, and C16 conjugated dienyl aliphatics are also common pheromone components. Indeed, field screening using C16 conjugated diene aliphatics demonstrated that males of five species of Sphingidae are attracted to bombykal family compounds, and these compounds were discovered in pheromone gland extracts of

Table 2.1 Sex pheromones and attractants of hawk moth species

Species	Components (%)	Identification criteria[a]	Attraction[b]	References
Sphinginae				
Agrius convolvuli	E11,E13-16:Ald (100)	GC-EAD, GC-MS	WT	Wakamura et al. (1996)
Dolbina tancrei	E9,Z11-15:Ald (94.3), Z9,Z11-15:Ald (5.7)	GC-EAD, GC-MS, GC, MTAD	FT	Uehara et al. (2013)
Manduca sexta	E10,Z12-16:Ald (65), E10,E12,Z14-16:Ald (35)	GC-MS	FT	Tumlinson et al. (1989) and Tumlinson et al. (1994)
Psilogramma increta	E11-16:Ald (100)	GC-EAD, GC-MS, GC, DMDS	No activity	Uehata et al. unpublished data
Sphinx constricta	E11,Z13-16:Ald (92), Z11,Z13-16:Ald (8)	GC-EAD, GC-MS, GC, MTAD	No activity	Uehara (2015)
Sphinx drupiferarum	E10,Z12-16:Ald	GC-MS	Not examined	Reed et al. (1987)
Sphinx vashti		EAD response to bombycal isomers	Not examined	Reed et al. (1987)
Macroglossinae				
Amphion floridensis	E10,Z12-16:Ald (100)		FT	Landolt et al. (1989)
Cephonodes hylas hylas	Z11-16:Ald (2), E10,Z12-16:Ald (5), E10,E12-16:Ald (93)	GC-EAD, GC-MS, GC	Low activity	Uehara (2015)
Daphnis nerii	Z11-16:Ald (25), E10,Z12-16:Ald (10), E10,E12-16:Ald (65)	GC-EAD, GC	Not examined	Uehara (2015)
Deilephila elpenor lewisii	E11-16:Ald (85), E10,E12-16:Ald (15)	GC-EAD, GC-MS, GC, MTAD, DMDS	FT	Bestmann et al. (1992) and Uehara et al. (2012)
Hemaris affinis	Z11-16:Ald (45), E10,Z12-16:Ald (20), E10,Z12-16:Ald (35)	GC-EAD, GC-MS, GC	FT	Uehara et al. (2015)
Hemaris diffinis	Z10, E12-16:Ald (100)		FT	Reed et al. (1987)
Hyles euphorbiae		EAD response to bombycal isomers	Not examined	Reed et al. (1987)
Hyles gallii		EAD response to bombycal isomers	Not examined	Reed et al. (1987)
Neogurelca himachala sangaica	E10,Z12-16:Ald (100)	GC-EAD, GC-MS, GC, MTAD	FT	Uehara et al. (2016)

(continued)

Table 2.1 (continued)

Species	Components (%)	Identification criteria[a]	Attraction[b]	References
Proserpinus flavofasciata	E10, E12-16:Ald (100)		FT	Reed et al. (1987)
Theretra japonica	E11-16:Ald (50), E10,Z12-16:Ald (5), E10,E12-16:Ald (45)	GC-EAD, GC	No activity	Uehara (2015)
Theretra oldenlandiae oldenlandiae	E11-16:Ald (40), E10,Z12-16:Ald (30), E10,E12-16:Ald (30)	GC-EAD, GC-MS, GC, MTAD, DMDS	FT	Uehara et al. (2012)
Smerinthinae				
Ambulyx schauffelbergeri	E10,Z12-16:Ald (5), E10,E12-16:Ald (20), E10,E12-16:OAc (75)	GC-EAD, GC	Not examined	Uehara (2015)
Callambulyx tatarinovii gabyae	E10,E12-16:Ald (100)	GC-EAD, GC-MS, GC	No activity	Uehara (2015)
Langia zenzeroides nawai	E10,E12-16:Ald (100)	GC-EAD, GC-MS, GC	No activity	Uehara (2015)
Marumba gaschkewitschii echephron	E10,Z12-16:Ald (30), E10,E12-16:Ald (70)	GC-EAD, GC-MS, GC	No activity	Uehara (2015)
Mimas christophi	Z11-16:Ald (24), E10,E12-16:Ald (1), E10,Z12-16:Ald (75)	GC-EAD, GC-MS, GC, MTAD	No activity	Uehara (2015)
Pachysphinx modesta	Z10, E12-16:OAc (100)		FT	Reed et al. (1987)
Phyllosphingia dissimilis dissimilis	E11-16:Ald (30), E10,E12-16:Ald (70)	GC-EAD, GC-MS, GC	No activity	Uehara (2015)
Smerinthus tokyonis	Z10,Z12-16:OAc (100)	GC-EAD, GC-MS, GC, MTAD	WT	Kozaki et al. unpublished data
Smerinthus planus planus	E10,E12-16:Ald (100)	GC-EAD, GC	No activity	Uehara (2015)
Smerinthus jamaicensis	Z10, E12-16:Ald (100)		FT	Reed et al. (1987)
Smerinthus cerisyi	Z10, E12-16:OAc (10), Z10, E12-16:Ald (90)	EAD response to bombycal isomers	FT	Reed et al. (1987)

[a]*GC* Gas chromatography, *GC-EAD* GC coupled with electroantennographic detector, *GC-MS* GC mass spectrometry, *MTAD* derivertization with 4-Methyl-1,2,4-triazoline-3,5-dione (MTAD), *DMDS* derivertization with dimethyl disulfide (DMDS)
[b]*WT* wind tunnel test, *FT* Field test

Sphinx drupiferarum, *Hyles gallii*, and *Amphion floridensis* using gas chromatographic analyses (Reed et al. 1987; Landolt et al. 1989).

However, some variations in pheromone components were reported. For example, the triene component E10,E12,Z14-16:Ald was identified from *M. sexta* (Tumlinson et al. 1989). The positional isomer (11*E*,13*E*)-11,13-hexadecadienal

(E11,E13-16:Ald) was identified from the sweet potato hornworm, *Agrius convolvuli* (Wakamura et al. 1996), and the one-carbon shortened components (9*E*,11*Z*)- and (9*Z*,11*Z*)-pentadecadienal (E9,Z11-15:Ald and Z9,Z11-15:Ald) were identified from *Dolbina tancrei* (Uehara et al. 2013). In addition, acetate type pheromones were detected in *Smerinthus tokyonis* (Kozaki et al., unpublished data), but the attractiveness of these compounds was only confirmed in the laboratory.

Considering the use of similar sex pheromone components in hawk moth species, how different species recognize respective potential mates is of interest. Comparisons of species that utilize similar sex pheromone components have revealed that a precise blend ratio of components is key to mate recognition. For example, although *Deilephila elpenor lewisii* and *Theretra oldenlandiae oldenlandiae* both use E11-16:Ald and (10*E*,12*E*)-10,12-hexadecadienal (E10,E12-16:Ald) as pheromone components, the blend ratios are 85:15 and 57:43, respectively (Uehara et al. 2012). Furthermore, *T. oldenlandiae oldenlandiae* uses a third component, E10,Z12-16:Ald, in addition to the two common components. Thus, as Tamaki (1977) proposed, the combination of components and their blend ratio confer species specificity to the sex pheromone communication systems in hawk moths.

A C16 conjugated diene structure is the common skeleton for hawk moth sex pheromones. Such pheromone components were also identified from other lepidopteran groups such as Bombycidae (Daimon et al. 2012), Crambidae (Klun et al. 1986; Raina et al. 1986; Honda et al. 1994), Saturniidae (McElfresh and Millar 1999a, b; McElfresh et al. 2001), and Noctuidae (Cork et al. 1988). Hawk moth species, such as *A. convolvuli* and *D. tancrei,* have a unique pheromone component that helps to achieve strict species specificity, but most hawk moth species rely on the blend ratio of more common components for mate recognition.

2.3 Sex Pheromone Biosynthesis

Elucidation of pheromone biosynthesis is important for understanding how species-specific pheromone components are derived from ubiquitous substances. In addition, mutations in genes that code for the enzymes involved in biosynthesis cause pheromonal changes, which may result in speciation. Thus, clarifying the molecular basis of sex pheromone biosynthesis can promote our understanding of the molecular basis of speciation.

Biosynthesis of pheromone components and regulation of the blend ratio have been studied in some moth species. In general, Type-I pheromones, such as bombykol, are synthesized from saturated fatty acids. In the case of the silkmoth, a C16 saturated fatty acid, palmitic acid, which is stored in the form of glycerolipid in the abdominal tip, is the starting substance for sex pheromone biosynthesis. Palmitic acid is conjugated to coenzyme A (fatty-acyl CoA or 16:CoA) and then modified by two successive enzymatic processes, desaturation and reduction. Namely, a single fatty acyl desaturase introduces two double bonds at a specified position of 16:CoA

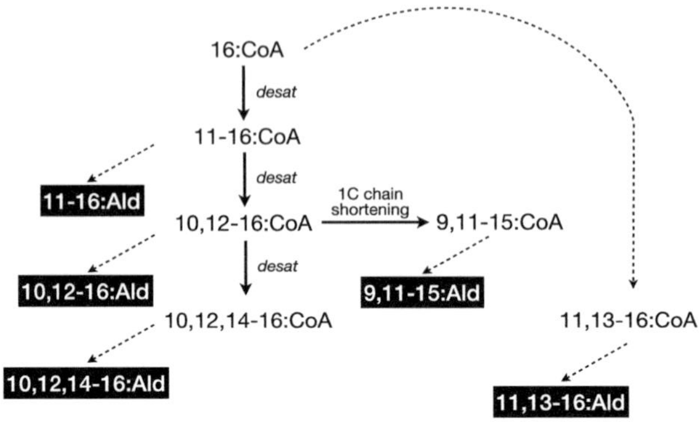

Fig. 2.1 Schematic diagram of sex pheromone biosynthetic pathways in hawk moths. Solid lines indicate a biochemically suggested reaction. Broken lines indicate a putative reaction

(Moto et al. 2004), and then a reductase catalyzes reduction of the fatty acyl moiety to the corresponding alcohol (Moto et al. 2003).

Hawk moths employ a biosynthetic pathway similar to that of *B. mori* for sex pheromone production (Fig. 2.1). The elucidation of the biosynthetic pathway for hawk moth pheromones was first attempted in *M. sexta* by observing the behavior of the precursors (Fang et al. 1992). Isotope-labeled palmitic acid applied to the female abdominal tip was incorporated into pheromone precursors, such as 11-16:CoA and 10,12-16:CoA, indicating that the saturated precursor has a double bond introduced at the 11th position or the 10th and 12th positions (Fang et al. 1995b; Svatoš et al. 1999). Molecular cloning of fatty acyl desaturase and its functional assay revealed that a single desaturase, Δ11-desaturase, first introduces a double bond at the 11th position of 16:CoA, and then at the 10th and 12th positions of 11-16:CoA (Matoušková et al. 2007). Moreover, transcriptome analysis of the sex pheromone gland demonstrated that desaturase genes are duplicated, and the counterpart desaturase introduces a double bond into the diene precursor to produce the triene pheromone component (Buček et al. 2015). After desaturation, fatty acyl-CoAs are transformed to the corresponding aldehydes, which are the sex pheromone components. However, the terminal steps of the biosynthesis of sex pheromones in *M. sexta* are still unclear. Although oxidase activity with broad substrate specificity was detected in the sex pheromone gland of *M. sexta*, no endogenous alcohols were found in the gland (Fang et al. 1995a; Luxová and Svatoš 2006).

The biosynthetic pathway for the hawk moth sex pheromones has mainly been studied in *M. sexta*. However, as many hawk moth species produce 10,12-hexadecadienyl aldehydes, employment of common biosynthetic pathways involving a bifunctional desaturase, which transforms C16:CoA to a diene precursor, and a putative reductase, which transforms the diene precursor to the corresponding alcohol, is suggested. Comparisons of pheromone components among hawk moth species have indicated that variations in these enzymes are reflected in the geometry of components and their blend ratio. Studies on *M. sexta* and *Ostrinia* spp. suggested that neo-functionalization and changes in desaturase substrate specificity can occur as a consequence of unequal cross-over and amino-acid substitution (Xue et al. 2007; Fujii et al. 2011; Rooney 2011; Buček et al. 2015), which may help to explain differences in sex pheromones among species.

Two hawk moth species, *A. convolvuli* and *D. tancrei*, produce unconventional pheromones. *A. convolvuli* uses E11,E13-16:Ald as a pheromone component. As noted above, the common motif in the sex pheromone, a 10,12-conjugated aliphatic, is produced by a bifunctional desaturase through an 11-monoene aliphatic. There are two possible routes by which E11,E13-16:Ald is biosynthesized. One possibility is that two desaturases, Δ11- and Δ13-desaturases, introduce two double bonds independently to the saturated fatty acid, 16:CoA. Alternatively, as in the biosynthetic pathway of sex pheromones in *M. sexta*, a bifunctional desaturase may be involved in the biosynthesis. An enzyme that introduces a double bond first at the 12th position then at the 11th and 13th positions is expected. In general, however, Δ12-desaturase is absent in Lepidoptera (Malcicka et al. 2018), making the first route more plausible.

D. tancrei uses sex pheromone components with an odd-numbered carbon chain, E9,Z11-15:Ald and Z9,Z11-15:Ald. Although such odd-numbered pheromone components have also been detected in other lepidopteran species, these pheromones are relatively rare because of the mechanism of Type-I sex pheromone biosynthesis. As previously described, Type-I pheromone components are usually derived from C16 fatty acid conjugated with CoA. 16:CoA may undergo chain shortening in 2-carbon units via a single cycle of reactions termed limited β-oxidation. Accordingly, even-numbered pheromones are common, whereas odd-numbered pheromone components must be specifically biosynthesized. According to biochemical studies conducted by Jurenka and Roelofs (1993), α-oxidation of even-numbered pheromone precursors is a likely reaction yielding these odd-numbered components.

To clarify how odd-numbered pheromone components are biosynthesized in *D. tancrei*, we investigated the composition of pheromone precursors in the female sex pheromone gland. A C15 saturated fatty acid, pentadecanoic acid (15:CoA), was detected, but the expected precursors for 9,11-15:Ald were not detected. Instead, precursors for 10,12-16:Ald, such as 11-16:CoA and 10,12-16:Ald, were detected (Uehara 2015). These results suggest that odd-numbered components of *D. tancrei* are biosynthesized from the same starting substance as that for bombykal through a similar biosynthetic pathway. Our tracer experiment using deuterium-labeled palmitic acid (16:CoA) supports this pathway (Uehara 2015). Accordingly, the key step for biosynthesizing 9,11-15:Ald may be a one-carbon shortening reaction after desaturation to 10,12-16:Ald. In the case of *D. tancrei*, the involvement of α-oxidation in sex pheromone biosynthesis is supported (Fig. 2.1).

Straight-chain shortening processes, including α-oxidation and β-oxidation, are expected to be involved in sex pheromone biosynthesis (Jurenka and Roelofs 1993). However, the molecular mechanisms remain unclear. One of the difficulties in elucidating the chain shortening reaction is that many enzymes, such as peroxidase, reductase, and acyl-transferase, are involved (Ding and Löfstedt 2015). As an ongoing project, we are analyzing the transcriptome of pheromone glands from female hawk moths to obtain the sequences of candidate genes responsible for α-oxidation.

In summary, most hawk moths employ a similar biosynthetic pathway, which includes desaturation and modification of the terminal functional groups, for sex pheromone production. Of note, the pathway is conserved even in species with an unusual pheromone component such as 9,11-15:Ald. Thus, changes in sex pheromone biosynthetic pathways, which may occur as a consequence of enzyme mutations, may be under strong phyletic constraints.

2.4 Sex Pheromone Recognition

To understand the sex pheromone communication system, we must examine pheromone recognition by male moths. *Manduca sexta* is one of the model insects suited for elucidating the mechanisms underlying pheromone recognition, and here we review it from peripheral reception toward central processing.

In general, pheromone molecules are first detected by olfactory receptor neurons (ORNs) in the peripheral sensory organ, the antenna. Olfactory receptors (ORs) and the olfactory receptor co-receptor (Orco) are co-expressed in the ORNs. These two kinds of receptors form a hetero-oligomer and act together as a pheromone receptor. Antennal transcriptome analysis for *M. sexta* revealed that several ORs and a single Orco (MsexOrco) are expressed (Grosse-Wilde et al. 2011). Some ORs, including MsexOR1 and MsexOR4, belong to the pheromone receptor clade (Koenig et al. 2015). Heterologous expression systems using *Xenopus* oocytes or cultured cells, such as HEK293 and CHO, have demonstrated that MsexOR1 co-expressed with MsexOrco responds to bombykal in a dose-dependent manner (Wicher et al. 2017). The response of MsexOR4 co-expressed with MsexOrco to bombykal and E10,E12,Z14-16:Ald, the minor pheromone component of *M. sexta*, was examined, but neither elicited a response. Although bombykal analogues, including geometric isomers, have been identified as sex pheromone components from hawk moth species, no receptors other than those for bombykal have been identified.

The axons of ORNs terminate in the glomeruli in the antennal lobe, which is the primary olfactory center in the brain. Sex pheromone signals are sent to a group of particularly enlarged glomeruli, the macroglomerular complex (MGC), and processed (Hansson et al. 1991, 1992; Kanzaki et al. 2003). The MGC in moths is generally composed of three subcompartments, the cumulus, toroid, and horseshoe (Hansson et al. 1991). Projection neurons, which are output neurons of glomeruli, send information about olfactory stimuli to higher processing centers in the protocerebrum (Kanzaki et al. 1989; Anton and Hansson 1995; Seki et al. 2005).

In *M. sexta*, the cumulus and toroid convey information on E10,E12,Z14-16:Ald and bombykal, respectively. Axons from each MGC project to the mushroom bodies and the inferior lateral protocerebrum (Lei et al. 2013). Additionally, synchrony of spiking in projection neurons from the same glomerulus encodes the behavioral response to an effective blend ratio of sex pheromone components (Martin et al. 2013). Although *A. convolvuli* uses a different sex pheromone component, E11,E13-16:Ald, the pheromone recognition system is very similar. The cumulus and toroid convey information on E11,E13-16:Ald and bombykal, respectively, and axons from each MGC project to the mushroom bodies and the inferior lateral protocerebrum (Nirazawa et al. 2017).

Neuroanatomical studies on other hawk moths are limited, but behavioral responses of *Neogurelca himachala sangaica* to bombykal and its geometric isomer revealed notable aspects of pheromone reception in hawk moths. The species employs a single-component sex pheromone system with bombykal (Uehara et al. 2016), a common pheromone component in hawk moths and other Lepidoptera. Therefore, how does the male moth recognize a conspecific female? In field experiments using a mixture of bombykal and its geometric isomer E10,E12-16:Ald as a lure, trap catches declined as the ratio of E10,E12-16:Ald to bombykal increased. E10,E12-16:Ald is a pheromone component of other sympatric hawk moths such as *D. elpenor lewisii, T. oldenlandiae oldenlandiae*, and *Hemaris affinis*. By electroantennogram analysis, the male antenna of *N. himachala sangaica* was found to respond to bombykal isomers to a similar degree as bombykal (Uehara et al. 2016). These results suggest that sex pheromones of other species are perceived by male *N. himachala sangaica* and act as a behavioral antagonist (Fig. 2.2).

It is possible that the pheromone system of *N. himachala sangaica* evolved from an ancestral pheromone that consisted of multiple components, like other hawk moth pheromones. The molecular phylogenetic tree of hawk moths was proposed

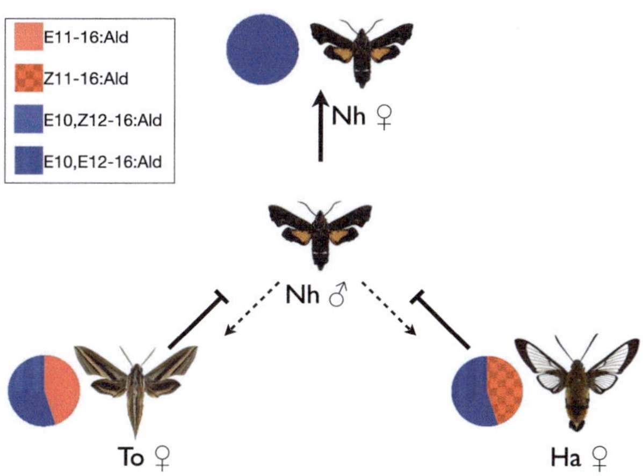

Fig. 2.2 Sex pheromone communication system of *Neogurelca himachala sangaica*. Male *N. himachala sangaica* moths are attracted to conspecific females by tracing the pheromone plume consisting of bombykal (E10,Z12-16:Ald) alone. However, the males are not attracted to sympatric species with a sex pheromone containing bombykal, due to behavioral antagonism of heterospecific components such as E10,E12-16:Ald. Nh: *N. himachala sangaica*, To: *Theretra oldenlandiae oldenlandiae*, Ha: *Hemaris affinis*. (Photos of hawk moths were provided by Dr. Hideshi Naka)

by Kawahara et al. (2009). According to the phylogenetic tree, the multi-component pheromone is likely the common ancestral trait. In Bombycidae, single-component pheromone systems have been reported in several species, and Namiki et al. (2014) described a possible evolutionary scenario based on a comparative neuroanatomy study of MGC in these species. Glomerular organization in the antennal lobe is associated with sex pheromone recognition (Namiki et al. 2014). Species with two pheromone components have two similar-sized glomeruli, whereas species with a single pheromone have glomeruli of the same type, but with one glomerulus extensively enlarged. These findings suggest that recognition of a mixture of two components requires two similarly sized glomeruli. When a moth evolves to use just one of the two components, the volume of either glomerulus may be enlarged to increase the sensitivity to this single component, and the other glomerulus may change its role to an inhibitory one (Namiki et al. 2014).

Although the receptor repertoire and glomerular organization of *N. himachala sangaica* are still unclear, we speculate that this species also evolved a single-component pheromone via a similar scenario. Thus, recognition of a heterospecific sex pheromone component by males may provide an alternative way to secure species specificity.

2.5 Sex Pheromone Communication System

In sex pheromone communication, it is essential that a species-specific channel be generated to avoid heterospecific mating. Although use of a unique component secures high species specificity, this rarely arises because of phyletic constraints on sex pheromone biosynthesis. Therefore, species-specific channels in the communication system of hawk moths are primarily achieved by the diversity of sex pheromones resulting from the specificity of pheromone components and the number of component combinations. The number of combinations can be calculated using a simple formula: $C(n, k) = n! / (k! (n - k)!)$, where C is the number of combinations, n is the total number of potential components, and k is the number of actual components. Theoretically, a species-specific pheromone channel can be generated by increasing the number of pheromone components (k). However, only one- to three-component pheromone systems are observed in hawk moths. Some restrictions on divergent pheromone components are expected. For example, females need to maintain many kinds of enzymes to generate different components, and males need to have the proper number of glomeruli to recognize many components. The number of sex pheromone components is likely constrained by the physiological costs of biosynthesis and recognition of sex pheromones. In this respect, single-component systems economize the physiological cost of biosynthesis and recognition. How divergence of the components and physiological costs are balanced, and how the physiological cost affects the fitness of a species need to be further addressed.

Traits in sex pheromone communicating systems other than the combination and blend ratio of pheromone components may contribute to the premating reproductive

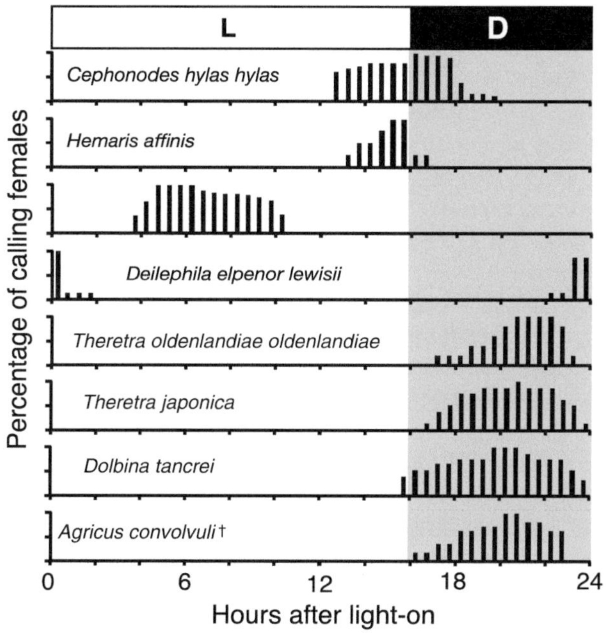

Fig. 2.3 Calling time of hawk moths. Calling behavior of female moths was observed every 1 h under 16 L:8D photoperiodic conditions. (†Data from Shimoda and Kiguchi (1995); other data from Uehara (2015))

isolation of closely related species. For example, the timing of sex pheromone release is an additional factor contributing to the isolation, as observed in *Ostrinia* species (Ishikawa et al. 1999). A characteristic feature of hawk moths is that the family includes both nocturnal and diurnal species, and they exhibit calling behavior at night and during the day, respectively (Fig. 2.3; Sasaki and Riddiford 1984; Itagaki and Conner 1988; Shimoda and Kiguchi 1995; Uehara 2015). That is, they communicate with sex pheromones to potential mates during the period they are active. If the calling time of species is temporally segregated, then similar pheromone components and blend ratios can be used by multiple species. We performed field trap experiments in which a synthetic pheromone lure was arranged at varying times at night and day for a week, but we never observed cross-attraction of a non-target species (Uehara 2015). This observation suggests that the timing of pheromone release is involved in premating reproductive isolation in addition to the components and blend ratio.

The contribution of visual cues to the sex pheromone communication system should be considered. In foraging behavior, day-flying hawk moth species navigate to nectar sources by two sensory modalities, flower scents and visual cues (Balkenius and Kelber 2006). Even nocturnal hawk moths, such as *M. sexta* (Sphinginae), respond to both types of cues (Raguso and Willis 2002, 2005; Goyret et al. 2007), but the relative importance of those cues differs between nocturnal and diurnal hawk moths (Balkenius et al. 2006). Nocturnal species primarily depend on odor, whereas diurnal species depend more on visual stimuli from nectar sources. In the pheromone communication of diurnal moths, such as the clear-wing moth *Synanthedon myopaeformis* (Secilidae) and the Japanese nine-spotted moth *Amata fortunei* (Arctiidae), visual cues aid in mate recognition (KonDo et al. 2012; Judd and Eby 2014). In the case of hawk moths, *H. affinis* and *N. himachala sangaica* were attracted to and captured by a sticky trap baited with a sex pheromone lure, whereas *Cephonodes hylas hylas* approached the trap but did not land on it (Uehara 2015). It is unclear whether this was caused by discordance of sex pheromone components or the absence of visual cues. A similar observation was reported in a field trap test: nectar feeders seldom landed on the sticky pheromone trap (Reed et al. 1987). At present, the visual contribution to the communication system is not fully understood. However, comparative studies between hawk moths with different daily activity (diurnal vs. nocturnal) and those with different foraging strategies (nectar vs. sap feeders vs. those that do not feed in adulthood) are good models.

Acknowledgments Some studies introduced in this chapter were corroborated by Dr. Tetsu Ando (Tokyo University of Agriculture and Technology), Dr. Hideshi Naka (Tottori University), Dr. Shigeru Matsuyama (University of Tsukuba), and Mr. Atsuya Kozaki (University of Tsukuba). This project was partly supported by Grants-in-Aid for JSPS Fellows and for Young Scientists (Start-up), JSPS KAKENHI Grant Number 15H06854, and the Sasakawa Scientific Research Grant from The Japan Science Society.

References

Ando T, Inomata S, Yamamoto M (2004) Lepidopteran sex pheromones. Top Curr Chem 239:51–96

Anton S, Hansson BS (1995) Sex pheromone and plant-associated odour processing in antennal lobe interneurons of male *Spodoptera littoralis* (Lepidoptera: Noctuidae). J Comp Physiol A 176:773–789

Balkenius A, Kelber A (2006) Colour preferences influences odour learning in the hawkmoth, *Macroglossum stellatarum*. Naturwissenschaften 93:255–258

Balkenius A, Rosen W, Kelber A (2006) The relative importance of olfaction and vision in a diurnal and a nocturnal hawkmoth. J Comp Physiol A 192:431–437

Bestmann HJ, Erler J, Garbe W et al (1992) Pheromone components of the female elephant hawkmoth, *Deilephila elpenor*, and the silver- striped hawk-moth, *Hippotion celerio*. Experientia 48:610–613

Buček A, Matoušková P, Vogel H et al (2015) Evolution of moth sex pheromone composition by a single amino acid substitution in a fatty acid desaturase. Proc Natl Acad Sci 112:12586–12591

Cork A, Chamberlain DJ, Beevor PS et al (1988) Components of female sex pheromone of spotted bollworm, *Earias vittella* F. (Lepidoptera: Noctuidae): identification and field evaluation in Pakistan. J Chem Ecol 14:929–945

Daimon T, Fujii T, Fujii T et al (2012) Reinvestigation of the sex pheromone of the wild silkmoth *Bombyx mandarina*: the effects of bombykal and bombykyl acetate. J Chem Ecol 38:1031–1035

Ding BJ, Löfstedt C (2015) Analysis of the *Agrotis segetum* pheromone gland transcriptome in the light of sex pheromone biosynthesis. BMC Genomics 16:711

Fang N, Tumlinson JH, Teal PE et al (1992) Fatty acyl pheromone precursors in the sex pheromone gland of female tobacco hornworm moths, *Manduca sexta* (L.). Insect Biochem Mol Biol 22:621–631

Fang N, Teal PE, Tumlinson JH (1995a) Characterization of oxidase (s) associated with the sex pheromone gland in *Manduca sexta* (L.) females. Arch Insect Biochem Physiol 29:243–257
Fang N, Teal PE, Doolittle RE et al (1995b) Biosynthesis of conjugated olefinic systems in the sex pheromone gland of female tobacco hornworm moths, *Manduca sexta* (L.). Insect Biochem Mol Biol 25:39–48
Fujii T, Ito K, Tatematsu M et al (2011) Sex pheromone desaturase functioning in a primitive *Ostrinia* moth is cryptically conserved in congeners' genomes. Proc Natl Acad Sci 108:7102–7106
Goyret J, Markwell PM, Raguso RA (2007) The effect of decoupling olfactory and visual stimuli on the foraging behavior of *Manduca sexta*. J Exp Biol 210(8):1398–1405
Grosse-Wilde E, Kuebler LS, Bucks S et al (2011) Antennal transcriptome of *Manduca sexta*. Proc Natl Acad Sci 108:7449–7454
Hall DR, Farman D, Domínguez JC, Pajares JA (2017) Female sex pheromone of the cone moth, *Dioryctria mendacella*: investigation of synergism between Type I and Type II pheromone components. J Chem Ecol 43:433–442
Hansson BS, Christensen TA, Hildebrand JG (1991) Functionally distinct subdivisions of the macroglomerular complex in the antennal lobe of the male sphinx moth *Manduca sexta*. J Comp Neurol 312:264–278
Hansson BS, Ljungberg H, Hallberg E et al (1992) Functional specialization of olfactory glomeruli in a moth. Science 256:1313–1315
Honda H, Himeno K, Yoshiyasu Y (1994) Chemotaxonomy of the cotton leaf-roller (Lepidoptera: Pyralidae) in Japan with special reference to differences in sex pheromones. Appl Entomol Zool 29:323–330
Honda H, Yamasaki R, Sumiuchi Y et al (2015) Hybrid sex pheromones of the hibiscus flower-bud borer, *Rehimena surusalis*. J Chem Ecol 41:1043–1049
Ishikawa Y, Takanashi T, Kim CG et al (1999) *Ostrinia* spp. in Japan: their host plants and sex pheromones. Entomol Exp Appl 91:237–244
Itagaki H, Conner WE (1988) Calling behavior of *Manduca sexta* (L.) (Lepidoptera: Sphingidae) with notes on the morphology of the female sex pheromone gland. Ann Entomol Soc Am 81:798–807
Johnson SD, Moré M, Amorim FW et al (2017) The long and the short of it: a global analysis of hawkmoth pollination niches and interaction networks. Funct Ecol 31:101–115
Judd GJ, Eby C (2014) Spectral discrimination by *Synanthedon myopaeformis* (Lepidoptera: Sesiidae) when orienting to traps baited with sex pheromone or feeding attractants. Can Entomol 146:8–25
Jurenka RA, Roelofs WL (1993) Biosynthesis and endocrine regulation of fatty acid derived sex pheromones in moths. In: Stanley DW, Nelson DR (eds) Insect lipids: chemistry, biochemistry and biology. University of Nebraska Press, Lincoln, pp 353–388
Kanzaki R, Arbas EA, Strausfeld NJ, Hildebrand JG (1989) Physiology and morphology of projection neurons in the antennal lobe of the male moth *Manduca sexta*. J Comp Physiol A 165:427–453
Kanzaki R, Soo K, Seki Y, Wada S (2003) Projections to higher olfactory centers from subdivisions of the antennal lobe macroglomerular complex of the male silkmoth. Chem Senses 28:113–130
Kawahara AY, Mignault AA, Regier JC et al (2009) Phylogeny and biogeography of hawkmoths (Lepidoptera: Sphingidae): evidence from five nuclear genes. PLoS One 4:e5719
Klun JA, Leonhardt BA, Schwarz M et al (1986) Female sex pheromone of the pickleworm, *Diaphania nitidalis* (Lepidoptera: Pyralidae). J Chem Ecol 12:239–249
Koenig C, Hirsh A, Bucks S et al (2015) A reference gene set for chemosensory receptor genes of *Manduca sexta*. Insect Biochem Mol Biol 66:51–63
KonDo Y, Naka H, Tsuchida K (2012) Pheromones and body coloration affect mate recognition in the Japanese nine-spotted moth *Amata fortunei* (Lepidoptera: Arctiidae). J Ethol 30:301–308
Landolt PJ, Tumlinson JH, Brennan MM (1989) Attraction of *Amphion floridensis* (Lepidoptera: Sphingidae) to bombykal, (E, Z)-10,12-hexadecadienal. Fla Entomol 72:324–327

Lei H, Chiu HY, Hildebrand JG (2013) Responses of protocerebral neurons in *Manduca sexta* to sex-pheromone mixtures. J Comp Physiol A 199:997–1014
Löfstedt C, Wahlberg N, Millar JM (2016) Evolutionary patterns of pheromone diversity in Lepidoptera. In: Allison JD, Cards RT (eds) Pheromone communication in moths: evolution, behavior and application. University of California Press, Berkeley, pp 43–78
Luxová A, Svatoš A (2006) Substrate specificity of membrane-bound alcohol oxidase from the tobacco hornworm moth (*Manduca sexta*) female pheromone glands. J Mol Catal B Enzym 38:37–42
Malcicka M, Visser B, Ellers J (2018) An evolutionary perspective on linoleic acid synthesis in animals. Evol Biol 45:15–26
Martin JP, Lei H, Riffell JA et al (2013) Synchronous firing of antennal-lobe projection neurons encodes the behaviorally effective ratio of sex-pheromone components in male *Manduca sexta*. J Comp Physiol A 199:963–979
Matoušková P, Pichová I, Svatoš A (2007) Functional characterization of a desaturase from the tobacco hornworm moth (*Manduca sexta*) with bifunctional Z11- and 10,12-desaturase activity. Insect Biochem Mol Biol 37:601–610
McElfresh JS, Millar JG (1999a) Geographic variation in sex pheromone blend of *Hemileuca electra* from southern California. J Chem Ecol 25:2505–2525
McElfresh JS, Millar JG (1999b) Sex pheromone of the common sheep moth, *Hemileuca eglanterina*, from the San Gabriel Mountains on California. J Chem Ecol 25:687–709
McElfresh JS, Hammond AM, Millar JG (2001) Sex pheromone components of the buck moth *Hemileuca maia*. J Chem Ecol 27:1409–1422
Millar JG (2000) Polyene hydrocarbons and epoxides: a second major class of lepidopteran sex attractant pheromones. Annu Rev Entomol 45:575–604
Millar JG, Grant GG, McElfresh JS et al (2005) (3*Z*,6*Z*,9*Z*,12*Z*,15*Z*)-pentacosapentaene, a key pheromone component of the fir coneworm moth, *Dioryctria abietivorella*. J Chem Ecol 31:1229–1234
Moto KI, Yoshiga T, Yamamoto M et al (2003) Pheromone gland-specific fatty-acyl reductase of the silkmoth, *Bombyx mori*. Proc Natl Acad Sci 100:9156–9161
Moto KI, Suzuki MG, Hull JJ et al (2004) Involvement of a bifunctional fatty-acyl desaturase in the biosynthesis of the silkmoth, *Bombyx mori*, sex pheromone. Proc Natl Acad Sci 101:8631–8636
Namiki S, Daimon T, Iwatsuki C et al (2014) Antennal lobe organization and pheromone usage in bombycid moths. Biol Lett 10:20140096
Nieukerken EJ, Kaila L, Kitching I et al (2011) Order Lepidoptera Linnaeus, 1758. In: Zhang Z-Q (ed) Animal biodiversity: an outline of higher-level classification and survey of taxonomic richness. Zootaxa, vol 3148, pp 212–221
Nirazawa T, Fujii T, Seki Y et al (2017) Morphology and physiology of antennal lobe projection neurons in the hawkmoth *Agrius convolvuli*. J Insect Physiol 98:214–222
Pittaway AR (1993) The hawkmoths of the western Palaearctic. Harley Books, Colchester, pp 156–157
Raguso RA, Willis MA (2002) Synergy between visual and olfactory cues in nectar feeding by naive hawkmoths, *Manduca sexta*. Anim Behav 64:685–695
Raguso RA, Willis MA (2005) Synergy between visual and olfactory cues in nectar feeding by wild hawkmoths, *Manduca sexta*. Anim Behav 69:407–418
Raina AK, Klun JA, Schwarz M et al (1986) Female sex pheromone of the melonworm, *Diaphania hyalinata* (Lepidoptera: Pyralidae), and analysis of male responses to pheromone in a flight tunnel. J Chem Ecol 12:229–237
Reed DW, Underhill EW, Giblin EM (1987) Attraction of sphingid moths (Lepidoptera: Sphingidae) to 10,12-hexadecadienyl aldehydes and acetates: evidence of pheromone components. J Chem Ecol 13:931–942
Rooney AP (2011) Pheromone emergencies and drifting moth genomes. Proc Natl Acad Sci 108(20):8069–8070
Sasaki M, Riddiford LM (1984) Regulation of reproductive behaviour and egg maturation in the tobacco hawk moth, *Manduca sexta*. Physiol Entomol 9:315–327

Seki Y, Aonuma H, Kanzaki R (2005) Pheromone processing center in the protocerebrum of *Bombyx mori* revealed by nitric oxide-induced anti-cGMP immunocytochemistry. J Comp Neurol 481:340–351

Shimoda M, Kiguchi K (1995) The sweet potato hornworm, *Agrius convolvuli*, as a new experimental insect: behavior of adult moths in a rearing cage. Jpn J Appl Entomol Zool 39:321–328

Starratt AN, Dahm KH, Allen N et al (1979) Bombykal, a sex pheromone of the sphinx moth *Manduca sexta*. Zeitschrift für Naturforschung C 34:9–12

Svatoš A, Kalinová B, Boland W (1999) Stereochemistry of lepidopteran sex pheromone biosynthesis: a comparison of fatty acid-CoA Δ11-(Z)-desaturases in *Bombyx mori* and *Manduca sexta* female moths. Insect Biochem Mol Biol 29:225–232

Tamaki Y (1977) Complexity, diversity, and specificity of behavior: modifying chemicals in Lepidoptera and Diptera. In: Shorey HH, Mckelvey JJ Jr (eds) Chemical control of insect behavior. Wiley, New York, pp 253–285

Truman JW, Riddiford LM (2002) Endocrine insights into the evolution of metamorphosis in insects. Annu Rev Entomol 47:467–500

Tumlinson JH, Brennan MM, Doolittle RE et al (1989) Identification of a pheromone blend attractive to *Manduca sexta* (L.) males in a wind tunnel. Arch Insect Biochem Physiol 10:255–271

Tumlinson JH, Mitchell ER, Doolittle RE et al (1994) Field tests of synthetic *Manduca sexta* sex pheromone. J Chem Ecol 20:579–591

Uehara T (2015) Biology and chemistry of pheromonal communication in Japanese hawk moth. PhD dissertation, University of Tsukuba

Uehara T, Naka H, Matsuyama S et al (2012) Identification and field evaluation of sex pheromones in two hawk moths *Deilephila elpenor lewisii* and *Theretra oldenlandiae* (Lepidoptera; Sphingidae). Appl Entomol Zool 47:227–232

Uehara T, Naka H, Matsuyama S et al (2013) Identification of conjugated pentadecadienals as sex pheromone components of the sphingid moth, *Dolbina tancrei*. J Chem Ecol 39:1441–1447

Uehara T, Naka H, Matsuyama S et al (2015) Identification of the sex pheromone of the diurnal hawk moth, *Hemaris affinis*. J Chem Ecol 41:9–14

Uehara T, Kitahara H, Naka H et al (2016) Single-component pheromone consisting of bombykal in a diurnal hawk moth, *Neogurelca himachala sangaica*. J Chem Ecol 42:517–522

Wakamura S, Yasuda T, Watanabe M et al (1996) Sex pheromone of the sweetpotato hornworm, *Agrius convolvuli* (L.) (Lepidoptera: Sphingidae): identification of a major component and its activity in a wind tunnel. Appl Entomol Zool 31:171–174

Wang H, Ando N, Kanzaki R (2008) Active control of free flight manoeuvres in a hawkmoth, *Agrius convolvuli*. J Exp Biol 211:423–432

Wicher D, Morinaga S, Halty-de Leon L et al (2017) Identification and characterization of the bombykal receptor in the hawkmoth *Manduca sexta*. J Exp Biol 220:1781–1786

Xue B, Rooney AP, Kajikawa M et al (2007) Novel sex pheromone desaturases in the genomes of corn borers generated through gene duplication and retroposon fusion. Proc Natl Acad Sci 104:4467–4472

Yan Q, Li HD, Chen Y et al (2018a) Identification and field evaluation of the sex pheromone of *Orthaga achatina* (Lepidoptera: Pyralidae). J Chem Ecol 44:886–893

Yan Q, Nghia NTN, Khanh CNQ et al (2018b) Unsaturated cuticular hydrocarbon components of the sex pheromone of eggplant fruit borer, *Leucinodes orbonalis* Guenée (Lepidoptera: Crambidae). J Chem Ecol 44:631–636

Chapter 3
Sex Pheromones of Mealybugs: Implications for Evolution and Application

Jun Tabata

Abstract Mealybugs, named after the abundant powdery wax covering their body surface, are soft-bodied scale insects that feed on plant sap, and most closely related to aphids and whiteflies. Mealybugs, as well as other scale insects, are characterized by their unusual shapes, which may be difficult to be recognized as insects. They exhibit marked sexual dimorphism; females are neotenic and almost immotile with no wings and retrogressed legs, whereas males are winged, but extremely tiny and short-lived. Volatile pheromones released by females are therefore essential for males to find conspecific mates efficiently. Consequently, mealybug pheromones have structurally radiated with a high species specificity, and thus offer a unique model to illustrate the diversity of odor-based sexual communication systems. Moreover, these aphrodisiac fragrances may be promising for pest management programs against notorious mealybugs that damage crops and fruits. This chapter compiles several studies on mealybugs and their pheromones, and discusses possible evolutionary scenarios as well as potential applications of pheromones.

Keywords Mealybug · Scale insect · Sexual dimorphism · Terpene · Pest management

3.1 Introduction

Mealybugs (Hemiptera: Pseudococcidae), named after the abundant powdery wax covering their body surface, are soft-bodied scale insects that feed on plant sap, and are most closely related to aphids and whiteflies. According to the systematic catalog compiled by Ben-Dov (1994), more than 1900 species in approximately 290 genera are listed in the group of mealybugs. Mealybugs are widely distributed across all

---

J. Tabata (✉)
National Agriculture and Food Research Organization, Tsukuba, Japan
e-mail: jtabata@affrc.go.jp

Y. Ishikawa (ed.), *Insect Sex Pheromone Research and Beyond*, Entomology Monographs, https://doi.org/10.1007/978-981-15-3082-1_3

zoogeographical regions, although they are abundant in the tropics and subtropics, and are considered to originate from these regions (Kawai 1980). Some species are notorious agricultural pests that damage major products, including rice, sugarcane, cassava, potato, and many fruits. These pest mealybugs destroy plants by directly depleting sap as well as by transmitting plant viruses. Furthermore, they excrete honeydew, on which microbes that severely reduce quality of crops and fruits propagate.

Studies on mealybug pheromones started in the early 1980s in an attempt to use these chemicals for plant protection. Although elucidation of the pheromone structures progressed slowly, studies on mealybug pheromones proceeded rapidly in the last two decades, probably due to the increased impact of mealybugs as agricultural pests. Pheromones released by females were found to be powerful and promising for regulating mate-finding behavior of males, and these message chemicals are commonly terpenoids with unique features and structural diversity. Thus, mealybug pheromones are not only a useful tool for pest management programs, but are also an intriguing model to study the diversification of chemical communication channels in insects.

Mealybugs have some characteristic features that may influence the evolution of their pheromone communication systems. One is their marked sexual dimorphism (Fig. 3.1). Females are neotenic and almost immotile with no wings and retrogressed legs. They produce protective powdery wax secretions, and adult females can live for several months until copulation and continue feeding and growing. In contrast, adult males are winged and motile, but they are tiny and fragile, weighing 0.5–1% of females, and have a short lifespan of at most a few days, during which they do not feed (Franco et al. 2009; Ross and Shuker 2009). Sex pheromones emitted by sedentary females serve as a key navigation tool for dwarf males, and thus are essential and considered to be under strong selection pressure to facilitate mating and reproduction.

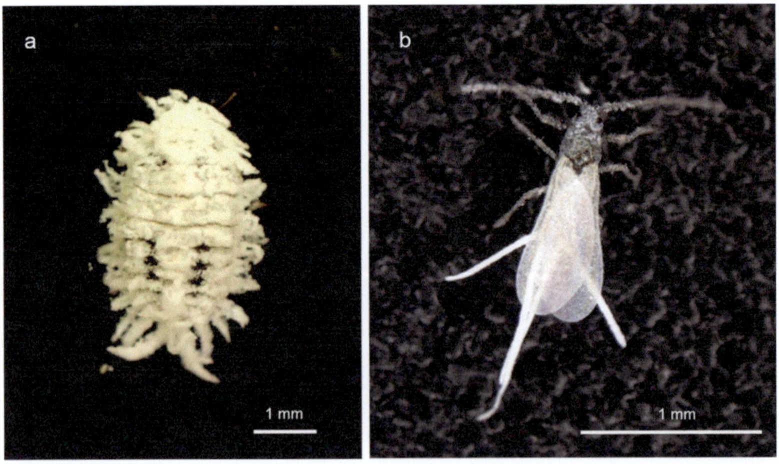

Fig. 3.1 Female (**a**) and male (**b**) adults of the cotton mealybug, *Phenacoccus solenopsis*, with different appearances and biology

Another unique feature of mealybugs is their bizarre genetic system, which is known as paternal genome elimination (PGE) (Nur 1980). In PGE systems, both sexes develop from fertilized eggs, but the paternal half of the genome in the male offspring is deactivated through heterochromatinization during early development. The deactivated set of the genome is possessed by all somatic cells of males but, due to breakdown during meiosis, it is not included in mature sperm, and thus not passed to the male offspring (Schraeder 1921). According to molecular phylogenetic studies, PGE evolved only once in scale insects (Cook et al. 2002), and the evolution of characteristic genetic systems of scale insects, including PGE systems in mealybugs, may be associated with their unique biology, which promotes conflicts over transmission between maternally and paternally inherited genes (intra-genomic conflict) (Ross et al. 2010b). Theoretically, fathers with PGE systems favor a female-biased sex ratio more than mothers because fathers share genes only with their daughters, whereas mothers share genes with all their offspring (Ross et al. 2010b). Given the high female-biased sex ratio in populations, females may have been forced to intensify their pheromone signals to attract males (limited mate resources), leading to promotion of the evolution of pheromone communications.

Mealybugs, as well as many other hemipteran insects, generally possess obligate symbiotic bacteria and genomic conflicts, which potentially influence the evolution of pheromones, may arise from interactions between host mealybugs and their endosymbionts. Mealybugs and other scale insects rely heavily on their symbiotic bacteria for the synthesis of essential amino acids and vitamins that are absent from their diet of plant sap (Ross et al. 2012). Two different proteobacteria, which belong to *Betaproteobacteria* (primary endosymbionts) and *Gammaproteobacteria* (secondary endosymbionts), are contained in specialized organs (bacteriomes) within the cytoplasm of the host mealybug cells, referred to as bacteriocytes, and secondary endosymbionts live inside the primary endosymbionts (von Dohlen et al. 2001). These endosymbionts, typical of endosymbionts in general, are vertically transmitted through the host maternal line via the cytoplasm of the eggs (Ross et al. 2012). This may cause conflict between the host and bacteria over the host sex ratio because maternally inherited bacteria favor a higher female-biased sex ratio than the host (Ross et al. 2010b). Such a conflict may mean that the interests of the endosymbionts are often aligned with the interests of genes in males under PGE genetic systems, which, as described above, also favor female-biased offspring sex ratios (Ross et al. 2010b). Thus, inter-genomic conflicts between mealybugs and endosymbionts also have the potential to enhance competitions among females in attracting males via pheromone signals.

As genomic conflicts among different genetic entities can cause a selection bias in the population sex ratio of mealybugs, their sex determination system and sex allocation pattern may also be influenced. Although the sex determination system of mealybugs is poorly understood, no sex chromosomes have ever been observed in mealybugs and autosomal genetic determination loci are also unlikely to be present (Brown and Nur 1964). Therefore, it has been assumed that sex is determined either by facultative imprinting on the gametes from the parents or by maternal effect proteins that are added to the eggs (Ross et al. 2010b). In either case, offspring sex determination may be, at least partially, influenced by parental conditions. Indeed, sex ratios of offspring have been demonstrated to be distorted by their maternal conditions in the citrus mealybug, *Planococcus citri*; population density, temperature, and the maternal age at the time of mating and oviposition affect sex allocation patterns, although non-biased or only slightly female-biased sex ratios were observed in most studies (James 1937; Nelson-Rees 1960; Varndell and Godfray 1996; Ross et al. 2010a, 2011). It is important to accumulate data on sex ratios, particularly in wild populations, for further discussion on the genomic conflicts and sex allocation patterns, and their influence on mealybug pheromone communication systems.

3.2 Diversity of Chemical Structures in Mealybug Pheromones

3.2.1 Common Features in Mealybug Pheromone Chemistry

Twenty-one pheromone compounds have been characterized to date from 18 mealybug species. The structures of all these pheromones are strictly species specific. This is in contrast to pheromone compounds of moths, which mostly use fatty alcohols or their derivatives with an unbranched carbon chain of an even number. These pheromone components are often shared with other species, but their combinations and blend ratios confer species specificity to the signal, avoiding encounters with the wrong mates (Roelofs and Cardé 1974; Cardé and Baker 1984; Löfstedt 1993). Unlike moths, mealybugs are considered to have evolved a unique biosynthesis and perception system in each species to utilize species-specific chemicals as pheromones.

Although there is structural diversity, several common features are found in mealybug pheromones. For example, all mealybug pheromones characterized to date are terpenes. More specifically, most of them are monoterpenes that are composed of two isoprene units coupled by irregular connections, rather than a general "head-to-tail" (1′–4) connection. In addition, mealybug pheromones are mostly esters of these terpene alcohols and carboxylic acids. The acids include acetic acid (C2), propionic acid (C3), butyric acid (C4), isobutyric acid (C4), and isoprene-analogous acids (C5), among which acetic acid and C5 acids are prevalent, and found in 7 and 9 of 21 pheromone compounds, respectively.

It is also a notable feature in mealybugs that pheromone activity is generally exhibited by a single component. Binary-component pheromones are reported in three species, the pink hibiscus mealybug *Maconellicoccus hirsutus* (Zhang et al. 2004), *Dysmicoccus grassii* (De Alfonso et al. 2012), and the Madeira mealybug *Phenacoccus madeirensis* (Ho et al. 2009); however, a combination of two components (*R*)-(2,2-dimethyl-3-isopropylidenecyclobutyl)methyl (*S*)-2-methylbutyrate and (*R*)-lavandulyl (*S*)-2-methylbutyrate (Zhang et al. 2004) is only required by *M. hirsutus* for male attraction. Females of *D. grassii* release a 6:1 mixture of (*R*)-lavandulyl propionate and

(*R*)-lavandulyl acetate, either of which singly attracts conspecific males (De Alfonso et al. 2012). Similarly, a 3:1 mixture of (1*R*,3*R*)-chrysanthemyl (*R*)-2-methylbutanoate and (*R*)-lavandulyl (*R*)-2-methylbutanoate was found in the sex pheromone of *Ph. madeirensis*, but the latter was only weakly attractive and had no synergistic effects (Ho et al. 2009).

One of the exceptions to the common features mentioned above is the hemiterpene that consists of one isoprene unit in the sex pheromone of the Matsumoto mealybug, *Crisicoccus matsumotoi*; isoprenyl 5-methylhexanoate (Tabata et al. 2012). This compound is an ester of a hemiterpene alcohol (isoprenol; 3-methyl-3-buten-1-ol) and a methyl-branched C7-carboxylic acid (5-methylhexanoic acid). It is currently unknown whether hemiterpene pheromones are used in other mealybugs or are specific to *C. matsumotoi*. Studies on pheromones of other members of the genus *Crisicoccus* as well as their phylogeny will provide essential information to illustrate the entire picture of terpenoid pheromones in mealybugs.

Another exception is a monoterpene aldehyde, rather than an ester, in the pheromone of the pineapple mealybug, *Dysmicoccus brevipes*. This compound, (1*S*,2*S*)-(1,2-dimethyl-3-methylenecyclopentyl)acetaldehyde (Mori 2016; Mori and Tabata 2017; Tabata et al. 2017a), is essential for mate-finding and copulation in sexual reproduction of *D. brevipes* (Tabata et al. 2017a). Aldehydes are frequently released from hemipteran insects as components of attractant pheromones, alarm pheromones, or defensive chemicals. However, *D. brevipes* is currently the only species that is known to use an aldehyde as the pheromone among mealybugs and related scale insects.

3.2.2 Lavandulol-Related Structures

Lavandulol, 2-isopropenyl-5-methyl-4-hexenol, is a characteristic component of lavender fragrance and is composed of two isoprene units with the tail of one isoprene connected in an irregular manner to the second carbon of the next unit (1′–2 linkage) (e.g., Tabata et al. 2015). All acyclic mealybug pheromones include a lavandulol-related structure in their monoterpene alcohol moieties; five of the alcohols are lavandulol and three others have the same skeleton as lavandulol.

(*S*)-Lavandulol and (*S*)-lavandulyl senecioate are found in airborne volatiles released by virgin females of the vine mealybug, *Planococcus ficus*. However, the former has neither attractive nor synergistic effects, and thus only the latter is concluded to be the pheromone compound in *Pl. ficus* (Hinkens et al. 2001). Moreover, Millar et al. (2002) confirmed that synthetic lavandulol antagonized the attraction of *Pl. ficus* males at higher doses. Synthetic lavandulyl senecioate in a racemic mixture exhibits activity comparable to a pure (*S*)-enantiomer, indicating that the (*R*)-enantiomer does not influence the attraction of males. Of note, another ester of lavandulol, (*S*)-lavandulyl isovalerate, is found in a mass-reared culture of *Pl. ficus* originated from an Israeli population (Zada et al. 2003). Moreover, several pheromone-response types were found in *Pl. ficus* males; (*S*)-lavandulyl isovalerate is attractive to some males but repellant to others, and these response patterns are at least partially inherited in Israeli populations (Kol-Maimon et al. 2010, 2014b).

As described above, lavandulyl esters were also discovered as minor components of the pheromones of *M. hirsutus* and *Ph. madeirensis*. (*R*)-Lavandulyl (*S*)-2-methylbutyrate is essential for the pheromonal activity for males of *M. hirsutus*, whereas the diastereomer, (*R*)-lavandulyl (*R*)-2-methylbutyrate, had no obvious attractiveness to males of *Ph. madeirensis*. However, (*R*)-lavandulyl (*R*)-2-methylbutyrate of *Ph. madeirensis* is reported to antagonize the activity of (*R*)-lavandulyl (*S*)-2-methylbutyrate in attracting *M. hirsutus* males (Zhang et al. 2006).

Double bond-positional isomers of lavandulol have recently been found in alcohol moieties of three mealybug pheromones. The passionvine mealybug, *Planococcus minor*, uses *trans*-2-isopropyl-5-methyl-2,4-hexadienyl acetate (Ho et al. 2007), which is called "β-lavandulyl acetate" in this article, as the pheromone. Its geometric isomer, *cis*-β-lavandulyl acetate, was reported to antagonize attraction of *Pl. minor* (Ho et al. 2007). The pheromone of the congener, *Pl. kraunhiae* (the Japanese mealybug), was found to be 2-isopropyliden-5-methyl-4-hexenyl butyrate (γ-lavandulyl butyrate) (Sugie et al. 2008). A species of a different genus, *Dysmicoccus neobrevipes* (the grey pineapple mealybug), also uses a lavandulyl analog, (*S*)-*trans*-2-isopropyl-5-methyl-3,5-hexadienyl acetate (δ-lavandulyl acetate). In *D. neobrevipes*, however, unnatural stereoisomers of δ-lavandulyl acetate were demonstrated to have no antagonistic effects on male-attraction by the natural isomer (Tabata and Ichiki 2015; Tabata and Ohno 2015).

The pheromone of the Comstock mealybug, *Pseudococcus comstocki*, is 3-ace toxy-2,6-dimethyl-1,5-heptadiene, the alcohol moiety of which is an unusual analog of lavandulol in that it lacks the primary carbon atom of lavandulol (Bierl-Leonhardt et al. 1980; Negishi et al. 1980). The absolute configuration of this secondary alcohol was confirmed to be (*R*) by its chiral syntheses with Sharpless asymmetric epoxidation of a tri-substituted olefin alcohol, followed by rearrangement and opening of the enantiomeric epoxides (Mori and Ueda 1981). The racemic synthetic pheromone was demonstrated to be as attractive to males as the pure (*R*)-enantiomer (Bierl-Leonhardt et al. 1982). This C9 alcohol is currently the only known example of a secondary alcohol included in mealybug pheromones. Further studies are warranted to clarify whether such structures are exceptional in mealybug pheromones.

3.2.3 Cyclopropane Structures

Cyclic monoterpenes have been found in 11 mealybug pheromones, among which three compounds possess a cyclopropane ring. These three are esters of 2,2-dimethyl-3-(2-methylprop-1-enyl)cyclopropyl)methanol (chrysanthemyl alcohol); namely, (1*R*,3*R*)-chrysanthemyl (*R*)-2-methylbutanoate in the pheromone of *Ph. madeirensis* (Ho et al. 2009), (1*R*,3*R*)-chrysanthemyl (*R*)-2-acetoxy-3-methylbutanoate in the citrophilous mealybug *Pseudococcus calceolariae* (El-Sayed et al. 2010; Unelius et al. 2011), and (1*S*,3*R*)-chrysanthemyl tiglate in the striped mealybug *Ferrisia virgate* (Tabata and Ichiki 2017). Of note, connections of isoprene units that consist of the chrysanthemyl structure include an irregular, lavandulol-type 1′–2 linkage.

The acid form of chrysanthemyl alcohol, chrysanthemic acid (or chrysanthemum acid) is well known as a key structure in naturally occurring pyrethroid insecticides, which were first discovered from pyrethrum plants (Khambay and Jewess 2010). Coincidentally, insecticidal natural products and mealybug pheromones share a common motif. This structure possesses two asymmetric carbons and has four possible stereoisomers. Among them, the (1*R*,3*R*)-*trans* configuration is the most prevalent in natural pyrethroids and generally exhibits higher insecticidal activity than the *cis* counterpart. In mealybug pheromones, those of *Ph. madeirensis* and *Ps. calceolariae* possess the (1*R*,3*R*)-*trans* configuration, whereas that of *F. virgate* possesses the (1*S*,3*R*)-*cis* configuration. In these three mealybugs, other unnatural stereoisomers do not exhibit obvious inhibitory effects on pheromone activity (Ho et al. 2011; Unelius et al. 2011; Tabata and Ichiki 2017).

3.2.4 Cyclobutane Structures

Two monoterpene structures with a cyclobutane ring are found in the alcohol part of mealybug pheromones: (2,2-dimethyl-3-isopropenylcyclobutyl)methanol and (2,2-dimethyl-3-isopropylidenecyclobutyl)methanol. These two alcohols have the same carbon skeleton with a difference only in the double bond position. The former was first discovered from the pheromone of *Pl. citri*, in a form of (1*R*,3*R*)-acetate (Bierl-Leonhardt et al. 1981), and is called planococcyl alcohol or planococcol in some reports. Another ester of planococcyl alcohol, (1*R*,3*R*)-planococcyl 3-methyl-3-butenoate, was subsequently reported as the pheromone of the citriculus mealybug, *Pseudococcus cryptus* (Arai et al. 2003). The other cyclobutanoid monoterpene was first found in the alcohol moiety of the pheromone of *M. hirsutus*, and was named maconelliol (Zhang et al. 2004). As previously mentioned, the pheromone of *M. hirsutus* is a 5:1 mixture of (*R*)-maconelliyl (*S*)-2-methylbutanoate and (*R*)-lavandulyl (*S*)-2-methylbutanoate. More recently, maconelliyl senecioate was identified from the pheromone of the cotton mealybug, *Phenacoccus solenopsis* (Tabata and Ichiki 2016).

Similar to cyclopropanoid (chrysanthemyl) pheromones, both planococcyl and maconelliyl structures include the same non-head-to-tail (1′–2) linkage of isoprene units like lavandulol. Moreover, the second carbon of the alcohol moiety of planococcyl and maconelliyl pheromones is commonly in the (*R*)-configuration. Thus, the four mealybugs that use these cyclobutanoid pheromones are expected to share a mechanism for biosynthesis of unusual cyclobutane structures with a unique connection of isoprene units (Tabata and Ichiki 2016).

3.2.5 Cyclopentane Structures

Four compounds that possess a cyclopentyl or cyclopentenyl ring are known in pheromones of three *Pseudococcus* and one *Dysmicoccus* mealybugs. In contrast to the limited variations in skeletons of cyclopropyl (chrysanthemyl esters) and cyclobutyl (planococcyl or maconelliyl esters) pheromones, the core structures of these cyclopentyl or cyclopentenyl pheromones are highly variable and are likely to be built with unique connections of isoprene units.

The pheromone of the grape mealybug, *Pseudococcus maritimus*, was clarified to be *trans*-(3,4,5,5-tetramethyl-2-cyclopentenyl)methyl isobutyrate (Figadère et al. 2007). The alcohol moiety is known as α-necrodol, which was first reported as one of the isomers isolated from the defensive secretions of a carrion beetle (*Necrodes surinamensis*) (Roach et al. 1990). A comparative analysis using a chiral stationary phase column-installed gas chromatograph and enantioselective synthetics via lipase-catalyzed kinetic resolution of an intermediate demonstrated that the natural pheromone is an 85:15 mixture of the (1*R*,4*R*)- and (1*S*,4*S*)-enantiomers (Zou et al. 2010). Two possible routes for generation of the necrodol skeleton have been proposed: (i) cyclization of geranyl diphosphate arising from a typical head-to-tail (1′–4) linkage to a bornane intermediate, followed by rearrangement to an isocamphane skeleton and cleavage of one ring of the bicyclic structure, and (ii) assembly of two isoprene units by 2 → 3′ and 4 → 2′ connections (Figadère et al. 2007).

The longtailed mealybug, *Pseudococcus longispinus*, also produces an unusual cyclopentenoid pheromone, 2-(1,5,5-trimethytl-2-cyclopentenyl)ethyl acetate (Millar et al. 2009). The absolute configuration was suggested to be (*S*) through enantioselective syntheses and field trap bioassays for male attraction (Ramesh et al. 2013, 2015). The structure of this compound suggests that it is assembled by a standard head-to-tail connection to form a geranyl-type intermediate, followed by a 3′–3 connection to form the 1,1,2,2-tetraalkycyclopentane core, followed by functional group introduction and/or modification (Millar et al. 2009).

The structure of the pheromone of the obscure mealybug, *Pseudococcus viburni*, was determined to be (2,3,4,4-tetramethylcyclopentyl)methyl acetate by Millar et al. (2005b), and a subsequent vibrational circular dichroism analysis (Figadère et al. 2008) and enantioselective synthesis (Hashimoto et al. 2008) revealed its absolute configuration to be (1*S*,2*S*,3*R*). More recently, another cyclopentyl structure, (1,2-dimethyl-3-methylenecyclopentyl)acetaldehyde, was discovered from *D. brevipes* (Tabata et al. 2017a), and its absolute configuration was demonstrated to be (1*S*,2*S*) by enantioselective synthesis (Mori and Tabata 2017). These two pheromones of *Ps. viburni* and *D. brevipes* are particularly irregular because they are composed of isoprene units, presumably with unprecedented 2′–2 and 3′–4 linkages (Millar et al. 2005b), and 2′–3 and 4′–4 linkages (Tabata et al. 2017a), respectively.

3.3 Evolution of Mealybug Pheromones

3.3.1 *Biosynthetic Background of Pheromone*

As monoterpene skeletons of mealybug pheromones are highly unique and have not been reported from any of their host plants, including well-studied crops, such as citrus fruit and grapes, these pheromones are considered to be synthesized de novo, rather than derived from some modifications of terpenes sequestered from the host plant (Millar et al. 2005a). However, little is known about backgrounds of pheromone biosynthesis in mealybugs. Moreover, the tissues or organs that produce and release pheromones have not been characterized, although metathoracic legs were proposed as a probable site for pheromone emission based on the results of body-sectional extraction, followed by behavioral bioassays and anatomical inspections by scanning electron microscopy (Waterworth et al. 2012). However, some implications on pheromone biosynthesis can be taken from numerous studies about terpene biosynthesis on other insects and plants.

Terpenes are generally biosynthesized from the universal terpene building blocks, isopentenyl diphosphate (IPP) and dimethylallyl diphosphate (DMAPP), which are predominantly derived from the mevalonate pathway (Breitmaier 2006). For syntheses of general monoterpenes with the head-to-tail linkage, IPP and DMAPP are coupled to form structurally diverse C10-precursor molecules. The diversity of these molecules is attributed to the functions of different prenyltransferases that catalyze condensations of IPP and DMAPP. For example, geranyl diphosphate synthase, a member of the *trans*-prenyltransferases, generates geranyl diphosphate to be converted to geraniol, and its geometric isomer, nerol, is produced via neryl diphosphate arising from a reaction of another member of the *cis*-prenyltransferases (neryl diphosphate synthase). For biosynthesis of prevalent monoterpenes, C10 precursors from IPP and DMAPP condensations are further catalyzed by another group of enzymes called monoterpene synthases.

Relatively little is known about the biosynthesis of irregular monoterpenes that include non-head-to-tail linkages. A lavandulyl diphosphate synthase that catalyzes a 1′–2 condensation of two DMAPP molecules to produce a lavandulol precursor was isolated from a lavender plant, and found to belong to the *cis*-prenyltransferase family (Demissie et al. 2013). On the other hand, a chrysanthemyl diphosphate synthase isolated from a pyrethrum plant that couples two DMAPPs via a c1′–2–3 linkage was reported to belong to the *trans*-prenyltransferase family (Rivera et al. 2001). Of note, chrysanthemyl diphosphate synthase was reported to generate lavandulyl diphosphate from two DMAPPs, in addition to the major product, chrysanthemyl diphosphate (Yang et al. 2014). Expression of a prenyltransferase (geranyl diphosphate synthase) gene, which builds a key structure of monoterpene pheromones, has been suggested to be regulated by juvenile hormone III, like other mevalonate pathway genes involved in pheromone biosynthesis in the bark beetle *Ips pini* (Gilg et al. 2005); however, hormonal regulation of monoterpene biosynthesis remains largely unclear in other insects.

3.3.2 Pheromone Perception in Males

As female pheromone structures clearly differ among mealybug species, males were considered to have strict perception systems to respond to conspecific pheromone compounds in each species. However, variations in male responses to pheromones were reported in *Pl. ficus* (Kol-Maimon et al. 2010). As previously mentioned, two lavandulyl esters are contained in volatiles released from females of this species, lavandulyl senecioate (LS) and lavandulyl isovalerate (LI). LS is prevalent in *Pl. ficus*, whereas LI was identified as a second component in a mass-reared colony, which originated from an Israeli population (Zada et al. 2003). Males of *Pl. ficus* display ambiguous responses to LI; trap experiments in an Israeli vineyard revealed that the addition of LI to LS significantly reduced the trap catches, indicating repellant effects of LI (Zada et al. 2003), but some laboratory-reared males were attracted to LI and LS (Kol-Maimon et al. 2010). Subsequent behavioral bioassays in the laboratory demonstrated that nine types of male responses (pherotypes), which are distinguishable by the responses (attraction, repulsion, and indifference) to LS and LI, are present in Israeli populations (Kol-Maimon et al. 2010). Moreover, these pherotypes were found to be heritable (Kol-Maimon et al. 2010, 2014b). Five of the nine pherotypes are observed in Portuguese populations, but none of the Portuguese males were attracted to LI (Kol-Maimon et al. 2010).

Although pheromone structures of mealybugs are species-specific, a few studies have reported the occurrence of cross-attraction between closely related species. Two *Planococcus* species, *Pl. citri* and *Pl. ficus*, may occur sympatrically on the same host plant and can hybridize to produce hybrid offspring, which are fertile but markedly female-biased (Rotundo and Tremblay 1982). Of note, some *Pl. citri* males were attracted to the *Pl. ficus* pheromone, but not vice versa (Kol-Maimon et al. 2014a). In addition, genotyping by nuclear ITS2 (internal transcribed spacer 2 in ribosomal DNA) and mitochondrial *CO1* (cytochrome c oxidase subunit 1) gene sequences revealed the presence of hybrid females in *Pl. citri* populations, but not in *Pl. ficus* populations (Kol-Maimon et al. 2014a). Two hypotheses were proposed to explain the occurrence of the *Pl. ficus* pherotype and genotype traits in *Pl. citri* populations: (i) the occurrence of ancient and/or contemporary gene flow between the two species, and (ii) an ancient sympatric speciation, by which *Pl. ficus* emerged from *Pl. citri*, has led to the present situation (Kol-Maimon et al. 2014a). In either event, the cross-attraction and hybridization between *Pl. citri* and *Pl. ficus* are important to discuss the divergence of their pheromone systems because the pheromones of the two species are substantially different with a cyclobutanoid in the former and an acyclic lavandulyl compound in the latter species.

3.3.3 Saltational Shifts of Pheromone Structures

Most mealybug pheromones have a common structural feature, carboxylic esters of monoterpene alcohols with irregular non-head-to-tail connections of isoprene units, and are therefore considered to be conserved among mealybug species as a whole

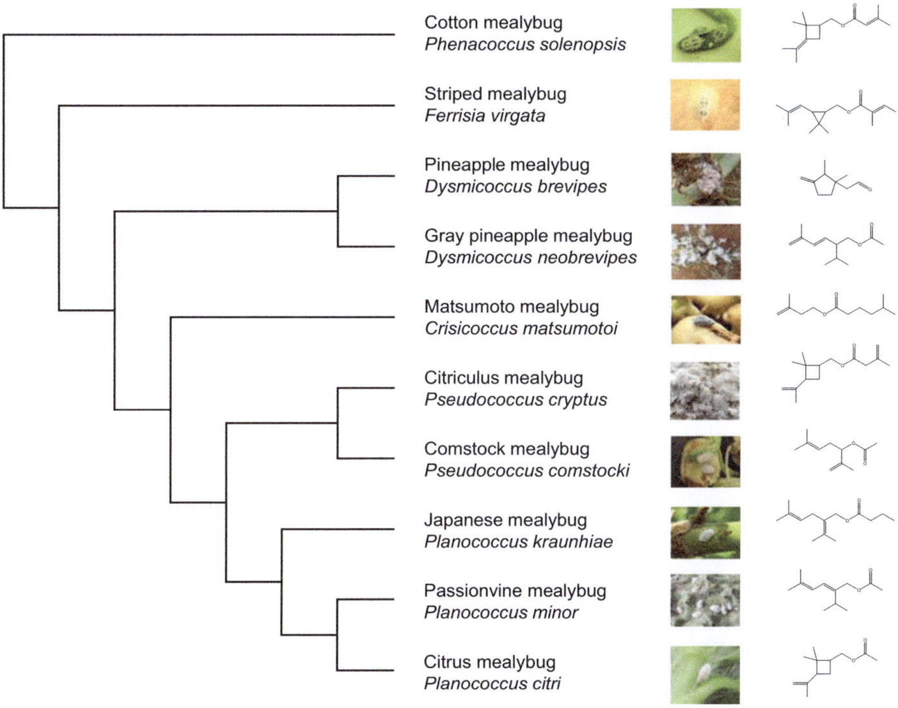

Fig. 3.2 Discordance between pheromone-structural similarities and phylogenetic relationship of mealybugs inferred from partial sequences of the mitochondrial *CO1* gene

(Millar et al. 2005a; Zou and Millar 2015). However, the structural similarities of the pheromones between each mealybug taxon are often inconsistent with the phylogenetic relationships (Fig. 3.2). For example, characteristic and similar cyclobutane structures (i.e., planococcol and maconelliol) are found in four species of different genera—*Planococcus citri*, *Pseudococcus cryptus*, *Phenacoccus solenopsis*, and *Maconellicoccus hirsutus*—indicating that the process to build the cyclobutane monoterpenes has been acquired (or lost) multiple times during mealybug speciation. Similarly, chrysanthemyl esters, which have never been found in insects other than mealybugs, were discovered as pheromones of three species of different genera, *Phenacoccus madeirensis*, *Pseudococcus calceolariae*, and *Ferrisia virgate*. This pattern also implies multiple independent evolutions that have enabled the generation of chrysanthemyl structures. These three mealybugs all produce only one enantiomer of their own chrysanthemyl pheromones, with no contamination of the other isomer, indicating that each species biosynthesizes its pheromone with strict stereoselectivity (Tabata and Ichiki 2017). Discordance between phylogeny and pheromone chemistry suggests that positive selection has worked on females to generate signals that can be clearly discriminated from those of closely related taxa during evolutionary radiation of mealybug species (Tabata et al. 2017a).

Such saltational shifts of pheromones are found in several insect taxa, including flies, beetles, and moths (Symonds and Elgar 2008; Symonds et al. 2009). In *Bactrocera* fruit flies (Diptera: Tephritidae), which have mechanisms other than pheromones to ensure reproductive isolation, male-produced sex pheromone compositions were demonstrated to have evolved through rapid saltational changes associated with speciation, followed by gradual divergence thereafter (Symonds et al. 2009). The aggregation pheromone blends of bark beetles of two genera, *Dendroctonus* and *Ips* spp. (Coleoptera: Curculionidae), are notably different between the genera, but have a less clear phylogenetic pattern between the species within each genus, suggesting that, within certain phylogenetic constraints, the aggregation pheromones have evolved with saltational shifts, resulting in sibling species being pheromonally different from one another (Symonds and Elgar 2004). In *Ostrinia* moths (Lepidoptera: Crambidae), saltational shifts of pheromone blends, which are composed of similar unsaturated fatty compounds, have been promoted by positive selection acting on orthologous genes encoding pheromone biosynthesis enzymes, leading to the avoidance of interspecific interference among sibling species (Tabata and Ishikawa 2016). As mealybugs are sedentary insects that cannot readily change their habitats, which often play a role in assortative mating and reproductive isolation (Symonds et al. 2009), interference among pheromone signals is likely to be much more serious than for motile insects, such as flies, beetles, or moths, and this may have driven the evolution of species-specific pheromone structures that ensure unique chemical channels (Tabata et al. 2017a).

Another unique feature in the evolution of mealybug pheromones is the decay of pheromone signaling as a consequence of asexual reproduction, which was observed in *D. brevipes*. This species is a known parthenogenetic species, and many previous studies in different areas revealed that this pest to pineapple production reproduces asexually by obligate apomictic thelytokous parthenogenesis (e.g., Ito 1938; Carter 1942; Bertin et al. 2013). However, a recent study demonstrated that two different lineages of *D. brevipes* with sexual and asexual reproduction systems coexist in a Japanese population. Molecular phylogenetic analyses indicated that these two lineages are genetically isolated and are likely to represent diverged lineages, yet they were estimated to have diverged relatively recently (ca. 1.3 million years ago) and are very closely related with no distinguishable morphological features (Tabata et al. 2016). Virgin adult females of the sexual lineage of *D. brevipes*, as well as those of other mealybug species with the general sexual reproduction system, emit pheromones and increase emission during the first 10 days of adulthood. On the other hand, females of the asexual sister lineage have completely lost pheromone production (Tabata et al. 2017a). In addition, no pheromone candidates that elicit male-antennal responses were detected in volatiles from the asexual females. As other *Dysmicoccus* spp. reproduce sexually, and obligate parthenogenesis is found only in *D. brevipes* (Nur 1971), the common ancestor of the two lineages of *D. brevipes* may have reproduced sexually (Tabata et al. 2016). Thus, as an evolutionary consequence of acquiring parthenogenetic reproduction, the asexual mealybugs are likely to have abandoned production of the sex pheromone, which had been required to attract the males of their ancestors (Tabata et al. 2017a).

Given the costs required to produce and maintain most characteristics, useless traits are considered to be disadvantageous and reduction of these traits should be favored by natural selection (Regal 1977). Sexual traits, such as pheromone signals, of asexual females generally exhibit large shifts compared with females of sexual relatives, often resulting in disappearance (van der Kooi and Schwander 2014). The evidence of the complete loss of the pheromone production in the asexual lineage of *D. brevipes* within a period of relatively rapid evolution implies the potential costs of pheromone production for mealybug females. Moreover, the pheromone is produced only during a limited period, even by the sexually reproducing mealybugs, suggesting that they reduce the cost of continuous emission of pheromones; mealybug females generally release their pheromones during a certain time of the photoperiod, but thereafter rapidly cease releasing pheromones, and most individuals stop producing pheromones and do not reactivate emission after copulation (Levi-Zada et al. 2014; Tabata et al. 2017a; Tabata and Teshiba 2018). Most pheromones are considered to be biosynthesized de novo and require nutrient resources for their production (Foster 2009). It is thus preferable to limit emission of the pheromones to necessity. Unnecessary emission is detrimental, particularly for mealybugs, which feed on nitrogen-deficient plant sap; nitrogen is not included in mealybug pheromones, but is necessary for the operation of biosynthetic cascades. Furthermore, pheromones are often intercepted and utilized as kairomones by natural enemies, and some mealybug pheromones have been reported to attract predators and parasitoids (Dunkleblum 1999; Franco et al. 2008, 2009; Tsueda 2014). These potential negative aspects may have driven the loss of pheromone production in the asexual *D. brevipes* after the acquisition of parthenogenetic reproduction. Approximately 10% of mealybug species reproduce either partially or completely by parthenogenesis (Nur 1971), and the species with parthenogenetic reproduction systems do not form a unique phylogenic cluster (Ross et al. 2010b), indicating that asexual reproduction systems, which appeared to be followed by abortion of pheromone production, have evolved multiple times during mealybug species radiation.

3.4 Application of Pheromones to Mealybug Pest Management

3.4.1 Pheromone Traps

Monitoring the occurrence of pests to predict damage to crops may aid in deciding when to spray insecticides in a more effective and efficient manner (Gut et al. 2004). Particularly for mealybugs, insecticidal control is often ineffective due to their cryptic behavior, waxy body cover, and clumped spatial distribution pattern (Franco et al. 2009). Traps baited with pheromones may be more useful for monitoring than other sampling methods that can be time-consuming and require technical expertise. As mealybug pheromones are structurally species-specific, pheromone traps provide reliable monitoring of the occurrence of a specific pest in each crop at each location.

The utility of pheromone traps was demonstrated in *Pl. kraunhiae* infesting persimmon orchards (Sawamura et al. 2015). Pheromone-trap captures illustrated seasonal fluctuations in mealybug occurrences with three to four generations a year in Japan, and the first trap of wild mealybugs precisely matched with the first emergence of adults from overwintering-generation individuals maintained in a Stevenson screen. Moreover, the estimated timing of the emergence of the first-generation nymphs, which was calculated from the cumulative effective temperature starting from the date of the pheromone trapping of adults in the overwintering generation, was consistent with the actual timing of nymph emergence. Thus, using the data of effective cumulative temperature and pheromone trapping, the timing of mealybug nymph emergence from eggs, when insecticide spraying is the most effective because mealybugs frequently wander around trees for dispersal, can be precisely predicted (Sawamura et al. 2015). Similarly, the pheromone traps were confirmed to precisely reflect the population sizes measured by more labor-intensive visual sampling methods in South African vineyards infested by *Pl. ficus* (Walton et al. 2004). The lures made of rubber septa impregnated with 100 μg of racemic lavandulyl senecioate, the synthetic *Pl. ficus* pheromone, were attractive to males for 10 weeks or more with an effective range of at least 50 m (Walton et al. 2004). Bahder et al. (2013) reported that only one pheromone trap per 12.14 ha was sufficient to provide the flight phenology data of *Ps. maritimus* in a Washington state vineyard for making informed management decisions, although four to eight traps per 12.14 ha were needed to achieve a sampling precision efficiency of 30% throughout the season.

Pheromone traps are also useful for the detection and monitoring of quarantine pests. Several mealybug species have recently expanded their distribution in association with increasing movement of plant materials by humans (Gullan et al. 2003) and climate change (Gutierrez et al. 2008; Jara et al. 2013). For example, *Ph. solenopsis* has been an emerging invasive pest since the 2000s. This species is originally described from North America, but was first recognized in 2005 in the cotton fields of India and Pakistan. Since then, it is reported to have damaged more than 200 plant species from 24 countries in tropical and subtropical regions (Fand and Suroshe 2015). As pheromone traps for mealybugs generally exhibit powerful species-specific activity, they will be a useful tool for quarantine worldwide. This feature of species specificity is particularly noteworthy for mealybugs, which are tiny with a similar appearance regardless of species and are difficult to be identified based on morphological characters.

3.4.2 Mating Disruption

As sex pheromones generally elicit marked behavioral and/or physiological responses in conspecific mates in minute amounts, the pheromone-mediated mate-finding systems utilized by pest insects have long been recognized to be of potential value in their management, even before the identification of any pest pheromones

(Wright 1964a, b; Cardé 1990; Tabata 2018). Mating disruption is one of the strategies to suppress the population growth of pest insects by mimicking their sex pheromones, and interfering with mate-finding and copulation. Most studies on mating disruption conducted since the early 1960s aimed at controlling moths, and the majority of mating disruptants currently available for farmers are designed for moth pests (Tabata et al. 2017b; Tabata 2018). Moths are generally nocturnal, and most species use a common mating system that involves long-distance attraction of males with pheromones emitted by females. Moreover, chemical constituents of moth pheromones are relatively simple and similar aliphatic compounds, suitable for the industrial synthesis. Thus, mating disruption has been adopted for controlling moths across more than 750,000 ha of agricultural fields and forests (Miller and Gut 2015).

Mealybugs are also potential targets of mating disruption because the pheromones released by immotile females are important to facilitate mating and reproduction by serving as a key navigation tool for males who have no mouth to take in nutrition or water and can search for mates for only a short time. The short active period of males, which is estimated at <35 h after emergence (Zada et al. 2008; Mendel et al. 2012), makes the use of mating disruption strategy for mealybug control more promising. In addition to the suppression of mating, delayed mating under the influence of mating disruptants may reduce the fertility of mated females (Ross et al. 2011); the number of eggs deposited by *Pl. kraunhiae* females decreased rapidly as they aged (Tabata and Teshiba 2018). Of note, pheromone emission and associated sexual attractiveness levels of females increase with aging until copulation, even though their reproductive performance is rapidly declining: senescent females continue releasing substantial amounts of pheromones and maintain their sexual appeal when they have almost completely lost fertility (Tabata and Teshiba 2018). This suggests a potential sexual conflict within a pheromone communication system in mealybugs, by which females benefit at the expense of the males through "dishonest" signals of fertility. In theory, females increase their sexual signaling efforts when they remain unmated in response to the elevated risk of reproductive failure (Simmons 2015; Umbers et al. 2015). As mating disruption essentially deprives adult mealybugs of copulation opportunities, "old maids" inevitably increase in populations in the presence of mating disruptants. According to the dataset on *Pl. kraunhiae* (Tabata and Teshiba 2018), the older, less fertile females are expected to release more pheromones and to demonstrate greater sexual attractiveness than the younger, fresh females. Thus, the potential sexual conflict in pheromone communication may enhance mating disruption effects in mealybugs.

The first trial of mating disruption targeting *Pl. ficus* mealybugs infesting California vineyards was reported by Walton et al. (2006). This species was first described from the specimen on figs in Hyeres and Nice in France (Ben-Dov 1994) and is considered to have originated from the Mediterranean basin. However, it currently invades vineyards in a broad range of regions, including California, Mexico, Brazil, Argentina, and South Africa. It is one of the most destructive mealybugs in grape vineyards because it affects fruit production by excreting honeydew, which propagates sooty mold and putrefactive bacteria and markedly reduces the quality of table and wine grapes (Chiotta et al. 2010). The active constituent of the mating disruptant used for *Pl. ficus* management is a racemic mixture of lavandulyl senecioate, which is as attractive as the pure (*S*)-enantiomer of the natural form (Millar et al. 2002; Zada et al. 2008). Walton et al. (2006) used the sprayable microencapsulated formulation of the racemic lavandulyl senecioate, applied with an air-blast sprayer, three and four times in 2003 and 2004, respectively, across 5–12 ha of commercial vineyards in California: the microencapsulated pheromone was mixed with water and applied at a rate of 10.7 g/ha across each of the vineyards of the treated plots in May, June, and July in 2003 (32.1 g/ha in total), and in April, May, June, and July in 2004 (53.6 g/ha in total). Consequently, a significant reduction of season-long trap catches of adult males, population densities, and fruit damages was observed in the treated plots in comparison with the control plots (Walton et al. 2006). However, mealybug densities were only 12% and 31% lower in the treated plots than in control plots in 2003 and 2004, respectively. The observed limitation of mating disruption efficacy may be attributable to the relatively short effective lifetime of the microencapsulated pheromone formulation.

Another formulation type of mating disruptants, reservoir dispensers made of solid matrix membranes (4 × 9 cm; CheckMate; Suterra, Oregon, USA) loaded with relatively high amounts (100–150 mg) of racemic lavandulyl senecioate, were used to suppress *Pl. ficus* populations in two commercial vineyards in Sardinia, Italy (Cocco et al. 2014). Only one set of the dispensers were placed in small plots in the vineyards in late-April to mid-May in 2008 and 2009 before adult males of the overwintering generation emerged. By calculation, the treated plots were filled with 62.5 g/ha and 93.8 g/ha of the synthetic pheromone in 2008 and 2009, respectively. As a result, the number of males captured in pheromone traps, the mealybug population densities, and the proportion of wild pregnant females decreased. Subsequently, Sharon et al. (2016) applied the same reservoir dispenser-type disruptant (CheckMate) to control *Pl. ficus* for two successive years in nine commercial vineyards in Israel. The dispensers, each of which was loaded with 150 mg of racemic lavandulyl senecioate, were set in April with 620 dispensers per hectare (93.0 g/ha). A significant reduction of the mealybug population was observed in vines in treated plots with low initial infestation levels after only 1 year of treatment. In plots with high initial infestation levels, a gradual reduction of the mealybug population size was observed after two successive years of treatment.

Further investigation to control mealybug populations by mating disruption is currently underway for *Pl. kraunhiae*, which is of East Asian origin, and is often found in Japan, Korea, Taiwan, and China (Kawai 1980; Ben-Dov 1994). This species is particularly destructive to Japanese orchards and damages many fruits, including persimmons, grapes, pears, citrus, and figs (Ueno 1963; Shibao and Tanaka 2000). The development of a mating disruption method for management of this pest has long been desired (Teshiba 2013) because this pest frequently resurges following the successive application of insecticides (Morishita 2005). The sex pheromone of *Pl. kraunhiae* (γ-lavandulyl butyrate) can be easily derived and synthesized from lavandulol via isomerization reactions with double-bond migration (Tabata 2013; Kinsho et al. 2015), which has enabled the practical application of mating disruption to *Pl. kraunhiae*.

The potential efficacy of mating disruption against *Pl. kraunhiae* has been demonstrated by Teshiba et al. (2009) using pheromone dispensers made of rubber septa, which were impregnated with γ-lavandulyl butyrate. They performed a field experiment in a commercial persimmon orchard; 167 sets of pheromone dispensers, each of which contained 30 mg of γ-lavandulyl butyrate, were placed in a 500 m^2 plot (5 g in total) in April before adult males of the overwintering generation had emerged. No males were attracted to the monitoring pheromone traps in the treated plot, indicating strong interference of mate-finding behavior. In addition, wild pregnant females were captured at a rate of only 1.4% in the treated plot, whereas the rate was 51.0% in the neighboring control plot. The population sizes, which were similar in the initial stage (0.02–0.03 mealybugs per twig), were smaller in the treated plot (0–0.03) than in the control plot (0.31–1.30) at the end of the experimental period. Consequently, the proportion of fruit damaged by mealybugs at harvest was only 0.2% in the treated plot, which was less than 1/20 of that in the control plot (4.2%).

In general, the cost of synthetic pheromones is a limiting factor for mating disruption (Rodriguez-Saona et al. 2009). Particularly, several mealybug pheromones are structurally complicated and are difficult to synthesize at industrial scales. Thus, field-scale application is not practical for many mealybug pheromones. However, the invention of a relatively simple synthetic method of lavandulol-related pheromones, including those of *Pl. ficus* and *Pl. kraunhiae*, has enabled the practical use of these pheromones as mating disruptants (Tabata et al. 2017b).

3.4.3 Parasitoid Recruitment: Possible Compensation for Limitations of Mating Disruption

Mealybug pheromones are often intercepted and used as kairomones by natural enemies, including predatory beetles and parasitic wasps (Dunkelblum 1999). In particular, several encyrtid wasps were demonstrated to be attracted to mealybug pheromones. For example, *Anagyrus* sp., which is closely related to *A. pseudococci* (Hymenoptera: Encyrtidae), a solitary koinobiont endoparasitoid of mealybugs, uses the sex pheromone of *Pl. ficus* (lavandulyl senecioate) as a kairomone and was strongly attracted to this compound in field surveys conducted in Mediterranean Europe (Franco et al. 2008). Moreover, Franco et al. (2011) observed that the *Pl. ficus* pheromone increases parasitism by *Anagyrus* sp. near *pseudococci* to a different mealybug species (*Pl. citri*). Similarly, the sex pheromone of *Pl. kraunhiae* (γ-lavandulyl butyrate) was reported to attract *A. fujikona*, another *Anagyrus* wasp specialized to attack *Pl. kraunhiae* (Tsueda 2014). Thus, mating disruption based on synthetic mealybug pheromones may theoretically affect the behavior of natural enemies, including parasitic wasps. However, no negative influences have hitherto been reported on kairomone-based host-locating behavior of parasitic wasps; in the study of mating disruption against *Pl. ficus* in Sardinia (Cocco et al. 2014), mating disruption treatments did not decrease the parasitism rate of *Anagyrus* wasps, which was 1.5-fold higher in the treated plot than in the control plot.

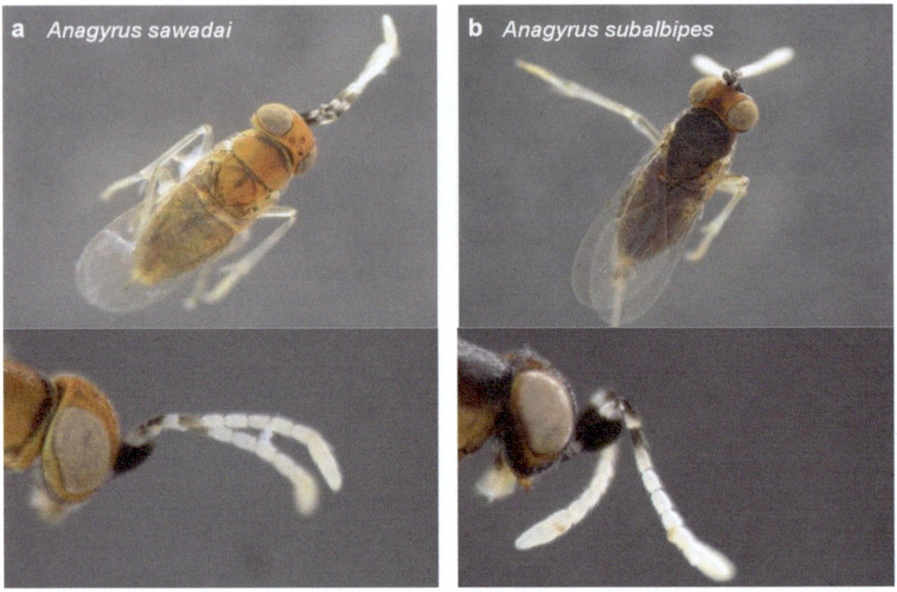

Fig. 3.3 Cyclolavandulyl butyrate (CLB) attracts two different *Anagyrus* wasps, *A. sawadai* (**a**) and *A.subalbipes* (**b**), both of which do not parasitize *Planococcus kraunhiae* in nature, but attack it in the presence of CLB

During field studies of the synthetic *Pl. kraunhiae* pheromone, its cyclization product, cyclolavandulyl butyrate, was discovered as another attractant for several mealybug parasitic wasps, including *Anagyrus sawadai* and *Anagyrus subalbipes* (Tabata et al. 2011; Fig. 3.3). Cyclolavandulyl butyrate was a byproduct during the acid-promoted reaction of double-bond migration from lavandulol to γ-lavandulol. Of note, *A. fujikona*, which is a major and specific parasitoid of *Pl. kraunhiae* that responds to the *Pl. kraunhiae* pheromone (γ-lavandulyl butyrate), is never attracted to cyclolavandulyl butyrate. Moreover, *A. sawadai* and *A. subalbipes*, which are strongly attracted to cyclolavandulyl butyrate, do not respond to γ-lavandulyl butyrate and are not likely to parasitize *Pl. kraunhiae* under natural conditions (Teshiba et al. 2012). Thus, cyclolavandulyl butyrate attracts "non-natural enemies" for *Pl. kraunhiae*. However, these "non-natural enemies" attacked *Pl. kraunhiae* in persimmon orchards when they recruited by cyclolavandulyl butyrate from the surrounding environment (Teshiba et al. 2012). In addition, the field application of cyclolavandulyl butyrate successfully suppressed *Pl. kraunhiae* population growth in persimmon orchards by enhancing the parasitism activity of wasps (Teshiba and Tabata 2017). This is a unique example demonstrating that a "non-natural enemy", which does not typically attack the pest in nature, can be enrolled in biological control programs by using its attractant. Industrial synthesis of cyclolavandulyl butyrate has already been established (Kinsho and Ishibashi 2015), although it remains unclear whether cyclolavandulyl butyrate is present as a natural product and plays a role in the ecosystems of mealybugs and their natural enemies. Thus, the use of parasitic wasps in combination with their attractants, including cyclolavandulyl butyrate, may be promising for management programs for many mealybug species.

Parasitoid recruitment by attractants may compensate for the limitations of mating disruption in pest management programs. In general, pest density is a strong limiting factor in pest control by semiochemicals (Gut et al. 2004), and the efficacy of mating disruption is markedly reduced in high-density populations. Moreover, mating disruption is not applicable to mealybugs with asexual reproduction systems (Tabata et al. 2017a). Incorporation of parasitoids may mitigate these constraints; population growth rates of parasitoids can increase in the presence of a large number of host insects, and the activity and performance of parasitoids are therefore expected to be greater in higher density pest populations. Thus, the combination of mating disruption and parasitoid recruitment is promising due to their complementary actions in mealybug pest management programs.

Acknowledgement The author thanks many co-researchers involved in mealybug management projects for their contributions to this manuscript through valuable discussions.

References

Arai T, Sugie H, Hiradate S et al (2003) Identification of a sex pheromone component of *Pseudococcus cryptus*. J Chem Ecol 29:2213–2223. https://doi.org/10.1023/A:1026214112242

Bahder BW, Naidu RA, Daane KM et al (2013) Pheromone-based monitoring of *Pseudococcus maritimus* (Hemiptera: Pseudococcidae) populations in concord grape vineyards. J Econ Entomol 106:482–490. https://doi.org/10.1603/EC12138

Ben-Dov Y (1994) A systematic catalogue of the mealybugs of the world (Insecta: Homoptera: Coccoidae: Pseudococcidae and Putoidae) with data on geographical distribution, host plants, biology and economic importance. Intercept, Andover

Bertin A, Bortoli LC, Botton M et al (2013) Host plant effects on the development, survival, and reproduction of *Dysmicoccus brevipes* (Hemiptera: Pseudococcidae) on grapevines. Ann Entomol Soc Am 106:604–609. https://doi.org/10.1603/AN13030

Bierl-Leonhardt BA, Moreno DS, Schwarz M et al (1980) Identification of the pheromone of the Comstock mealybug. Life Sci 27:399–402. https://doi.org/10.1016/0024-3205(80)90188-5

Bierl-Leonhardt BA, Moreno DS, Schwarz M et al (1981) Isolation, identification, and synthesis of the sex pheromone of the citrus mealybug, *Planococcus citri* (Risso). Tetrahedron Lett 22:389–392. https://doi.org/10.1016/0040-4039(81)80107-4

Bierl-Leonhardt BA, Moreno DS, Schwarz M et al (1982) Isolation, identification, synthesis, and bioassay of the pheromone of the comstock mealybug and some analogs. J Chem Ecol 8:689–699. https://doi.org/10.1007/BF00988310

Breitmaier E (2006) Terpenes. Wiley-VCH, Weinheim

Brown SW, Nur U (1964) Heterochromatic chromosomes in Coccoids. Science 145:130–136. https://doi.org/10.1126/science.145.3628.130

Cardé RT (1990) Principles of mating disruption. In: Ridgeway R, Silverstein RM, Inscoe M (eds) Behavior-modifying chemicals for insect management. Marcel Dekker, New York, pp 47–71

Cardé RT, Baker TC (1984) Sexual communication with pheromones. In: Bell WJ, Cardé RT (eds) Chemical ecology of insects. Chapman and Hall, New York, pp 355–383

Carter W (1942) The geographical distribution of mealybug wilt with notes on some other insect pests of pineapple. J Econ Entomol 35:10–15. https://doi.org/10.1093/jee/35.1.10
Chiotta ML, Ponsone ML, Torres AM et al (2010) Influence of *Planococcus ficus* on *Aspergillus* section *Nigri* and ochratoxin A incidence in vineyards from Argentina. Lett Appl Microbiol 51:212–218. https://doi.org/10.1111/j.1472-765X.2010.02884.x
Cocco A, Lentini A, Serra G (2014) Mating disruption of *Planococcus ficus* (Hemiptera: Pseudococcidae) in vineyards using reservoir pheromone dispensers. J Insect Sci 14:144. https://doi.org/10.1093/jisesa/ieu006
Cook LG, Gullan PJ, Trueman HE (2002) A preliminary phylogeny of the scale insects (Hemiptera: Sternorrhyncha: Coccoidea) based on nuclear small-subunit ribosomal DNA. Mol Phylogenet Evol 25:43–52. https://doi.org/10.1016/S1055-7903(02)00248-8
De Alfonso I, Hernandez E, Velazquez Y et al (2012) Identification of the sex pheromone of the mealybug *Dysmicoccus grassii* Leonardi. J Agric Food Chem 60:11959–11964. https://doi.org/10.1021/jf304065d
Demissie ZA, Erland LAE, Rheault MR et al (2013) The biosynthetic origin of irregular monoterpenes in *Lavandula*: isolation and biochemical characterization of a novel *cis* prenyl diphosphate synthase gene—lavandulyl diphosphate synthase. J Biol Chem 288:6333–6341. https://doi.org/10.1074/jbc.M112.431171
Dunkelblum E (1999) Scale insects. In: Hardie J, Minks AK (eds) Pheromones of non-lepidopteran insects associated with agricultural plants. CAB International, Oxfordshire, pp 251–276
El-Sayed AM, Unelius CR, Twidle A et al (2010) Chrysanthemyl 2-acetoxy-3-methylbutanoate: the sex pheromone of the citrophilous mealybug, *Pseudococcus calceolariae*. Tetrahedron Lett 51:1075–1078. https://doi.org/10.1016/j.tetlet.2009.12.106
Fand BB, Suroshe SS (2015) The invasive mealybug *Phenacoccus solenopsis* Tinsley, a threat to tropical and subtropical agricultural and horticultural production systems–a review. Crop Prot 69:34–43. https://doi.org/10.1016/j.cropro.2014.12.001
Figadère BA, Devlin FJ, Millara JG et al (2008) Determination of the absolute configuration of the sex pheromone of the obscure mealybug by vibrational circular dichroism analysis. Chem Commun 2008(9):1106–1108. https://doi.org/10.1039/B717440C
Figadère BA, McElfresh JS, Borchardt D et al (2007) *trans*-α-Necrodyl isobutyrate, the sex pheromone of the grape mealybug, *Pseudococcus maritimus*. Tetrahedron Lett 48:8434–8437. https://doi.org/10.1016/j.tetlet.2007.09.155
Foster S (2009) Sugar feeding via trehalose haemolymph concentration affects sex pheromone production in mated *Heliothis virescens* moths. J Exp Biol 212:2789–2794. https://doi.org/10.1242/jeb.030676
Franco JC, Silva EB, Cortegano E et al (2008) Kairomonal response of the parasitoid *Anagyrus* spec. nov. near *pseudococci* to the sex pheromone of the vine mealybug. Entomol Exp Appl 126:122–130. https://doi.org/10.1111/j.1570-7458.2007.00643.x
Franco JC, Zada A, Mendel Z (2009) Novel approaches for the management of mealybug pests. In: Ishaaya I, Horowitz AR (eds) Biorational control of arthropod pests. Springer, Dordrecht, pp 233–278
Franco JC, Silva EB, Fortuna T et al (2011) Vine mealybug sex pheromone increases citrus mealybug parasitism by *Anagyrus* sp. near *pseudococci* (Girault). Biol Control 58:230–238. https://doi.org/10.1016/j.biocontrol.2011.06.008
Gilg AB, Bearfield JC, Tittiger C et al (2005) Isolation and functional expression of an animal geranyl diphosphate synthase and its role in bark beetle pheromone biosynthesis. Proc Natl Acad Sci U S A 102:9760–9765. https://doi.org/10.1073/pnas.0503277102
Gullan PJ, Downie DA, Steffan SA (2003) A new pest species of the mealybug genus *Ferrisia* Fullaway (Hemiptera: Pseudococcidae) from the United States. Ann Entomol Soc Am 96:723–737. https://doi.org/10.1603/0013-8746(2003)096[0723:ANPSOT]2.0.CO;2
Gut LJ, Stelinski LL, Thomson DR et al (2004) Behaviour-modifying chemicals: prospects and constraints in IPM. In: Koul O, Dhaliwal GS, Cuperus GW (eds) Integrated pest management—potential, constraints and challenges. CABI Publishing, Oxford, pp 73–122

Gutierrez AP, Ponti L, d'Oultremont T et al (2008) Climate change effects on poikilotherm tritrophic interactions. Clim Chang 87:167–192. https://doi.org/10.1007/s10584-007-9379-4

Hashimoto K, Morita A, Kuwahara S (2008) Enantioselective synthesis of a mealybug pheromone with an irregular monoterpenoid skeleton. J Org Chem 73:6913–6915. https://doi.org/10.1021/jo801147c

Hinkens DM, McElfresh JS, Millar JG (2001) Identification and synthesis of the sex pheromone of the vine mealybug, *Planococcus ficus*. Tetrahedron Lett 42:1619–1621. https://doi.org/10.1016/S0040-4039(00)02347-9

Ho H-Y, Hung C-C, Chuang T-H et al (2007) Identification and synthesis of the sex pheromone of the passionvine mealybug, *Planococcus minor* (Maskell). J Chem Ecol 33:1986–1996. https://doi.org/10.1007/s10886-007-9361-7

Ho H-Y, Su Y-T, Ko C-H et al (2009) Identification and synthesis of the sex pheromone of the Madeira mealybug, *Phenacoccus madeirensis* Green. J Chem Ecol 35:724–732. https://doi.org/10.1007/s10886-009-9649-x

Ho H-Y, Ko CH, Cheng CC et al (2011) Chirality and bioactivity of the sex pheromone of Madeira mealybug (Hemiptera: Pseudococcidae). J Econ Entomol 104:823–826. https://doi.org/10.1603/EC10374

Ito K (1938) Studies on the life history of the pineapple mealybug, *Pseudococcus brevipes* (Ckll.). J Econ Entomol 31:291–298. https://doi.org/10.1093/jee/31.2.291

James HC (1937) Sex ratios and the status of the male in Pseudococcinae (Hem. Coccidae). Bull Entomol Res 28:429–461. https://doi.org/10.1017/S0007485300038906

Jara V, Meza FJ, Zaviezo T et al (2013) Climate change impacts on invasive potential of pink hibiscus mealybug, *Maconellicoccus hirsutus* (Green), in Chile. Clim Chang 117:305–317. https://doi.org/10.1007/s10584-012-0542-1

Kawai S (1980) Scale insects of Japan in colors. Zenkoku-Noson-Kyouiku-Kyoukai, Tokyo

Khambay BPS, Jewess PJ (2010) Pyrethroids. In: Gilbert LI, Gill SS (eds) Insect control biological and synthetic agents. Elsevier, Oxford, pp 1–29

Kinsho T, Ishibashi N (2015) Method to produce cyclolavandulol and its derivetives. Japan Patent P2015-78179A

Kinsho T, Ishibashi N, Yamashita M et al (2015) Method to produce 2-isopropyliden-5-methyl-4-hexenyl butyrate. Japan Patent P2015-110553A

Kol-Maimon H, Levi-Zada A, Franco JC et al (2010) Male behaviors reveal multiple pherotypes within vine mealybug *Planococcus ficus* (Signoret) (Hemiptera; Pseudococcidae) populations. Naturwissenschaften 97:1047–1057. https://doi.org/10.1007/s00114-010-0726-3

Kol-Maimon H, Ghanim M, Franco JC et al (2014a) Evidence for gene flow between two sympatric mealybug species (Insecta; Coccoidea; Pseudococcidae). PLoS One 9:e88433. https://doi.org/10.1371/journal.pone.0088433

Kol-Maimon H, Mendel Z, Franco JC et al (2014b) Paternal inheritance in mealybugs (Hemiptera: Coccoidea: Pseudococcidae). Naturwissenschaften 101:791–802. https://doi.org/10.1007/s00114-014-1218-7

Levi-Zada A, Fefer D, David M et al (2014) Diel periodicity of pheromone release by females of *Planococcus citri* and *Planococcus ficus* and the temporal flight activity of their conspecific males. Naturwissenschaften 101:671–678. https://doi.org/10.1007/s00114-014-1206-y

Löfstedt C (1993) Moth pheromone genetics and evolution. Philos Trans R Soc Lond B 340:167–177

Mendel Z, Protasov A, Jasrotia P et al (2012) Sexual maturation and aging of adult male mealybug (Hemiptera: Pseudococcidae). Bull Entomol Res 102:385–394. https://doi.org/10.1017/S0007485311000605

Millar JG, Daane KM, McElfresh JS et al (2002) Development and optimization of methods for using sex pheromone for monitoring the mealybug *Planococcus ficus* (Homoptera: Pseudococcidae) in California vineyards. J Econ Entomol 95:706–714. https://doi.org/10.1603/0022-0493-95.4.706

Millar JG, Daan KM, McElfresh JS et al (2005a) Chemistry and applications of mealybug sex pheromones. In: Petroski RJ, Tellez MR, Behle RW (eds) Semiochemicals in pest and weed control. American Chemical Society, Washington, DC, pp 11–27

Millar JG, Midland SL, McElfresh JS et al (2005b) (2,3,4,4-Tetramethylcyclopentyl)methyl acetate, a sex pheromone from the obscure mealybug: first example of a new structural class of monoterpenes. J Chem Ecol 31:2999–3005. https://doi.org/10.1007/s10886-005-9320-0

Millar JG, Moreira JA, McElfresh JS et al (2009) Sex pheromone of the longtailed mealybug: a new class of monoterpene structure. Org Lett 11:2683–2685. https://doi.org/10.1021/ol802164v

Miller JR, Gut LJ (2015) Mating disruption for the 21st century: matching technology with mechanism. Environ Entomol 44:427–453. https://doi.org/10.1093/ee/nvv052

Mori K (2016) Pheromone synthesis. Part 260: synthesis of (±)-(*anti*-1,2-dimethyl-3-methylenecyclopentyl)acetaldehyde, the racemate of the female-produced sex pheromone of the pineapple mealybug (*Dysmicoccus brevipes*), and its *syn*-isomer. Tetrahedron 72:6578–6588. https://doi.org/10.1016/j.tet.2016.08.072

Mori K, Tabata J (2017) Pheromone synthesis. Part 262: determination of the absolute configuration of the female sex pheromone [(1*S*,2*S*)-(−)-(1,2-dimethyl-3-methylenecyclopentyl) acetaldehyde] of the pineapple mealybug (*Dysmicoccus brevipes*) by synthesis coupled with X-ray analysis. Tetrahedron 73:6530–6541. https://doi.org/10.1016/j.tet.2017.09.046

Mori K, Ueda H (1981) Synthesis of the optically active forms of 2,6-dimethyl-1,5-heptadien-3-ol acetate, the pheromone of the comstock mealybug. Tetrahedron 37:2581–2583. https://doi.org/10.1016/S0040-4020(01)98960-4

Morishita M (2005) Resurgence of Japanese mealybug, *Planococcus kraunhiae* (Kuwana), in persimmon induced by a synthetic pyrethroid cypermethrin. Annu Rep Kansai Plant Prot Soc 47:125–126. https://doi.org/10.4165/kapps.47.125

Negishi T, Uchida M, Tamaki Y et al (1980) Sex pheromone of the comstock mealybug, *Pseudococcus comstocki* Kuwana: isolation and identification. Appl Entomol Zool 15:328–333. https://doi.org/10.1303/aez.15.328

Nelson-Rees WA (1960) A study of sex predetermination in the mealy bug *Planococcus citri* (Risso). J Exp Zool 144:111–137. https://doi.org/10.1002/jez.1401440203

Nur U (1971) Parthenogenesis in coccids (Homoptera). Am Zool 11:301–308. https://doi.org/10.1093/icb/11.2.301

Nur U (1980) Evolution of unusual chromosome systems in scale insects (Coccoidea: Homoptera). In: Blackman RL, Hewitt GM, Ashburner M (eds) Insect cytogenetics. Blackwell, Oxford, pp 97–118

Ramesh R, Swaroop PS, Gonnade RG et al (2013) Syntheses and determination of absolute configurations and biological activities of the enantiomers of the longtailed mealybug pheromone. J Org Chem 78:6281–6284. https://doi.org/10.1021/jo400491n

Ramesh R, Bell V, Twidle AM et al (2015) Enantiospecific synthesis of both enantiomers of the longtailed mealybug pheromone and their evaluation in a New Zealand vineyard. J Org Chem 80:7785–7789. https://doi.org/10.1021/acs.joc.5b01131

Regal PJ (1977) Evolutionary loss of useless features: is it molecular noise suppression? Am Nat 111:123–133. https://doi.org/10.1086/283143

Rivera SB, Swedlund BD, King GJ et al (2001) Chrysanthemyl diphosphate synthase: isolation of the gene and characterization of the recombinant non-head-to-tail monoterpene synthase from *Chrysanthemum cinerariaefolium*. Proc Natl Acad Sci U S A 98:4373–4378. https://doi.org/10.1073/pnas.071543598

Roach B, Eisner T, Meinwald J (1990) Defense mechanisms of arthropods. 83. α- and β-Necrodol, novel terpenes from a carrion beetle (*Necrodes surinamensis*, Silphidae, Coleoptera). J Org Chem 55:4047–4051. https://doi.org/10.1021/jo00300a020

Rodriguez-Saona CR, Polk DF, Barry JD (2009) Optimization of pheromone deployment for effective mating disruption of oriental beetle (Coleoptera: Scarabaeidae) in commercial blueberries. J Econ Entomol 102:659–669. https://doi.org/10.1603/029.102.0226

Roelofs WL, Cardé RT (1974) Sex pheromones in the reproductive isolation of lepidopterous species. In: Birch MC (ed) Pheromones. North-Holland Publishing Company, Amsterdam, pp 96–114

Ross L, Shuker DM (2009) Scale insects. Curr Biol 19:R184–R186. https://doi.org/10.1016/j.cub.2008.12.023

Ross L, Langenhof MBW, Pen I et al (2010a) Sex allocation in a species with paternal genome elimination: the roles of crowding and female age in the mealybug *Planococcus citri*. Evol Ecol Res 12:89–104

Ross L, Pen I, Shuker DM (2010b) Genomic conflict in scale insects: the causes and consequences of bizarre genetic systems. Biol Rev 85:807–828. https://doi.org/10.1111/j.1469-185X.2010.00127.x

Ross L, Dealey EJ, Beukeboom LW et al (2011) Temperature, age of mating and starvation determine the role of maternal effects on sex allocation in the mealybug *Planococcus citri*. Behav Ecol Sociobiol 65:909–919. https://doi.org/10.1007/s00265-010-1091-0

Ross L, Shuker DM, Normark BB et al (2012) The role of endosymbionts in the evolution of haploid-male genetic systems in scale insects (Coccoidea). Ecol Evol 2:1071–1081. https://doi.org/10.1002/ece3.222

Rotundo G, Tremblay E (1982) Hybridization and sex pheromone response in two closely related mealybug species (Homoptera: Pseudococcidae). Syst Entomol 7:475–478. https://doi.org/10.1111/j.1365-3113.1982.tb00459.x

Sawamura N, Narai Y, Teshiba M et al (2015) Forecasting the occurrence of young Japanese mealybug larva *Planococcus kraunhiae* (Kuwana) (Hemiptera: Pseudococcidae) using sex pheromone traps and total effective temperature for persimmon. Jpn J Appl Entomol Zool 59:183–189. https://doi.org/10.1303/jjaez.2015.183

Schraeder F (1921) The chromosomes of *Pseudococcus nipae*. Biol Bull 40:259–270. https://doi.org/10.2307/1536736

Sharon R, Zahavi T, Sokolsky T et al (2016) Mating disruption method against the vine mealybug, *Planococcus ficus*: effect of sequential treatment on infested vines. Entomol Exp Appl 161:65–69. https://doi.org/10.1111/eea.12487

Shibao M, Tanaka H (2000) Seasonal occurrence of Japanese mealybug, *Planococcus kraunhiae* (Kuwana), on the fig, *Ficus carica* L., and control of the pest by insecticides. Jpn J Appl Entomol Zool Chugoku Branch 42:1–6

Simmons LW (2015) Sexual signalling by females: do unmated females increase their signalling effort? Biol Lett 11:20150298. https://doi.org/10.1098/rsbl.2015.0298

Sugie H, Teshiba M, Narai Y et al (2008) Identification of a sex pheromone component of the Japanese mealybug, *Planococcus kraunhiae* (Kuwana). Appl Entomol Zool 43:369–375. https://doi.org/10.1303/aez.2008.369

Symonds MR, Elgar MA (2004) The mode of pheromone evolution: evidence from bark beetles. Proc R Soc B Biol Sci 271:839. https://doi.org/10.1098/rspb.2003.2647

Symonds MR, Elgar MA (2008) The evolution of pheromone diversity. Trends Ecol Evol 23:220–228. https://doi.org/10.1016/j.tree.2007.11.009

Symonds MRE, Moussalli A, Elgar MA (2009) The evolution of sex pheromones in an ecologically diverse genus of flies. Biol J Linn Soc 97:594–603. https://doi.org/10.1111/j.1095-8312.2009.01245.x

Tabata J (2013) A convenient route for synthesis of 2-isopropyliden-5-methyl-4-hexen-1-yl butyrate, the sex pheromone of *Planococcus kraunhiae* (Hemiptera: Pseudococcidae), using a β,γ- to α,β-double-bond migration in an unsaturated aldehyde. Appl Entomol Zool 48:229–232. https://doi.org/10.1007/s13355-013-0173-7

Tabata J (2018) Mating disruption: concepts and keys for effective application. In: Tabata J (ed) Chemical ecology of insects: applications and associations with plants and microbes. CRC Press/Taylor and Francis, Boca Raton, pp 197–233

Tabata J, Ichiki RT (2015) A new lavandulol-related monoterpene in the sex pheromone of the grey pineapple mealybug, *Dysmicoccus neobrevipes*. J Chem Ecol 41:194–201. https://doi.org/10.1007/s10886-015-0545-2

Tabata J, Ichiki RT (2016) Sex pheromone of the cotton mealybug, *Phenacoccus solenopsis*, with an unusual cyclobutane structure. J Chem Ecol 42:1193–1200. https://doi.org/10.1007/s10886-016-0783-y

Tabata J, Ichiki RT (2017) (1*S*,3*R*)-*cis*-Chrysanthemyl tiglate: sex pheromone of the striped mealybug, *Ferrisia virgata*. J Chem Ecol 43:745–752. https://doi.org/10.1007/s10886-017-0879-z

Tabata J, Ishikawa Y (2016) Divergence of the sex pheromone systems in "oriental" *Ostrinia* species. In: Allison JD, Cardé RT (eds) Pheromone communication in moths. University of California Press, Oakland, pp 245–257

Tabata J, Ohno S (2015) Enantioselective synthesis of the sex pheromone of the grey pineapple mealybug, *Dysmicoccus neobrevipes* (Hemiptera: Pseudococcidae), for determination of the absolute configuration. Appl Entomol Zool 50:341–346. https://doi.org/10.1007/s13355-015-0337-8

Tabata J, Teshiba M (2018) Sexual attractiveness and reproductive performance in ageing females of a coccoid insect. Biol Lett 14:20180262. https://doi.org/10.1098/rsbl.2018.0262

Tabata J, Teshiba M, Hiradate S et al (2011) Cyclolavandulyl butyrate: an attractant for a mealybug parasitoid, *Anagyrus sawadai* (Hymenoptera: Encyrtidae). Appl Entomol Zool 46:117–123. https://doi.org/10.1007/s13355-010-0012-z

Tabata J, Narai Y, Sawamura N et al (2012) A new class of mealybug pheromones: a hemiterpene ester in the sex pheromone of *Crisicoccus matsumotoi*. Naturwissenschaften 99:567–574. https://doi.org/10.1007/s00114-012-0935-z

Tabata J, Teshiba M, Shimizu N et al (2015) Mealybug mating disruption by a sex pheromone derived from lavender essential oil. J Essent Oil Res 27:232–237. https://doi.org/10.1080/10412905.2015.1007219

Tabata J, Ichiki RT, Tanaka H et al (2016) Sexual versus asexual reproduction: distinct outcomes in relative abundance of parthenogenetic mealybugs following recent colonization. PLoS One 11:e0156587. https://doi.org/10.1371/journal.pone.0156587

Tabata J, Ichiki RT, Moromizato C et al (2017a) Sex pheromone of a coccoid insect with sexual and asexual lineages: fate of an ancestrally essential sexual signal in parthenogenetic females. J R Soc Interface 14:20170027. https://doi.org/10.1098/rsif.2017.0027

Tabata J, Yasui H, Tsujii N et al (2017b) Mating disruption: progress and perspective for application to non-lepidopteran insects. Jpn J Appl Entomol Zool 61:63–71. https://doi.org/10.1303/jjaez.2017.63

Teshiba M (2013) Integrated management of *Planococcus kraunhiae* Kuwana (Homoptera: Pseudococcidae) injuring Japanese persimmons. Jpn J Appl Entomol Zool 57:129–135. https://doi.org/10.1303/jjaez.2013.129

Teshiba M, Tabata J (2017) Suppression of population growth of the Japanese mealybug, *Planococcus kraunhiae* (Hemiptera: Pseudococcidae), by using an attractant for indigenous parasitoids in persimmon orchards. Appl Entomol Zool 52:153–158. https://doi.org/10.1007/s13355-016-0452-1

Teshiba M, Shimizu N, Sawamura N et al (2009) Use of a sex pheromone to disrupt the mating of *Planococcus kraunhiae* (Kuwana) (Hemiptera: Pseudococcidae). Jpn J Appl Entomol Zool 53:173–180. https://doi.org/10.1303/jjaez.2009.173

Teshiba M, Sugie H, Tsutsumi T et al (2012) A new approach for mealybug management: recruiting an indigenous, but 'non-natural' enemy for biological control using an attractant. Entomol Exp Appl 142:211–215. https://doi.org/10.1111/j.1570-7458.2011.01214.x

Tsueda H (2014) Response of two parasitoid wasps to the sex pheromone of Japanese mealybug, *Planococcus kraunhiae*. Jpn J Appl Entomol Zool 58:147–152. https://doi.org/10.1303/jjaez.2014.147

Ueno H (1963) Studies on the scale insects injury to the Japanese persimmon, *Diospyros kaki* L. I. On the overwintered larvae of the mealy bug, *Pseudococcus kraunhiae* Kuwana. Jpn J Appl Entomol Zool 7:85–91. https://doi.org/10.1303/jjaez.7.85

Umbers KDL, Symonds MRE, Kokko H (2015) The mothematics of female pheromone signaling: strategies for aging virgins. Am Nat 185:417–432. https://doi.org/10.1086/679614

Unelius CR, El-Sayed AM, Twidle A et al (2011) The absolute configuration of the sex pheromone of the citrophilous mealybug, *Pseudococcus calceolariae*. J Chem Ecol 37:166–172. https://doi.org/10.1007/s10886-010-9904-1

van der Kooi CJ, Schwander T (2014) On the fate of sexual traits under asexuality. Biol Rev Camb Philos Soc 89:805–819. https://doi.org/10.1111/brv.12078

Varndell NP, Godfray HCJ (1996) Facultative adjustment of the sex ratio in an insect *(Planococcus citri*, Pseudococcidae) with paternal genome loss. Evolution 50:2100–2105. https://doi.org/10.1111/j.1558-5646.1996.tb03597.x

von Dohlen CD, Kohler S, Alsop ST et al (2001) Mealybug beta-proteobacterial endosymbionts contain gamma-proteobacterial symbionts. Nature 412:433–436. https://doi.org/10.1038/35086563

Walton VM, Daane KM, Pringle KL (2004) Monitoring *Planococcus ficus* in South African vineyards with sex pheromone-baited traps. Crop Prot 23:1089–1096. https://doi.org/10.1016/j.cropro.2004.03.016

Walton VM, Daane KM, Bentley WJ et al (2006) Pheromone-based mating disruption of *Planococcus ficus* (Hemiptera: Pseudococcidae) in California vineyards. J Econ Entomol 99:1280–1290. https://doi.org/10.1603/0022-0493-99.4.1280

Waterworth RA, Redak RA, Millar JG (2012) Probable site of sex pheromone emission in female vine and obscure mealybugs (Hemiptera: Pseudococcidae). J Insect Behav 25:287. https://doi.org/10.1007/s10905-011-9297-1

Wright RH (1964a) After pesticides–what? Nature 204:121–125. https://doi.org/10.1038/204121a0

Wright RH (1964b) Insect control by nontoxic means. Science 144:487. https://www.jstor.org/stable/1713119

Yang T, Gao L, Hu H et al (2014) Chrysanthemyl diphosphate synthase operates in planta as a bifunctional enzyme with chrysanthemol synthase activity. J Biol Chem 289:36325–36335. https://doi.org/10.1074/jbc.M114.623348

Zada A, Dunkelblum E, Assael F et al (2003) Sex pheromone of the vine mealybug, *Planococcus ficus* in Israel: occurrence of a second component in a mass-reared population. J Chem Ecol 29:977–988. https://doi.org/10.1023/A:1022944119077

Zada A, Dunkelblum E, Assael F et al (2008) Attraction of *Planococcus ficus* males to racemic and chiral pheromone baits: flight activity and bait longevity. J Appl Entomol 132:480–489. https://doi.org/10.1111/j.1439-0418.2008.01277.x

Zhang A, Amalin D, Shirali S et al (2004) Sex pheromone of the pink hibiscus mealybug, *Maconellicoccus hirsutus*, contains an unusual cyclobutanoid monoterpene. Proc Natl Acad Sci U S A 101:9601–9606. https://doi.org/10.1073/pnas.0401298101

Zhang A, Wang S, Vitullo J et al (2006) Olfactory discrimination among sex pheromone stereoisomers: chirality recognition by pink hibiscus mealybug males. Chem Senses 31:621–626. https://doi.org/10.1093/chemse/bjl001

Zou Y, Millar JG (2015) Chemistry of the pheromones of scale and mealybug insects. Nat Prod Rep 32:1067–1113. https://doi.org/10.1039/c4np00143e

Zou Y, Daane KM, Bentley WJ et al (2010) Synthesis and bioassay of racemic and chiral *trans*-α-necrodyl isobutyrate, the sex pheromone of the grape mealybug *Pseudococcus maritimus*. J Agric Food Chem 58:4977–4982. https://doi.org/10.1021/jf904452v

Chapter 4
Hybrid Sex Pheromone Communication Systems in Seed Beetles

Kenji Shimomura and Kanju Ohsawa

Abstract Seed beetles of the subfamily Bruchinae, of the genus *Callosobruchus* in particular, are economically important pests of stored legumes. Two types of sex pheromones are involved in the mating processes of seed beetles. One is the volatile sex attractant pheromone that provides specific information about the presence of a mating partner, and the other is the contact sex pheromone that not only provides information for the recognition of species and sex, but also induces courtship behavior. Among *Callosobruchus* seed beetles, sex pheromones of 5 species, *C. maculatus, C. chinensis*, *C. analis*, *C. subinnotatus*, and *C. rhodesianus*, have been investigated with the intent of utilizing them for pest control. In this chapter, we summarize the biology of 5 species, and the chemistry of their sex attractant and contact sex pheromones. This is followed by the relationship between the structure and mating specificity of the pheromones, and suggestions for future research.

Keywords *Callosobruchus* · Volatile sex pheromones · Contact sex pheromones

4.1 Introduction

Grain legumes provide an important protein source worldwide (Day 2013). Infestation of legumes by stored product pests is a serious problem causing a huge loss in product. *Callosobruchus* seed beetles are among the most serious pests of stored products (Rees 2004). The genus *Callosobruchus* (Coleoptera: Chrysomelidae: Bruchinae) includes at least 20 species (Borowiec 1987). These species are distributed worldwide, particularly in tropical and subtropical regions (Rees 2004). Most are pests of stored legumes (Tuda et al. 2006), suggesting that they have adapted to

K. Shimomura (✉)
Department of Chemistry for Life Sciences and Agriculture, Tokyo University of Agriculture, Tokyo, Japan
e-mail: k3shimom@nodai.ac.jp

K. Ohsawa
Tokyo University of Agriculture, Tokyo, Japan

Y. Ishikawa (ed.), *Insect Sex Pheromone Research and Beyond*, Entomology Monographs, https://doi.org/10.1007/978-981-15-3082-1_4

the toxic chemicals in the legumes through evolutional processes (Tuda 2007). Extensive field sampling led to the discovery of a new species that feeds on non-economic legumes (Tuda et al. 2005).

Seed beetles are 3–7 mm in length. Females lay eggs on the seed coat or pod. Newly hatched larvae burrow directly into the seed, and larvae grow within it through their larval life. Just before pupation, larvae burrow a tunnel for emergence, which reaches near the surface of the seed, leaving the seed coat intact. Pupation takes place inside the cavity in the seed, and the adult emerges through the tunnel, leaving a circular hole on the seed (Southgate 1979). Adults do not feed on dried seeds; they may consume nectar and pollen from flowers (Zannou et al. 2003). The adult lifespan is variable; it may be >100 days when adults have access to flowers and nectar, or < 10 days when adults are confined under tropical conditions. As emerged adults are mature and ready to mate, they propagate quickly. Under optimal conditions (30–35 °C, 70–90% RH), the life cycle completes in as fast as 22–25 days (Southgate 1979).

Two types of sex pheromones are involved in the mating processes of the seed beetles. One is the volatile sex attractant pheromone that provides specific information about the presence of a mating partner, and the other is the contact sex pheromone that not only provides information for the recognition of species and sex, but also induces courtship behavior (Tanaka et al. 1981). Among *Callosobruchus* seed beetles, the sex pheromones of 5 species, *C. maculatus, C. chinensis*, *C. analis*, *C. subinnotatus*, and *C. rhodesianus*, have been investigated with the intent of utilizing them for pest control. In this chapter, we summarize the biology of 5 species, and the chemistry of their sex attractant and contact sex pheromones. This is followed by the relationship between the structure and mating specificity of the pheromones, and suggestions for future research.

4.2 Cowpea Weevil, *Callosobruchus maculatus* (F)

The presence of a sex attractant pheromone produced by mature *C. maculatus* females was first demonstrated by Rup and Sharma (1978). Qi and Burkholder (1982) characterized the behavioral mating traits in laboratory assays. When a female and male were placed together in a filter-paper-lined Petri dish, the male immediately drummed the surface of the filter paper with his antennae to search for the female. Copulation attempts mostly began within 1 min and, during the first moment, the male stroked the abdomen of the female with his antennae. When the female remained passive, the male mounted her for copulation. Copulation continued for 6.8 ± 4.2 min.

The secretion site of the sex attractant pheromone was investigated by the ablation of body parts. Female heads and thoraces elicited responses from 15 and 20% of males, respectively, whereas female abdomens elicited responses from 75% of males. Further investigation revealed that all males tested responded to the female ventral abdominal tip (Qi and Burkholder 1982). Ramaswamy et al. (1995) also reported that the solvent extracts from the female pygidium elicited the highest electroantennogram (EAG) responses from the male antennae. A more detailed study revealed that the intersegmental membrane region connecting tergite VIII and sternite VIII, which connects the pygidium to the ovipositor, likely produces the sex attractant pheromone (Pierre et al. 1997).

Changes in male behavioral and EAG responses were observed to assess the rhythmicity and dynamics of the production of the sex attractant pheromone (Qi and Burkholder 1982; Shu et al. 1996). The beetles were sexually active in the photophase, and virgin females emitted more sex attractant pheromone during the first 2 h of photophase. The release of pheromone was observed only in the first week of adult life. The production of the pheromone by virgin females increased from day 1 to day 3, remained at a high level on day 4, and then declined until day 7. Mating significantly reduced the production of the sex attractant pheromone by females; however, the reduction only lasted for 2 days and remating was observed from 2 days after the first mating. The presence of host beans did not affect the production of the sex attractant pheromone by the females.

In this species, two distinct adult forms, flight and flightless, which differ in their morphology, physiology, and behavior, are known to exist (Utida 1954). The flightless females are sexually mature at emergence; maturation is independent from exogenous factors. In contrast, maturation and associated emission of the sex attractant pheromone are delayed in the females of the flight form (Lextrait et al. 1995). The flightless form was suggested to only develop in stored beans, whereas the flight form is present in the field (Lextrait et al. 1994). In the flight form, environmental factors, such as the presence of the host plant (bean), stimulate the release of the sex attractant pheromone. Air temperature also affects the age when the females start to release the sex attractant pheromone (Lextrait et al. 1995). Lextrait et al. (1994) compared locomotor responses of the males of flight and flightless forms in a tubular olfactometer. The flightless form exhibited similar responses regardless of the stimulation by the sex attractant pheromone, i.e., either delivered as short pulses or as a continuous flow. In contrast, the flight form displayed markedly reduced responses when the stimulus was continuous.

The structures of the female-produced sex attractant pheromone were first described by Cork et al. (1991), although experimental data were absent. They reported that females of *C. maculatus* produced (*Z*)-3-methyl-2-heptenoic acid (**1**) and (*Z*)-3-methyl-3-heptenoic acid (**2**). Later, Phillips et al. (1996) identified 5 short-chain fatty acids including **1** and **2** as sex attractant pheromones: **1**, **2**, (*E*)-3-methyl-2-heptenoic acid (**3**), (*E*)-3-methyl-3-heptenoic acid (**4**), and 3-methyleneheptanoic acid (**5**). In laboratory bioassays using a single vial and wind tunnel, the five single compounds exhibited activity. In the single-vial bioassay, each compound elicited responses from >75% of males within 5 s. On comparison of the biological activity of 5 single compounds and a mixture of all compounds, the activity of **2** and **4** was higher than that of the others. In the wind-tunnel bioassay, 200 ng each of the 5 acids alone elicited male upwind flight; however, the beetles approached and contacted the lure only when **2**, **4**, **5**, or a mixture of all 5 acids (40 ng each) was used as a lure (Phillips et al. 1996).

The cuticular lipids of the adult *C. maculatus* were first characterized by Baker and Nelson (1981). There was 11 μg of lipids per adult, which represented 0.21% of the body weight, and hydrocarbons amounted to 9.7 μg (87.9% of cuticular lipids). Although four series of alkanes, i.e., n-alkanes, internal branched monomethylalkanes, terminal branched monomethylalkanes, and internal branched dimethyl alkanes were detected, no unsaturated alkenes were contained in the hydrocarbons. No significant difference in hydrocarbon composition was found between males and females.

The contact sex pheromone of *C. maculatus* was analyzed by Nojima et al. (2007). Cuticular lipids were extracted from the sheets of filter paper collected from the cages that housed females. The behavioral assay was conducted using a glass rod as a female dummy (Fig. 4.1). The bioassay-guided purification of the crude extracts revealed that a mixture of acidic and neutral fractions from the acid-base partition of the crude extracts elicited copulation attempts from males. After subsequent purification, the active acidic compounds were identified as a mixture of dicarboxylic acids, 2,6-dimethyloctane-1,8-dioic acid (**6**) and azelaic acid (nonanedioic acid, **7**). The active neutral compounds were identified as a mixture of C_{27}–C_{35} straight-chain and methyl-branched hydrocarbons. Bioassay of copulation-stimulating activity using synthetic standards of **6** and **7** revealed that 10 μg of **6** elicited full copulation behavior from the males, and 43% of males responded to **7** at a higher dose of 100 μg. Although hydrocarbons alone were only slightly active, the addition of hydrocarbons to the dicarboxylic acids significantly increased the activity. Thus, the structure of the hydrocarbons may not be important. For example, the addition of non-native hydrocarbons, such as hexadecane, octadecane, or liquid paraffin, to the dicarboxylic acids also increased the activity to almost the same degree as the natural hydrocarbons.

Compound **6** has two chiral centers, thus 4 stereoisomers are present. The stereoisomeric composition of **6** was determined by the 2D-HPLC-Ohrui-Akasaka method as (2*R*,6*S*): (2*S*,6*R*): (2*S*,6*S*): (2*R*,6*R*) = 43: 38: 18: trace (Yajima et al. 2006). In the behavioral bioassay, although all synthetic stereoisomers presented with the natural hydrocarbons elicited copulation behavior from the males, the more abundant isomer in the natural product was slightly more active (Yajima et al. 2006).

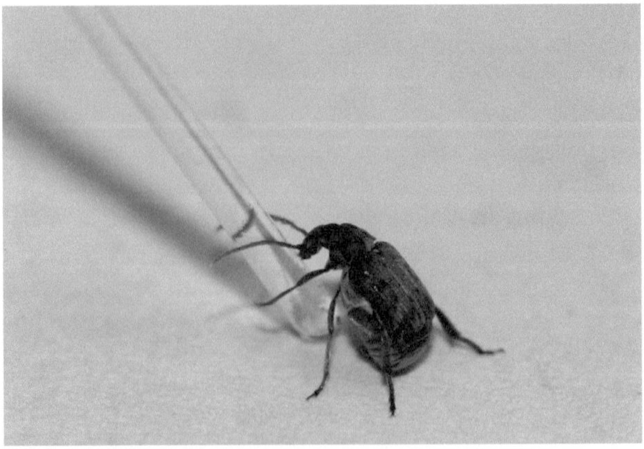

Fig. 4.1 Typical copulation behavior of male *Callosobruchus maculatus* toward a glass rod, the tip of which was treated with cuticular lipids extracted from the sheets of filter paper lining the container that housed *C. maculatus* females

4.3 Southern Cowpea Weevil, *Callosobruchus chinensis* (L.)

The presence of sex attractant pheromone in *C. chinensis*, also known as the azuki bean weevil, was first reported by Honda and Yamamoto (1976). The typical sequence of mating behavior is as follows: When a male detected the presence of a female, he raised his antennae and ran toward her in a zigzag pattern (sex attraction). As the male contacted the female, he lowered his antennae and tried to bend his head. If the female took several steps forward, the male followed her with his antennae lowered. When the female stopped, the male bent his head, and extended the tip of his abdomen toward the female and extruded his genital organ (copulation release). The female raised her abdomen slightly and accepted copulation.

The ether extract of the filter paper shelter in the Petri dish that housed virgin females exhibited both sex attraction and copulation release activity. When the ether extract was maintained at 90 °C for 24 h, however, the attraction activity was selectively lost. These results suggested that different types of sex pheromones, one volatile and the other non-volatile, are involved in the attraction and copulation-release activity, respectively (Tanaka et al. 1981).

The contact sex pheromone was extracted from the filter paper in the Petri dishes that housed the females. The behavioral assay was conducted using a glass rod as a female dummy. Based on the bioassay-guided purification of the crude extracts, the neutral-basic fraction and acidic fraction did not exhibit copulation-release activity when tested separately; however, original copulation activity was observed when the two fractions were recombined, indicating synergism between the two. (*E*)-3,7-dimethyl-2-octene-1,8-dioic acid (**8**) was identified as an active compound in the acidic fraction and named callosobruchusic acid (Tanaka et al. 1981, 1982). Several hydrocarbons, 3-methylpentacosane, 11-methylheptacosane, 3-methylheptacosane, 11-methylnonacosane, 13-methylnonacosane, 11,15-dimethylnonacosane, 13-methylhentriacontane, 9,13-dimethylhentriacontane, and 11,15-dimethyltritriacontane, were identified as synergists. The geometric isomer of **8** was found to be inactive; only the (*E*)-form, **8**, was active when combined with natural hydrocarbons (Tanaka et al. 1982).

The stereoisomeric composition of **8** was determined by the 2D-HPLC-Ohrui-Akasaka method to be $R{:}S = 3.3\text{–}3.4{:}\ 1$ (Yajima et al. 2007), indicating that the natural pheromone (**8**) is enantiomerically impure. To clarify the relationship between stereochemistry and the pheromone activity, optically active dicarboxylic acids were synthesized (Mori et al. 1983). The ED_{50} of the isomers when presented with 12 μg of natural hydrocarbons was 6.5 ng for (*S*)-**8** and 11.4 ng for (*R*)-**8**. Therefore, both enantiomers are biologically active, and *C. chinensis* females produce an approximately three-fold larger amount of the less active enantiomer (*R*)-**8**.

The sex attractant pheromone of *C. chinensis* was identified by Shimomura et al. (2008). GC-electroantennographic detector (EAD) analysis of the headspace volatile collected from virgin females revealed two strong and one weak EAD responses. A larger amount of the pheromone was extracted from corrugated filter papers in Petri dishes that housed virgin females. After subsequent purification steps and structural analysis, two compounds that elicited strong EAD responses were identified as isomers of homosesquiterpene aldehyde, (2*Z*,6*E*)-7-ethyl-3,11-dimethyl-2,6,10-dodecatrienal (**9**) and (2*E*,6*E*)-7-ethyl-3,11-dimethyl-2,6,10-dodecatrienal (**10**). The ratio of **9** and **10** in the pheromone extracts was estimated to be 1:1 by GC analysis; however, as approximately 60% of **9** (2*Z*,6*E*) and 30% of **10** (2*E*,6*E*) isomerized to **10** and **9** under our thermal conditions of GC, respectively, **9** was considered to be the dominant component of the pheromone. In the laboratory assay using an olfactometer, the attractancy of **10** was lower than that of **9**. Isomer **9** exhibited equal or higher activity than mixtures of **9** and **10** with different blend ratios, suggesting that there is neither synergism nor antagonism between the two isomers.

Chiluwal et al. (2017a) conducted a field bioassay using a semi-open greenhouse and laboratory-reared *C. chinensis* beetles. They compared the attractancy of pheromones with 2 different ratios of isomers; **9** (2*Z*,6*E*): **10** (2*E*,6*E*) = 4: 6 and 1: 9. The number of males captured by the traps baited with the 4:6 lure was significantly higher than that baited with the 1:9 lure.

Aiming at the development of alternative pest management measures, the anti-pheromonal activity of essential oils from *Illicium verum*, *Croton anisatum*, and *Gaultheria fragrantissima*, which exhibit repellency or insecticidal activity against *C. chinensis* (Park et al. 2016; Chiluwal et al. 2017b), were examined (Chiluwal et al. 2018a). These essential oils significantly reduced the activity of the sex attractant pheromone in laboratory bioassays using a Y-tube olfactometer. Furthermore, *trans*-anethole, a major component of *I. verum* and *C. anisatum*, and methyl salicylate, a major component of *G. fragrantissima*, antagonized the pheromone in the laboratory and semi-field bioassays. These anti-pheromonal effects may be due to either masking of the sex attractant pheromone or confusion among males in detecting pheromone plumes in the background of the essential oil plumes (Chiluwal et al. 2018a).

Ionizing radiation is commonly used to produce sterile insects in insect pest management programs (sterile insect technique, SIT). The effects of gamma radiation on the production of sex attractant pheromone of virgin females were evaluated by Chiluwal et al. (2018b). It is a prerequisite for successful sterilization that azuki bean beetles are sterilized without total disruption of the calling behavior. Irradiation by 300 Gy caused a 73.6 ± 13.4 and 62.5 ± 16.1% reduction in the production of two pheromone components, **9** and **10**, respectively.

4.4 *Callosobruchus analis* (F)

The sex attractant pheromone of *C. analis* was first reported by Cork et al. (1991). They collected headspace volatiles from virgin females of *C. analis* and analyzed them by GC-EAD. Only one compound elicited EAD responses from the antennae of male *C. analis*, which was identified as (*Z*)-3-methyl-2-heptenoic acid (**1**). This compound was not emitted by the males or mated females in the presence of the males. Although synthetic **1** and its geometric isomer, (*E*)-3-methyl-2-heptenoic acid (**3**), elicited EAG responses from male antennae of *C. analis*, only **1** exhibited attractancy in the pitfall bioassay. Compound **3** was inactive even in combination with **1**. Phillips et al. (1996), however, reported that although the female extract collected from the filter paper in the container that housed females induced sexual activity in the no-choice vial bioassay, synthetic **1** did not elicit responses from the males. Furthermore, the amount of **1** in the extract from the filter paper was below the detection limit. This inconsistency between the two studies has yet to be resolved.

The contact sex pheromone of *C. analis* was examined by Shimomura et al. (2010d). The males demonstrated copulation responses to freeze-killed conspecific females but not to freeze-killed males or solvent-washed females. Furthermore, the cuticular lipids collected by rubbing the body surface of the females with filter paper disks elicited copulation behavior from the males, confirming the presence of the contact sex pheromone in *C. analis*. To identify the pheromone, solvent-washed freeze-killed females were employed as controls for the bioassay. The acid-base partition of the crude extracts revealed that only the acidic fraction was sufficient to induce copulation attempts from the males. After subsequent purification and structural analysis, the active compounds were identified as 2,6-dimethyloctane-1,8-dioic acid (**6**) and (*E*)-3,7-dimethyl-2-octene-1,8-dioic acid (**8**). The stereoisomeric composition of **6** and **8** was determined by the 2D-HPLC-Ohrui-Akasaka method as 100% (2*S*,6*R*)-**6** and 100% (*S*)-**8**, indicating that both compounds were optically pure. The ratio of (2*S*,6*R*)-**6**: (*S*)-**8** was 1.8: 1.

Synthetic stereoisomers of **6** and **8** were evaluated for their copulation-release activity using the female dummy. A mixture of (2*S*,6*R*)-**6** and (*S*)-**8** at the natural ratio exhibited copulation-release activity in a dose-dependent manner, and 95% of males exhibited copulation behavior at 0.5 of the female equivalent. When (2*S*,6*R*)-**6** and (*S*)-**8** were tested alone, they elicited copulation behavior from 70% and 20% of males, respectively. Although the two diastereomers, (2*R*,6*R*)-**6** and (2*S*,6*S*)-**6**, elicited weak copulation behavior, no males exhibited copulation behavior to the enantiomer (2*R*,6*S*)-**6**. The racemic mixture of (2*S*,6*R*)-**6** and (2*R*,6*S*)-**6** also did not induce copulation activity.

In the subsequent study, copulation activity was evaluated using a glass rod as the female dummy. Males exhibited copulation behavior to the glass rod applied with the crude extract or synthetic mixture of (2*S*,6*R*)-**6** and (*S*)-**8** plus the neutral fraction. However, no copulation behavior was induced by the glass rod applied with the synthetic mixture alone or neutral fraction. Furthermore, copulation behavior was induced by the glass rod applied with the synthetic mixture of (2*S*,6*R*)-**6** and (*S*)-**8** plus the hydrocarbon fraction and 5% ether-eluted fraction of the neutral fraction. These results suggested that there is another synergist for the contact sex pheromone of *C. analis,* as in the case of *C. maculatus* and *C. chinensis* (Shimomura et al. 2010a).

4.5 *Callosobruchus subinnotatus* (Pic)

Based on the ultrastructure of the integumentary glands, the female pygidium was implicated as the source of sex attractant pheromone production in *C. subinnotatus* (Ramaswamy et al. 1995). Mbata et al. (1997) reported that a pair of *C. subinnota-*

tus placed in a glass vial started mating within 2–6 min after pairing during the scotophase, whereas it was 4–10 min after pairing during the photophase. The solvent extracts from pygidia of virgin females, among others, elicited the highest EAG responses from male antennae (Shu et al. 1998). These results confirmed the production of the sex attractant pheromone by the female, and the pygidium as the site of production.

Changes in male EAG responses were observed to study the rhythmicity and dynamics of the production of the female sex attractant pheromone (Shu et al. 1998). Production of the pheromone began before the female emerged from the bean. The amount of pheromone increased for 2 days after emergence to reach its maximum, and then decreased gradually. Mating suppressed the production of the sex attractant pheromone. Responses of male antennae to the extract from 1-day-old mated females were significantly lower than those to the extract from virgin females of the same age. Suppression of the production occurred immediately after copulation and lasted for 1 day. Two days after mating, females produced similar amounts of sex attractant pheromone as virgin females of the same age.

The identification of the sex attractant pheromone was performed by Shu et al. (1999). The whole-body extract of 2-day-old virgin females elicited strong EAG responses from the male antennae. The EAG active compounds were identified as two geometrical isomers, (*Z*)-3-methyl-2-heptenoic acid (**1**) and (*E*)-3-methyl-2-heptenoic acid (**3**). EAG responses of the male antennae to the two compounds, **1** and **3,** were almost the same in the log-linear ranges of doses. The EAG responses to the mixtures of **1** and **3** blended at varying ratios were significantly different; an equal mixture of **1** and **3** elicited a significantly higher EAG response than did 10% of **1** plus 90% of **3**, 90% of **1** plus 10% of **3**, or either isomer alone.

In the Y-tube olfactometer assay, **3** alone was not attractive to the male beetles. The mixtures of **1** and **3** at varying ratios, or **1** alone was more attractive than the control. Moreover, when the two isomers were compared individually against mixtures of two isomers, the males favored the mixtures over the individual isomers, suggesting that both **1** and **3** are components of the sex attractant pheromone.

To clarify how males of *C. subinnotatus* respond to the pheromone source in an environment, such as grain storage silos and warehouses, where air circulation is poor or absent, behavioral bioassays were conducted in the presence and absence of airflow using a laboratory olfactometer (Mbata et al. 2000). The percentage of males that reached the pheromone source was higher and the time spent to reach the pheromone sources was shorter under the conditions with directional airflow. However, males responded to the sex attractant pheromone source even in the absence of airflow. The males reached the pheromone source significantly faster when the mixtures of **1** and **3** were presented than when the compounds were presented alone. The number of males that reached the pheromone sources was also higher when the mixtures of **1** and **3** were used as the odor source.

4.6 *Callosobruchus rhodesianus* (Pic)

C. rhodesianus is mainly distributed in southern Africa, but has spread around the equator where it is sporadically reported (Southgate 1979; Giga and Smith 1983). Due to the morphological similarity, *C. rhodesianus* was once regarded as synonymous with *C. chinensis*, but Southgate (1958) concluded them to be separate species.

The sex attractant pheromone in *C. rhodesianus* was identified by Shimomura et al. (2010b). The headspace volatiles collected separately from virgin females and males were assayed using a Y-tube olfactometer in the laboratory. Only the volatiles collected from the virgin females were attractive to the males, suggesting the presence of the female-produced sex attractant pheromone. GC-EAD analysis using the male antennae revealed the presence of two EAD-active compounds. Through olfactometer-guided purification, only one compound was found to be sufficient for attractancy. Further purification steps and structural analysis identified the active compound to be (*E*)-7-ethyl-3,11-dimethyl-6,10-dodecadienal (2,3-dihydrohomofarnesal, **11**). The absolute configuration of **11** was determined to be (*S*)-**11**. Bioassay of synthetic enantiomers demonstrated that only (*S*)-**11** was attractive. Neither (*R*)-**11** nor the racemic **11** exhibited attractancy, indicating an inhibitory effect of (*R*)-isomer (Shimomura et al. 2010b). The other EAD-active compound was also isolated and the structure was determined to be (*E*)-6-ethyl-2,10-dimethyl-5,9-undecadienal. The absolute configuration and biological function of this compound, however, remain unclear (Shimomura et al. 2010c).

The presence of the contact sex pheromone in *C. rhodesianus* was first demonstrated by Shimomura et al. (2010d). The males of *C. rhodesianus* exhibited copulation responses to the freeze-killed conspecific females but not to the freeze-killed males or solvent-washed females. Furthermore, cuticular lipids collected by rubbing the body surface of the females with filter paper disks elicited copulation behavior from the males when applied to the solvent-washed female (female dummy), indicating the presence of the contact sex pheromone. Structural elucidation of the contact sex pheromone was performed by Shimomura et al. (2016). A glass rod was used as the female dummy in the bioassay. The rod applied with the crude lipid extracts of the female elicited copulation behavior from the males. Acid-base partition of the crude extracts revealed that only the neutral fraction was sufficient for eliciting male copulation activity. After subsequent purification, the active compounds were identified as 6,10,14-trimethyl-2-pentadecanone (hexahydrofarnesyl acetone, **13**) and 2-nonadecanone (**14**). Series of C_{27}–C_{33} hydrocarbons (n-alkanes, and a homologous series of mono- and dimethyl branched alkanes) were identified as synergists. As neither **13** nor **14** exhibited significant copulation-release activity when tested alone with the natural hydrocarbons, both **13** and **14** together with the hydrocarbons are needed to fully induce copulatory activity.

4.7 Specificity of Pheromone-Based Mate Recognition

Due to the increase in international trade, the chance of encountering native and non-native *Callosobruchus* seed beetles has significantly increased. For example, in a study conducted in the central region of Ghana, *C. maculatus*, *C. chinensis*, *C. analis*, *C. rhodesianus*, and *C. subinnotatus* emerged from the cowpea stocks from Burkina Faso, Ejura, Ghana, and Niger, which were available in the markets (Bawa et al. 2017). Mate recognition by the males is basically secured by the sex pheromones produced by the females; however, interspecific mating is observed under both laboratory and natural conditions. For example, Nojima et al. (2007) reported that males of *C. maculatus* and *C. chinensis* frequently exhibited copulation behavior with females of other species under laboratory conditions. In Afrotropical regions, *C. rhodesianus* was reported to have reproductive interference from *C. maculatus* (Giga and Canhão 1997).

Homologs of short-chain fatty acids have been identified from *C. maculatus*, *C. subinnotatus*, and *C. analis* as components of sex attractant pheromones (Table 4.1). A few components are shared by the three species. The sex attractant pheromone of *C. maculatus* consists of (*Z*)-3-methyl-2-heptenoic acid (**1**), (*Z*)-3-methyl-3-heptenoic acid (**2**), (*E*)-3-methyl-2-heptenoic acid (**3**), (*E*)-3-methyl-3-heptenoic acid (**4**), and 3-methyleneheptanoic acid (**5**), whereas that of *C. subinnotatus* consists of **1** and **3**. Mbata et al. (2000) examined the incidence of cross attraction between *C. maculatus* and *C. subinnotatus*, which use a mixture of **1** and **2,** and a mixture of **1** and **3** as the sex pheromone, respectively. Although the sex pheromone of *C. maculatus* (**1** and **2**) attracted only conspecific males, that of *C. subinnotatus* (**1** and **3**) attracted males of both species in the laboratory assays. These results suggested that compound **2** antagonizes the responses of *C. subinnotatus* to the sex pheromone of *C. maculatus* (Mbata et al. 2000).

Table 4.1 Chemical structures of the sex attractant pheromones of *Callosobruchus* seed beetles

	Structure	Species		Structure	Species
1	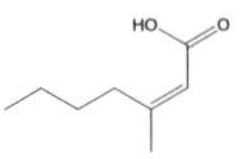 (*Z*)-3-methyl-2-heptenoic acid	*C. maculatus* *C. analis* *C. subinnotatus*	9	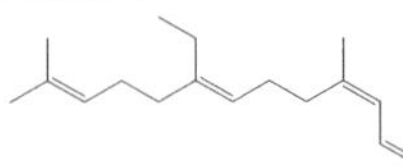 (2*Z*,6*E*)-7-ethyl-3,11-dimethyl-2,6,10-dodecatrienal	*C. chinensis*
2	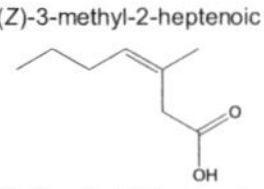 (*Z*)-3-methyl-3-heptenoic acid	*C. maculatus*	10	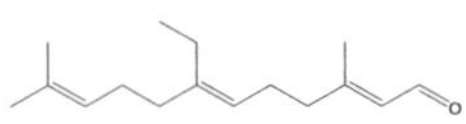 (2*E*,6*E*)-7-ethyl-3,11-dimethyl-2,6,10-dodecatrienal	*C. chinensis*
3	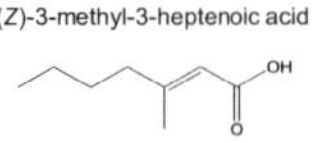 (*E*)-3-methyl-2-heptenoic acid	*C. maculatus* *C. subinnotatus*	11	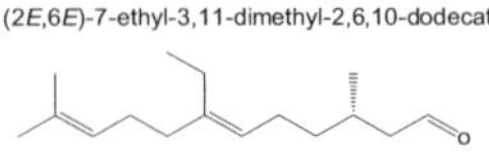 (*S*,*E*)-7-ethyl-3,11-dimethyl-6,10-dodecadienal	*C. rhodesianus*
4	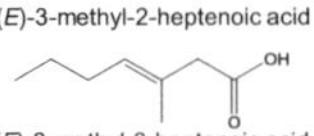 (*E*)-3-methyl-3-heptenoic acid	*C. maculatus*	12	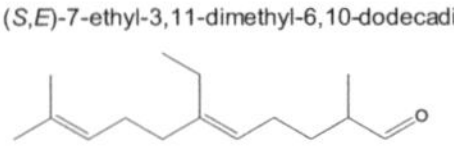 (*E*)-6-ethyl-2,10-dimethyl-5,9-undecadienal	*C. rhodesianus*
5	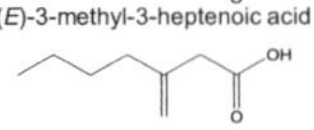 3-methyleneheptanoic acid	*C. maculatus*			

In contrast to the sex attractant pheromones of *C. maculatus*, *C. subinnotatus*, and *C. analis*, homosesquiterpene aldehydes have been identified as sex attractant pheromone components from *C. chinensis* [(2*Z*,6*E*)-7-ethyl-3,11-dimethyl-2,6,10-dodecatrienal (**9**) and (2*E*,6*E*)-7-ethyl-3,11-dimethyl-2,6,10-dodecatrienal (**10**)], and from *C. rhodesianus* [(*E*)-7-ethyl-3,11-dimethyl-6,10-dodecadienal (**11**)] (Table 4.1). The structures of the two types of sex attractant pheromones, short-chain fatty acids and homosesquiterpene aldehydes, are markedly different. Considering the incidence of mating between *C. maculatus* and *C. chinensis* (or *C. rhodesianus*) as described above, reproductive interference from *C. maculatus* may have been one of the factors that prompted the shift in sex attractant pheromones in *C. chinensis* and *C. rhodesianus*.

Following the attraction by the sex attractant pheromones, contact sex pheromones play an essential role in the final mate recognition in *Callosobruchus* species (Table 4.2). To assess the species specificity of contact sex pheromones, cross-species copulation assays were performed using freeze-killed females of *C. maculatus*, *C. chinensis*, *C. rhodesianus*, and *C. analis* (Shimomura et al. 2010d). The males of *C. maculatus* and *C. chinensis* exhibited copulation behavior to conspecific females and all heterologous females. In contrast, the males of *C. rhodesianus* and *C. analis* exhibited copulation behavior only to the conspecific female dummies (Fig. 4.2).

In *C. maculatus*, the main component of the contact sex pheromone is 2,6-dimethyloctane-1,8-dioic acid (**6**). The stereoisomeric composition is not regulated, and the activity is independent of stereochemistry. On the other hand, *C. chinensis* utilizes (*E*)-3,7-dimethyl-2-octene-1,8-dioic acid (**8**), which differs from **6** by one double bond, as the main component of the contact sex pheromone. The stereoisomeric composition of **8** in *C. chinensis* is also not regulated, and its activity

Table 4.2 Chemical structures of the contact sex pheromones and analogs of *Callosobruchus* seed beetles

	Structure	Species		Structure	Species
6	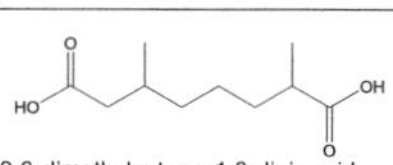 2,6-dimethyloctane-1,8-dioic acid	*C. maculatus*	13	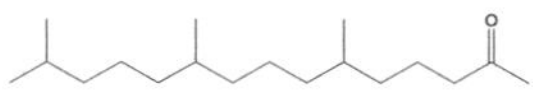 6,10,14-trimethyl-2-pentadecanone	*C. rhodesianus*
7	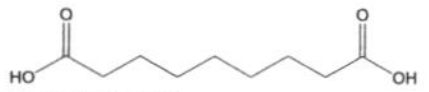 nonanedioic acid	*C. maculatus*	14	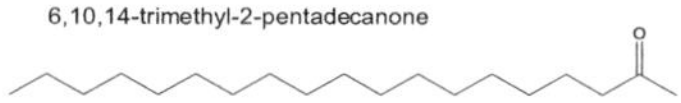 2-nonadecanone	*C. rhodesianus*
8	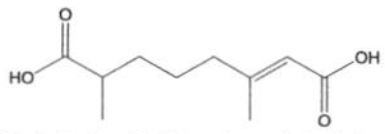 (*E*)-3,7-dimethyl-2-octene-1,8-dioic acid	*C. chinensis*	15	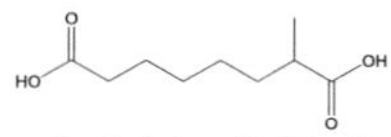 2-methyloctane-1,8-dioic acid	(Derivation) *C. rhodesianus*
(2*S*,6*R*)-6	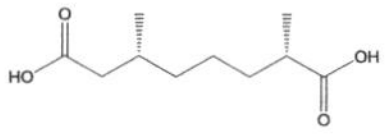 (2*S*,6*R*)-2,6-dimethyloctane-1,8-dioic acid	*C. analis*	16	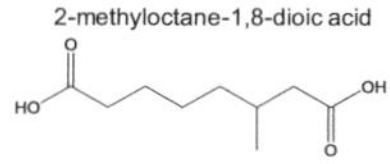 3-methyloctane-1,8-dioic acid	(Derivation) *C. rhodesianus*
(*S*)-8	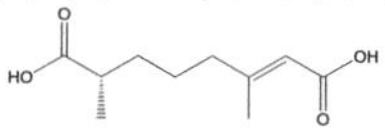 (*S*,*E*)-3,7-dimethyl-2-octene-1,8-dioic acid	*C. analis*			

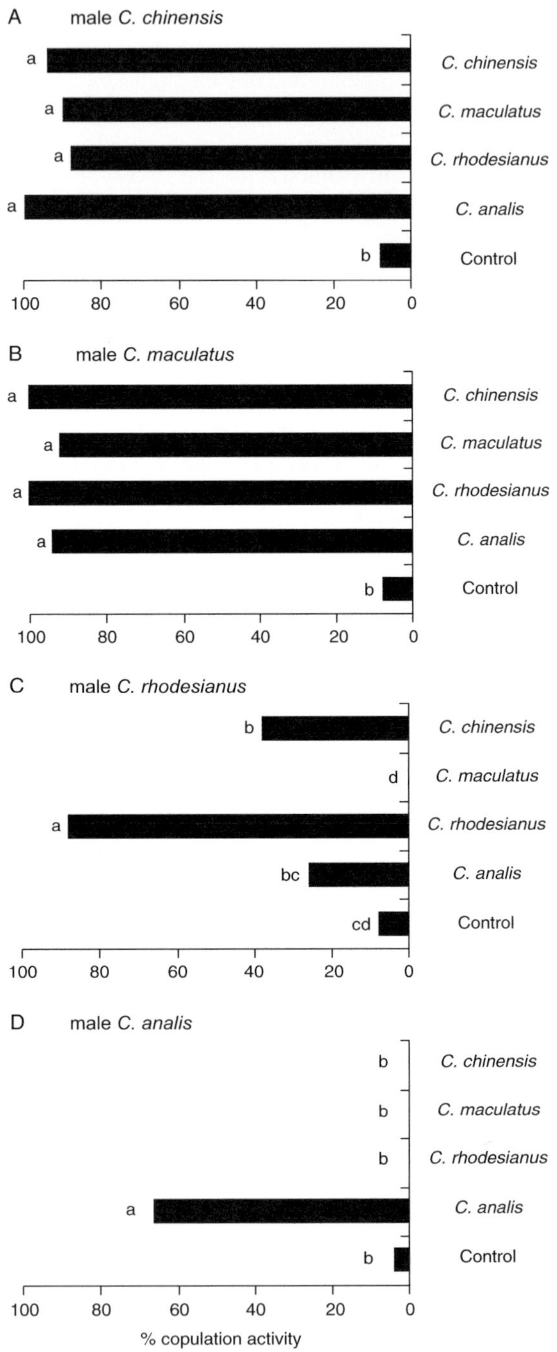

Fig. 4.2 Comparison of cross-species copulation behavior of male (**a**) *Callosobruchus chinensis*, (**b**) *C. maculatus*, (**c**) *C. rhodesianus*, and (**d**) *C. analis* toward females of their own species and three congeners (n = 50). Conspecific females washed with solvents (dummies) were used as the control. Use of the same letter within species indicates no significant difference (Fisher's exact probability test followed by a sequential Bonferroni test: $P > 0.05$) (Reproduced from Shimomura et al. (2010d) with permission)

is independent of stereochemistry. Thus, mate recognition systems based on contact sex pheromones in *C. maculatus* and *C. chinensis* are not strict.

In *C. analis*, the main components of the contact sex pheromone are **6** and **8**. Stereoisomeric compositions of the pheromone components are strictly regulated, and the enantiomer exhibits antagonistic effects. Therefore, mate recognition based on contact sex pheromone is specific in this species.

C. rhodesianus utilizes 6,10,14-trimethyl-2-pentadecanone (**13**) and 2-nonadecanone (**14**) as the main components of the contact sex pheromone. As the structures of these compounds are significantly different from dicarboxylic acids used as pheromone components in *C. maculatus* and *C. chinensis*, the mate recognition specificity of male *C. rhodesianus* is achieved by the difference in the main component of the contact sex pheromone. However, asymmetric cross-copulation behavior was observed between *C. maculatus* and *C. rhodesianus*. Although males of *C. rhodesianus* exhibited copulation behavior only with conspecific females, males of *C. maculatus* also exhibited copulation behavior with *C. rhodesianus* (Shimomura et al. 2010d). To clarify the cause of asymmetric cross-copulation behavior, compounds in the cuticular lipids of female *C. rhodesianus* that elicit copulation behavior from male *C. maculatus* were surveyed (Shimomura et al. 2017). The crude extracts collected from female *C. rhodesianus* induced copulation behavior from male *C. maculatus*. The acid-base partition of the crude extracts revealed that the combination of acidic and neutral fractions had heterospecific copulation-inducing activity. After subsequent purification and structural analysis, the active acidic compounds were identified as 2-methylsuberic acid (2-methyloctane-1,8-dioic acid, **15**), 3-methylsuberic acid (3-methyloctane-1,8-dioic acid, **16**), and azelaic acid (nonanedioic acid, **7**). The copulation-inducing activity was examined using synthetic standards and natural hydrocarbons from female *C. rhodesianus*. When acidic compounds were singly combined with natural hydrocarbons, **15** and **16**, but not **7**, elicited significant copulation activity from male *C. maculatus*. Compounds **15** and **16** are monomethyl analogs of the *C. maculatus* contact sex pheromone (2,6-dimethyloctane-1,8-dioic acid, **6**). The median effective dose (ED_{50}) of **7** for eliciting copulation activity from *C. maculatus* was 270-fold higher than that of **6** (Nojima et al. 2007), suggesting that the methyl side chain at the α- and/or β-position is important for eliciting copulation behavior from *C. maculatus*. Therefore, heterospecific copulation activity of male *C. maculatus* toward *C. rhodesianus* was induced by the presence of contact sex pheromone analogs in the cuticular wax of *C. rhodesianus* (Shimomura et al. 2017).

As reviewed above, *Callosobruchus* seed beetles use hybrid sex pheromone systems for mate recognition. In both sex attractant pheromones and contact sex pheromones, the structural shift of the components was not always gradual and instead saltational in some cases. Future studies on the biosynthetic pathways of pheromone components, in addition to a clarification of the molecular basis of pheromone reception, will provide greater insight into the evolution of sex pheromone communication systems in *Callosobruchus* seed beetles.

References

Baker JE, Nelson DR (1981) Cuticular hydrocarbons of adults of the cowpea weevil, *Callosobruchus maculatus*. J Chem Ecol 7:175–182. https://doi.org/10.1007/BF00988645

Bawa SA, Ofori EKS, Osae M (2017) Species diversity and relative abundance of *Callosobruchus* (Coleoptera: Chrysomelidae) in stored cowpea in four major agricultural produce markets in the central region, Ghana. J Stored Prod Res 72:117–120. https://doi.org/10.1016/J.JSPR.2017.04.007

Borowiec L (1987) The genera of seed-beetles (Coleoptera: Bruchidae). Pol Pismo Entomol 57:3–207

Chiluwal K, Kim J, Do BS, Maharjan R, Park CG (2017a) Attractiveness of male azuki bean beetle to the synthetic blends of 2*E*- and 2*Z*-homofarnesals. J Asia Pac Entomol 20:1183–1189. https://doi.org/10.1016/J.ASPEN.2017.09.003

Chiluwal K, Kim J, Do BS, Park CG (2017b) Essential oils from selected wooden species and their major components as repellents and oviposition deterrents of *Callosobruchus chinensis* (L.). J Asia Pac Entomol 20:1447–1453. https://doi.org/10.1016/J.ASPEN.2017.11.011

Chiluwal K, Kim J, Do BS, Roh GH, Park CG (2018a) Methyl salicylate and *trans* -anethole affect the pheromonal activity of homofarnesal, the female sex pheromone of azuki bean beetle. Entomol Res 48:354–361. https://doi.org/10.1111/1748-5967.12294

Chiluwal K, Kim J, Do BS, Roh GH, Park HJ, Park CG (2018b) Effect of gamma irradiation on fecundity, sterility, and female sex pheromone production of *Callosobruchus chinensis* (Coleoptera: Bruchidae). J Econ Entomol 112:156. https://doi.org/10.1093/jee/toy317

Cork A, Hall D, Blaney W, Simmonds MS (1991) Identification of a component of the female sex pheromone of *callosobruchus analis* (coleoptera: bruchidae). Tetrahedron Lett 32:129–132. https://doi.org/10.1016/S0040-4039(00)71236-6

Day L (2013) Proteins from land plants – potential resources for human nutrition and food security. Trends Food Sci Technol 32:25–42. https://doi.org/10.1016/J.TIFS.2013.05.005

Giga DP, Canhão J (1997) Interspecies interference of oviposition behaviour in the cowpea weevils *Callosobruchus rhodesianus* (Pic) and *Callosobruchus maculatus* (F.). Int J Trop Insect Sci 17:251–255. https://doi.org/10.1017/S1742758400016441

Giga DP, Smith RH (1983) Comparative life history studies of four *Callosobruchus* species infesting cowpeas with special reference to *Callosobruchus rhodesianus* (Pic) (Coleoptera: Bruchidae). J Stored Prod Res 19:189–198. https://doi.org/10.1016/0022-474X(83)90007-3

Honda H, Yamamoto I (1976) Evidence for and chemical nature of a sex pheromone present in azuki bean weevil, *Callosobruchus chinensis* L. Proc Symp Insect Pheromones their Appl (Nagaoka Tokyo) 164

Lextrait P, Biemont J-C, Pouzat J (1994) Comparison of walking locomotory reactions of two forms of *Callosobruchus maculatus* males subjected to female sex pheromone stimulation (Coleoptera: Bruchidae). J Chem Ecol 20:2917–2930. https://doi.org/10.1007/BF02098398

Lextrait P, Biémont J-C, Pouzat J (1995) Pheromone release by the two forms of *Callosobruchus maculatus* females: effects of age, temperature and host plant. Physiol Entomol 20:309–317. https://doi.org/10.1111/j.1365-3032.1995.tb00820.x

Mbata GN, Shu S, Ramaswamy SB (2000) Sex pheromones of *Callosobruchus subinnotatus* and *C. maculatus* (Coleoptera: Bruchidae): congeneric responses and role of air movement. Bull Entomol Res 90:147–154. https://doi.org/10.1017/S0007485300000250

Mbata GN, Shu S, Ramaswamy SB (1997) Rhythmicity of mating and oviposition in *Callosobruchus subinnotatus* (Pic) (Coleoptera: Bruchidae). J Insect Behav 10:409–423. https://doi.org/10.1007/BF02765607

Mori K, Ito T, Tanaka K, Honda H, Yamamoto I (1983) Synthesis and biological activity of optically active forms of (*E*)-3,7-dimethyl-2-octene-1,8-dioic acid (callosobruchusic acid) : a component of the copulation release pheromone (erectin) of the azuki bean weevil. Tetrahedron 39:2303–2306. https://doi.org/10.1016/S0040-4020(01)91957-X

Nojima S, Shimomura K, Honda H, Yamamoto I, Ohsawa K (2007) Contact sex pheromone components of the cowpea weevil, *Callosobruchus maculatus*. J Chem Ecol 33:923–933. https://doi.org/10.1007/s10886-007-9266-5

Park CG, Shin E, Kim J (2016) Insecticidal activities of essential oils, *Gaultheria fragrantissima* and *Illicium verum*, their components and analogs against *Callosobruchus chinensis* adults. J Asia Pac Entomol 19:269–273. https://doi.org/10.1016/J.ASPEN.2016.03.001

Phillips TW, Phillips JK, Webster FX, Tang R, Burkholder WE (1996) Identification of sex pheromones from cowpea weevil, *Callosobruchus maculatus*, and related studies with *C. analis* (coleoptera: Bruchidae). J Chem Ecol 22:2233–2249. https://doi.org/10.1007/BF02029543

Pierre D, Biémont J-C, Pouzat J, Lextrait P, Thibeaudeau C (1997) Location and ultrastructure of sex pheromone glands in female *Callosobruchus maculatus* (Fabricius) (Coleoptera: Bruchidae). Int J Insect Morphol Embryol 25:391–404. https://doi.org/10.1016/S0020-7322(96)00014-1

Qi Y-T, Burkholder WE (1982) Sex pheromone biology and behavior of the cowpea weevil *Callosobruchus maculatus* (Coleoptera: Bruchidae). J Chem Ecol 8:527–534. https://doi.org/10.1007/BF00987800

Ramaswamy SB, Shu S, Monroe WA, Mbata GN (1995) Ultrastructure and potential role of integumentary glandular cells in adult male and female *Callosobruchus subinnotatus* (Pic) and *C. maculatus* (Fabricius) (Coleoptera : Bruchidae). Int J Insect Morphol Embryol 24:51–61. https://doi.org/10.1016/0020-7322(94)P3968-Y

Rees D (2004) Beetles (Order: Coleoptera). In: Rees D (ed) Insects of stored products. CSIRO Publishing, Canberra, pp 11–120

Rup PJ, Sharma SP (1978) Behavioural response of males and females of *Callosobruchus maculatus* (F.) to the sex pheromones. Indian J Ecol 5:72–76

Shimomura K, Akasaka K, Yajima A, Mimura T, Yajima S, Ohsawa K (2010a) Contact sex pheromone components of the seed beetle, *Callosobruchus analis* (F.). J Chem Ecol 36:955–965. https://doi.org/10.1007/s10886-010-9841-z

Shimomura K, Koshino H, Yajima A, Matsumoto N, Kagohara Y, Kamada K, Yajima S, Ohsawa K (2010b) 2,3-Dihydrohomofarnesal: female sex attractant pheromone component of *Callosobruchus rhodesianus* (Pic). J Chem Ecol 36:824–833. https://doi.org/10.1007/s10886-010-9824-0

Shimomura K, Koshino H, Yajima A, Matsumoto N, Yajima S, Ohsawa K (2010c) A new sesquiterpenoid produced by female *Callosobruchus rhodesianus* (Pic): a possible component of the sex attractant pheromone. Tetrahedron Lett 51:6860–6862. https://doi.org/10.1016/J.TETLET.2010.10.100

Shimomura K, Matsui S, Ohsawa K, Yajima S (2016) Saltational evolution of contact sex pheromone compounds of *Callosobruchus rhodesianus* (Pic). Chemoecology 26:15–23. https://doi.org/10.1007/s00049-015-0204-7

Shimomura K, Matsui S, Ohsawa K, Yajima S (2017) Identification of cuticular compounds collected from *Callosobruchus rhodesianus* (Pic) eliciting heterospecific mating behavior with male *Callosobruchus maculatus* (F.). Chemoecology 27:65–73. https://doi.org/10.1007/s00049-017-0231-7

Shimomura K, Mimura T, Ishikawa S, Yajima S, Ohsawa K (2010d) Variation in mate recognition specificities among four *Callosobruchus* seed beetles. Entomol Exp Appl 135:315–322. https://doi.org/10.1111/j.1570-7458.2010.00994.x

Shimomura K, Nojima S, Yajima S, Ohsawa K (2008) Homofarnesals: female sex attractant pheromone components of the southern cowpea weevil, *Callosobruchus chinensis*. J Chem Ecol 34:467–477. https://doi.org/10.1007/s10886-008-9451-1

Shu S, Koepnick WL, Mbata GN, Cork A, Ramaswamy SB (1996) Sex pheromone production in *Callosobruchus maculatus* (Coleoptera: Bruchidae): electroantennographic and behavioral responses. J Stored Prod Res 32:21–30. https://doi.org/10.1016/0022-474X(95)00038-9

Shu S, Mbata GN, Cork A, Ramaswamy SB (1999) Sex pheromone of *Callosobruchus subinnotatus*. J Chem Ecol 25:2715–2727. https://doi.org/10.1023/A:1020899407497

Shu S, Mbata GN, Ramaswamy SB (1998) Female sex pheromone in *Callosobruchus subinnotatus* (Coleoptera: Bruchidae): production and male responses. Ann Entomol Soc Am 91:840–844. https://doi.org/10.1093/aesa/91.6.840

Southgate BJ (1958) Systematic notes on species of *Callosobruchus* of economic importance. Bull Entomol Res 49:591. https://doi.org/10.1017/S0007485300053864

Southgate BJ (1979) Biology of the Bruchidae. Annu Rev Entomol 24:449–473. https://doi.org/10.1146/annurev.en.24.010179.002313

Tanaka K, Ohsawa K, Honda H, Yamamoto I (1981) Copulation release pheromone, erectin, from the azuki bean weevil (*Callosobruchus chinensis* L.). J Pestic Sci 6:75–82. https://doi.org/10.1584/jpestics.6.75

Tanaka K, Ohsawa K, Honda H, Yamamoto I (1982) Synthesis of erectin, a copulation release pheromone of the azuki bean weevil, *Callosobruchus chinensis* L. J Pestic Sci 7:535–537. https://doi.org/10.1584/jpestics.7.535

Tuda M (2007) Applied evolutionary ecology of insects of the subfamily Bruchinae (Coleoptera: Chrysomelidae). Appl Entomol Zool 42:337–346. https://doi.org/10.1303/aez.2007.337

Tuda M, Chou L-Y, Niyomdham C, Buranapanichpan S, Tateishi Y (2005) Ecological factors associated with pest status in *Callosobruchus* (Coleoptera: Bruchidae): high host specificity of non-pests to Cajaninae (Fabaceae). J Stored Prod Res 41:31–45. https://doi.org/10.1016/J.JSPR.2003.09.003

Tuda M, Rönn J, Buranapanichpan S, Wasano N, Arnqvist G (2006) Evolutionary diversification of the bean beetle genus *Callosobruchus* (Coleoptera: Bruchidae): traits associated with stored-product pest status. Mol Ecol 15:3541–3551. https://doi.org/10.1111/j.1365-294X.2006.03030.x

Utida S (1954) "Phase" dimorphism observed in the laboratory population of the cowpea weevil, *Callosobruchus quadrimaculatus*. J Appl Zool 18:161–168. (in Japanese with English summary)

Yajima A, Akasaka K, Nakai T, Maehara H, Nukada T, Ohrui H, Yabuta G (2006) Direct determination of the stereoisomer constitution by 2D-HPLC and stereochemistry–pheromone activity relationship of the copulation release pheromone of the cowpea weevil, *Callosobruchus maculatus*. Tetrahedron 62:4590–4596. https://doi.org/10.1016/J.TET.2006.02.059

Yajima A, Akasaka K, Yamamoto M, Ohmori S, Nukada T, Yabuta G (2007) Direct determination of the stereoisomeric composition of Callosobruchusic acid, the copulation release pheromone of the azuki bean weevil, *Callosobruchus chinensis* L., by the 2D-Ohrui-Akasaka method. J Chem Ecol 33:1328–1335. https://doi.org/10.1007/s10886-007-9311-4

Zannou ET, Glitho IA, Huignard J, Monge JP (2003) Life history of flight morph females of *Callosobruchus maculatus* F.: evidence of a reproductive diapause. J Insect Physiol 49:575–582. https://doi.org/10.1016/S0022-1910(03)00029-5

Chapter 5
Pheromones in Longhorn Beetles with a Special Focus on Contact Pheromones

Midori Fukaya and Hiroe Yasui

Abstract Longhorn beetles (Coleoptera: Cerambycidae) utilize pheromones to find and recognize their mates. Long-range attractant pheromones have recently been chemically identified from a number of cerambycid species. In contrast, knowledge on the contact pheromones in cerambycids, which are essential for the recognition of mates, is limited. In this chapter, we first overview the volatile pheromones in Cerambycidae. We then describe the identification of contact pheromones in two cerambycid species, the yellow-spotted longicorn beetle *Psacothea hilaris* and the white-spotted longicorn beetle *Anoplophora malasiaca*. The contact pheromone of the former species consists of a single hydrocarbon, whereas that of the latter species is a markedly complex mixture of 15 components (8 hydrocarbons, 4 aliphatic ketones, and 3 lactones), the identification of which was achieved through research spanning a quarter of century. Lastly, difficulties and problems in the studies of contact pheromones in cerambycids are discussed.

Keywords Longhorn beetles · Volatile sex pheromones · Contact sex pheromones · *Psacothea hilaris* · *Anoplophora malasiaca*

M. Fukaya (✉)
College of Bioresource Sciences, Nihon University (NUBS), Fujisawa, Kanagawa, Japan
e-mail: fukaya.midori@nihon-u.ac.jp

H. Yasui
Central Region Agricultural Research Center, National Agriculture and Food Research Organization (NARO), Tsukuba, Ibaraki, Japan

Y. Ishikawa (ed.), *Insect Sex Pheromone Research and Beyond*, Entomology Monographs, https://doi.org/10.1007/978-981-15-3082-1_5

5.1 Introduction

Cerambycidae is a large family of longhorn beetles comprised of over 36,000 species in nine subfamilies distributed worldwide (e.g., Monné et al. 2017). Their larvae grow inside plants, mostly in arboreous plants. Many species are known as pests because their larvae bore into wood and cause extensive damage to either forest or urban trees (Haack 2017). With the increase of international trade, the larvae in wooden materials are often transported to non-native countries and cause problems as invasive pests (e.g., Eyre and Haack 2017; Meng et al. 2015; Sarto et al. 2018).

On the other hand, cerambycid beetles are considered to play essential roles in the forest ecosystem and indicate biodiversity of the forest (e.g., Gatti et al. 2018; Goldsmith 2007; Handley et al. 2015; Iwona et al. 2018). Some species are considered to be flagship or umbrella species in their fauna (e.g., Albert et al. 2012; Drag et al. 2015; Eckelt et al. 2018). Although many reports have been published by both naturalists and applied entomologists on cerambycid species, those on their basic biology are limited. Although the presence of pheromones in cerambycid species has long been presumed (e.g., Duffy 1953; Heintze 1925; Linsley 1959; Smyth 1934), those of only several species had been identified by the end of the twentieth century.

Chemical ecology of this family markedly progressed after the entry into the twenty-first century. Longhorn beetles use airborne chemicals, such as volatiles derived from plants and their feeding sites, and volatile pheromones to locate their hosts and mates, and chemicals, such as trail pheromones and contact pheromones, act in the short-range to help find and recognize their mates (e.g., Fukaya and Honda 1992; Galford 1977; Hoover et al. 2014; Wang et al. 2002; Zhang et al. 2003). Among these semiochemicals, volatile pheromones have been actively investigated and, to date, identified from more than 100 species (reviewed by Hanks and Millar 2016; Millar and Hanks 2017). Practical application of volatile pheromones and other semiochemicals for the management of pest species in this family is now underway (e.g., Barbour et al. 2019; Sweeney et al. 2017). In contrast, contact pheromones have been identified from only a small number of species, although they play essential roles in the mating systems in Cerambycidae (Ginzel 2010; Millar and Hanks 2017).

In this chapter, we first outline the typical mating behavior of cerambycid beetles; longhorn beetles use both chemical and non-chemical cues to find and copulate with their mates. We then provide an overview of the characteristics of volatile pheromones used by this group of beetles. After describing the first identification of a single cuticular hydrocarbon as a contact pheromone in the yellow-spotted longicorn beetle *Psacothea hilaris*, we focus on a unique contact sex pheromone of the white-spotted longicorn beetle *Anoplophora malasiaca*. The contact pheromone of this species consists of 15 components belonging to three chemical groups, i.e., hydrocarbons, ketones, and lactones. Among these components, essential components are novel lactones, and other components act synergistically to induce male mating behavior.

5.2 Mate Location and Function of Semiochemicals

Mating behaviors of cerambycid beetles are regulated by semiochemicals and, in some species, by non-chemical cues that act as synergists. A male and female typically meet and copulate on their larval host plants or at their feeding sites such as flower heads or sap production sites (e.g., Goldsmith 1987; Wang and Chen 2005). In general, female beetles are attracted by volatile pheromones emitted by males; however, in some species, volatiles pheromones are emitted by females (Fig. 5.1). Plant volatiles act synergistically on these long-range pheromones or singly act as attractants. Flying females (or males) usually do not land directly on the body of their conspecifics but in their vicinity. The visual factors of the mate enhance the effects of volatile attractants or regulate the orientation behavior of females (or males) in their vicinity (Fig. 5.1; Fukaya et al. 2004, 2005a, b; Iwabuchi 1985). If neither sex has specific long-range attractant pheromones, they may encounter each other by chance while wandering on the host plants (e.g., Fukaya 1992; Fukaya and Honda 1992; Luo et al. 2011).

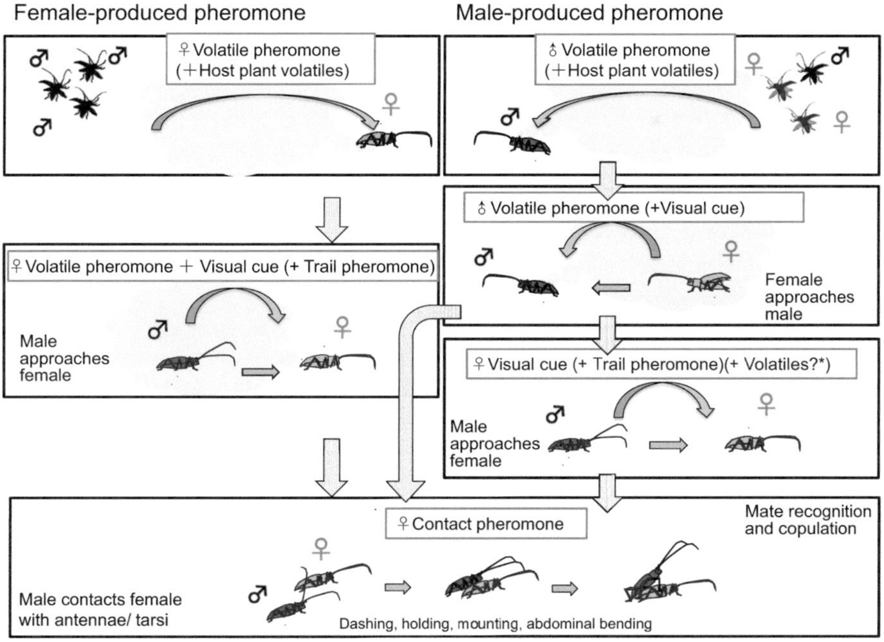

Fig. 5.1 Schema representing the typical patterns of mate location and precopulatory behavior in cerambycids. The female-produced pheromone strongly attracts males, whereas the male-produced pheromone attracts females and males. Plant volatiles (larval host odor and flower scent) or kairomones produced by bark beetles often act synergistically with the volatile pheromone. On the host or mating sites, the male approaches the female by visual cues and the female-produced pheromone. When a male contacts a female with his antennae, he dashes forward and mounts her

When males contact females by their antennae, they recognize their mates by the contact sex pheromone on the body of females and exhibit precopulatory behaviors; they dash and try to hold females, mount them, and bend their abdomen in an attempt to copulate (e.g., Akutsu 1983; Akutsu and Kuboki 1983; Fukaya and Honda 1992; Ginzel 2010).

5.3 Overview of the Volatile Pheromones in Cerambycidae

Most cerambycid species have volatile pheromones (Millar and Hanks 2017). To date, airborne pheromones that act in the long-range have been identified from species belonging to five subfamilies of Cerambycidae. The same compounds or compounds with the same chemical motif are shared as the pheromone components by species in the same genus, tribe, or subfamily of cerambycids (e.g., Mitchell et al. 2015; Ray et al. 2015; Silk et al. 2011). Some pheromone components in a subfamily are produced by the opposite sex in a different subfamily (e.g., Fettköther et al. 1995; Ray et al. 2012a) (Fig. 5.2).

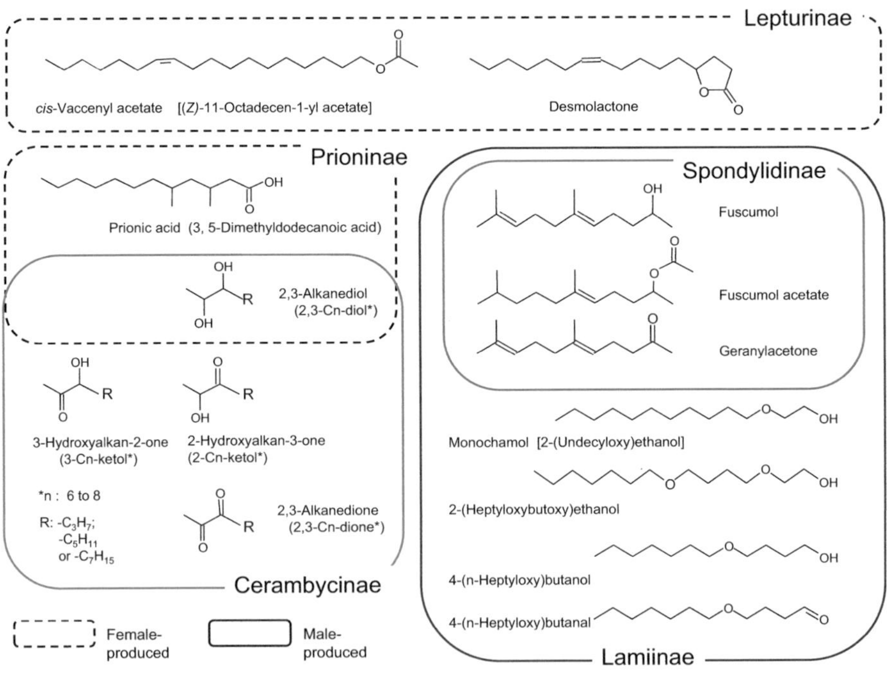

Fig. 5.2 Representative examples of volatile pheromone components identified from five subfamilies of Cerambycidae. Similar chemical motifs are noted across these subfamilies. Some components are shared in a genus, subfamily, or among different subfamilies. In addition to components illustrated here, common plant scent components, such as terpineol, hexenol, nerol, nonanone, and butanol, and many other motifs are shared in many cerambycines. See the text for details

5.3.1 Female-Produced Volatile Pheromones

Female-produced sex pheromones acting in the long-range have been identified from subfamilies Prioninae and Lepturinae. These female sex pheromones and sex attractants exhibit strong activity on males in field trapping tests.

Larvae of Prioninae (302 genera; Monné et al. 2017) usually feed on rotten wood, and their adults, which have black or brown bodies, are nocturnal (Bílý and Mehl 1989). In this subfamily, 3,5-dimethyldodecanoic acid (prionic acid) was identified as the female-produced sex attractant pheromone in *Prionus californicus* and some congeners (Rodstein et al. 2009, 2011). This compound attracted not only conspecific males, but also other species in tribe Prionini (Barbour et al. 2011; Wickham et al. 2016a). In addition to this compound, 2,3-alkanediols were identified as female-produced pheromone components; (2*R*,3*S*)-2,3-octanediol (2*R*,3*S*-C8-diol) in *Megopis costipennis* (Wickham et al. 2015a, b) and (2*R*,3*R*)-2,3-hexanediol (2*R*,3*R*-C6-diol) in *Tragosoma depsarium* (Ray et al. 2012a).

Lepturinae species (210 genera) are considered to be diurnal flower visitors, and their feeding and copulation are often observed on the flower heads. (4*R*,9*Z*)-Hexadec-9-en-4-olide [(*R*)-desmolactone] was identified as the female-produced sex pheromone in *Desmocerus californicus californicus* and *D. aureipennis aureipennis* (Ray et al. 2012b, 2014), and this compound also attracted males of several other *Desmocerus* species (Ray et al. 2014). A lepturinine species *Ortholeptura valida* uses *cis*-vaccenyl acetate, a well-known pheromone component of *Drosophila* males, as a major component of female-produced pheromone (Ray et al. 2011).

In addition to the two subfamilies described above, the presence of a long-range female-produced pheromone is reported in a small subfamily Necydalinae (2 genera); *Callisphyris apicicornis* (Curkovic and Ferrera 2012).

5.3.2 Male-Produced Volatile Pheromones

Male-produced long-range pheromones were identified from three subfamilies: Spondylidinae (32 genera), Cerambycinae (1757 genera), and Lamiinae (2964 genera). The male-produced pheromones attract both sexes, in most species, from a distance.

Species in Spondylidinae, a small subfamily, are nearly all nocturnal or crepuscular except for *Tetropium* species. Male-produced volatile pheromones containing fuscumol [(*E*)-6,10-dimethylundeca-5,9-dien-2-ol] and fuscumol acetate were identified from the genus *Tetropium*: *T. fuscum*, which is native to Europe, and *T. cinnamopterum cinnamopterum* and *T. abietis*, which are native to North America (Halloran et al. 2018; Silk et al. 2007; Sweeney et al. 2010). A mixture of (*S*)-fuscumol and fuscumol acetate attracted *T. schwarzianum* in screening trials in the field (Hanks and Millar 2013). Geranylacetone, an analog of (*S*)-fuscumol, was reported to be the pheromone component of *Asemum caseyi*. Males of *A. nitidum* produce both (*S*)-fuscumol and geranylacetone (Collignon et al. 2016; Halloran et al. 2018). A synthetic blend of host plant volatiles increased the effects of fuscumol and geranylacetone (Collignon et al. 2016).

Cerambycinae is a subfamily with marked diversity in morphology, life cycle, and habit. In Cerambycinae, typical components of male-produced pheromones are unbranched 6-, 8-, or 10 carbon chains with a ketone at position 2 and a hydroxyl group at position 3, or two hydroxyls at positions 2 and 3 (reviewed by Millar and Hanks 2017). In Cerambycidae family, the male-produced pheromone was first identified from *Xylotrechus pyrrhoderus*. It consists of 2*S*,3*S*-C8-diol and (*S*)-2-hydroxyoctan-3-one (2*S*-C8-ketol) (Iwabuchi et al. 1986; Sakai et al. 1984). 2,3-C8-diol and 2*S*-C8-ketol were also detected in the congeners (Iwabuchi et al. 1987; Kuwahara et al. 1987). C6 components were also identified: 3*R*-C6-ketol, 2*R* and 2*S*-3-C6-diol, and 2,3-C6-dione in *Hylotrupes bajulus*, and 3*R*-C6-ketol in *Anelaphus inflaticollis* and *Neoclytus* species (Fettköther et al. 1995; Lacey et al. 2004, 2007; Ray et al. 2009b, 2015). A compound with a new hydroxy-ketone structural motif [trichoferone: (2*R*,4*R*/*S*)-2-hydroxy-4-methyl-1-phenylhexan-3-one] was identified as the male-produced pheromone of *Trichoferus campestris* (Ray et al. 2019).

Cerambycinae species also use other types of compounds with different chemical motifs as the long-range pheromone. For example, *Rosalia funebris* in North America uses (*Z*)-3-decenyl (*E*)-2-hexenoate (Ray et al. 2009a), whereas its sibling *R. alpine* in Europe uses a novel alkylated pyrone (Kosi et al. 2017). *Aromia bungii,* native to East Asia, uses (*E*)-2-*cis*-6,7-epoxynonenal as a male-produced pheromone component (Fukaya et al. 2017; Xu et al. 2017; Yasui et al. 2019a; Zou et al. 2019). *Paranoplium gracile* uses monoterpene alcohol (Collignon et al. 2019), and the pheromone of *Megacyllene caryae* is a complex blend consisting of 2,3-C8-diols with several terpenoids and aromatic compounds (e.g., Lacey et al. 2008b; Handley et al. 2015). Components with more diverse structures are expected in the Cerambycinae pheromone. In many species of this subfamily, volatile pheromones were found to be or assumed to be produced by the glandular tissue in the prothorax and emitted from pores on the surface (e.g., Hoshino et al. 2015; Iwabuchi 1986; Lacey et al. 2007; Noldt et al. 1995).

Lamiinae is the largest subfamily in Cerambycidae. Hydroxyethers were identified as male-produced pheromone components in this subfamily. The pheromone of *A. glabripennis* consists of 4-(heptyloxy) butanol and 4-(heptyloxy)butanal (Zhang et al. 2002). In this species, (*E,E*)-α-farnesene, the minor component of the pheromone, and host plant volatiles act as synergists (Crook et al. 2014; Nehme et al. 2009). *A. chinensis* males also produce 4-(heptyloxy)butanol. This compound attracted both sexes when it was combined with α-pinene and ethanol, common plant volatiles (Hansen et al. 2015). Another hydroxyether, 2-(undecyloxy)ethanol ("monochamol") was identified from *Monochamus galloprovincialis* (Pajares et al. 2010). Monochamol has been identified as the pheromone component or demonstrated to act as an attractant for many *Monochamus* species (e.g., Fierke et al. 2012; Macias-Samano et al. 2012; Pajares et al. 2013), including the Japanese pine sawyer

M. alternatus, the vector of pine wilt disease (Teale et al. 2011). This compound acts as an attractant in four other genera of the same tribe Monochamini (Millar and Hanks 2017). Host volatiles, such as α-pinene and ethanol, and pheromone components of *Ips* and *Dendroctonus* (Scolytidae) bark beetles, such as ipsenol, ipsdienol, frontalin, endo- and exo-brevicomin, act as kairomones attracting many species of *Monochamus* adults, in addition to acting as synergists of monochamol (Allison et al. 2001, 2003, 2012, 2013; Macias-Samano et al. 2012; Miller et al. 2013; Pajares et al. 2004; Teale et al. 2011). Another type of component in the long-range pheromones of this subfamily is sesquiterpene derivatives. Fuscumol and fuscumol acetate, previously identified as pheromone components in Spondylidinae, have been reported to be the pheromone or the attractant in many Lamiinae species (e.g., Hanks et al. 2012; Hanks and Millar 2013; Hughes et al. 2013; Mitchell et al. 2011; Wong et al. 2012).

Male-produced pheromone components attract both sexes, although there is often a bias toward females. The sex ratio of the responders to the male-produced pheromone depends on species, and forms a spectrum from over 1 to below 0.5. For example, 0.8–0.5 in *Aromia bungii* (Xu et al. 2017), ≈0.7 in *T campestris* (Ray et al. 2019), and ≈0.5 in *Xylotrechus antilope antilope* (Molander et al. 2019). Regardless of the sex ratio of responders, the male-produced pheromone in Cerambycidae is often referred to as the "aggregation-sex pheromone" (cf. Hanks and Millar 2016). The pheromones released by males in this family attract females and lead to encounters by both sexes and mating, but it also attracts rival males. Thus, the male-produced pheromones of Cerambycidae may not always be beneficial to the pheromone releaser. This feature is consistent with the definition of the aggregation-sex pheromone (Cardé 2014).

5.4 Identification of the Contact Pheromone of *Psacothea hilaris*

The chemical component of contact pheromone in the cerambycid family was first identified by Fukaya et al. (1996) from the yellow-spotted longicorn beetle *Psacothea hilaris*, a pest of mulberry and fig trees. We review this study and point out remaining problems.

When a fresh female carcass (body) of *P. hilaris* was presented to a male of the same species, he exhibited a series of precopulatory behaviors: dashing, holding, and copulation attempt. The pheromonal activity to induce such responses was lost when the body was thoroughly washed with diethyl ether using Soxhlet extractor for 12 hr. The activity was restored when the extract of female body surface was applied to the washed body (Fukaya and Honda 1992).

The pheromonal activity was evaluated as follows. A test sample was applied to a gelatin capsule (dummy of a female) fixed on a filter paper disk and given to a male (Fig. 5.3). When the male bent his abdomen to the end of the dummy, the sample was evaluated as having pheromonal activity. For isolation of the active

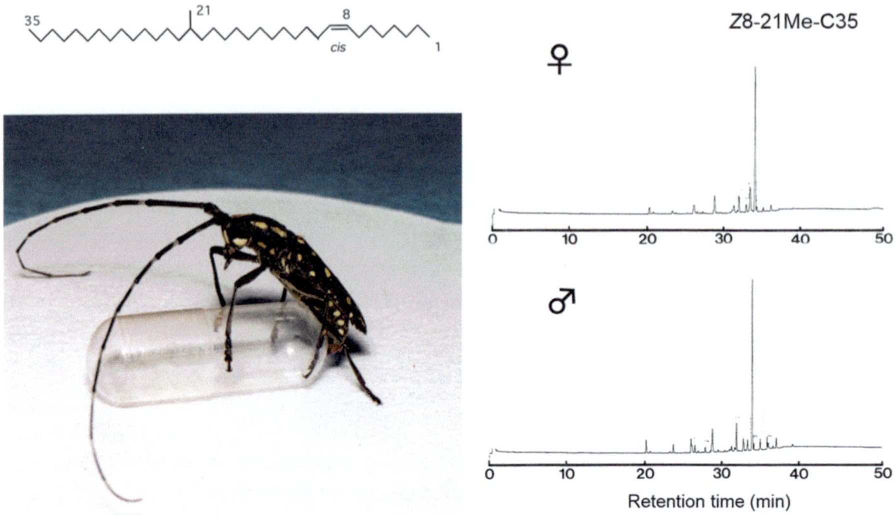

Fig. 5.3 Contact pheromone of the yellow-spotted longicorn beetle, *Psacothea hilaris* (Pascoe). Upper left: (*Z*)-21-methyl-8-pentatriacontene (Z8-21Me-C35), the contact pheromone component. Lower left: male copulatory attempt (abdominal bending) to the dummy treated with the synthetic contact-pheromone component. Right: gas chromatograms of female and male cuticular hydrocarbons (non-polar fraction). (From Fukaya et al. (1996), excluding the male GC profile)

compounds in the body wax, extraction from the elytra, rather than whole body, was advantageous because the extracts of the latter contained large amounts of contaminants that were difficult to remove (Fukaya and Honda 1992, 1995).

In order to separate and identify the active components, the crude extract was fractionated by silica-gel column chromatography and analyzed by gas chromatography-mass spectrometry (GC-MS). The pheromonal activity was detected in the non-polar and weakly polar fractions. The non-polar fraction (eluted by hexane) contained cuticular hydrocarbons (CHCs). CHCs were further separated by a silver nitrate-impregnated silica gel column into several fractions containing alkanes, monoenes, dienes, and trienes. The major component in the monoene fraction induced male copulatory attempts (Fukaya et al. 1996). No activity was observed with fractions containing alkanes, dienes, or trienes. The active compound was revealed to be (*Z*)-21-methyl-8-pentatriacontene (*Z*8-21Me-C35), a long-chain monomethyl alkene, by GC-MS and NMR analyses. *Z*8-21Me-C35 was a major component of CHCs, which comprised >60% of the total CHCs in the female extract (Fig. 5.3; Fukaya et al. 1996). The pheromonal activity of geometrical and chiral isomers of this major component was next evaluated. The (*Z*)-isomer, the natural component, exhibited higher activity than the (*E*)-isomer; however, no difference in activity was found between (*S*)- and (*R*)-enantiomers (Fukaya et al. 1997; Fukusaki et al. 1998).

The contact pheromone was assumed to be detected by the antenna, maxillary palpi, or tarsi in *P. hilaris* because when these sense organs were impaired by chemical treatment or ablation, male responses to a female or dummy treated with the pheromone or female extract were abolished (Fukaya and Honda 1992). Dai and

Honda (1990) identified ten types of sensilla on the antennal flagellum: sensilla chaetica (type-I, II, III), s. basiconica (type-I, II, III), s. trichodea, conical pegs, dome organs, and s. squamiformia. The sensilla responsible for perception of the contact pheromone have not yet been identified in this beetle, although s. trichodea is presumed to be the sensor of the contact chemical based on the presence of a monopore on its tip. Moreover, in another cerambycid *Phoracanta semipunctata*, sensilla trichodea were noted as the candidate receptor of cuticular surface components (Lopes et al. 2005). Sensilla trichodea are found on the male antenna of other cerambycids such as *Monochamus* species (Dyer and Seabrook 1975), *Xylotrechus grayii grayii*, and *Aromia bungii* (Chen et al. 2014; Di Palma et al. 2017).

Regarding the contact pheromone of *P. hilaris*, a few points remain to be clarified (Millar and Hanks 2017). One is the low activity of the major component. One female equivalent (FE) of the crude female extract induced copulatory attempts from >90% of the males, whereas 1 FE of a racemic mixture of *Z*8-21Me-C35 induced responses from only 34% of the males (Fukaya et al. 1996). Indeed, pheromonal activity was also detected in the weakly polar fractions (Fukaya et al. 1996; Fukaya 1992). When the weakly polar fractions were further separated by HPLC, pheromonal activity was not stably retained in any fraction (Fukaya et al. 1996). It is possible that unidentified compound(s) in the weakly polar fractions synergize the activity of *Z*8-21Me-C35.

Second, the absence of sexual dimorphism in terms of the pheromone component remains to be accounted for. No distinct difference in the profiles of CHCs is observed between the sexes: *Z*8-21Me-C35 is also the major CHC in males (Fig. 5.3). The dead body of a male and the crude extract of male elytra exhibited pheromonal activity toward males, although it was lower than that of the female counterparts (Fukaya and Honda 1996b). Male-male mounting was frequently observed on mulberry trees in a field cage (2 m × 2 m × 2 m), into which 20 females and males were each released. Larger males often mounted on smaller males and bent their abdomen toward smaller ones even though smaller ones struggled to escape (Fukaya 1992). Male-male mounting was also observed in *M alternatus* in a field cage experiment (Fauziah et al. 1987).

Third, two enantiomers of the pheromone *Z*8-21Me-C35 induced male precopulatory behaviors from "holding" to "abdominal bending", but did not induce "dashing" toward the female. In contrast, the female body, the extract of female elytra, and its non-polar fraction, elicited "dashing" and the following precopulatory behaviors (Fukaya and Honda 1995; Fukaya et al. 1996). As strong dash-eliciting activity was found in the ether extract of female prothorax rather than in the extract of elytra, unknown compound(s) may be involved in the induction of "dashing".

5.5 CHCs as Contact Pheromone Components in Other Cerambycids

Identification of contact pheromone components in Cerambycidae followed the identification in *P. hilaris*. Active components have been identified from species belonging to 4 subfamilies: Spondylidinae, Lamiinae, Cerambycinae, and Prioninae.

Moreover, candidate compounds have been reported for species in Lepturinae. In Lamiinae, to which *P. hilaris* belongs, 8 long-chained n-alkanes and methyl branched alkanes were revealed to be the components of the contact pheromone in *Anoplophora malasiaca* (Fukaya et al. 1999, 2000), as described later. On the other hand, in *A. glabripennis,* a species closely related to *A. malasiaca*, five aliphatic higher alkenes (*Z*9-C23, *Z*9-C25, *Z*7-C25, *Z*7-C27, *Z*9-C27) were identified as components of the contact pheromone (Zhang et al. 2003). These alkenes were more abundant in females than in males. No single component exhibited pheromonal activity; however, a mixture of all components induced a series of male precopulatory behaviors (Zhang et al. 2003). Of note, two congeners, *A. malasiaca* and *A. glabripennis*, utilize saturated and unsaturated hydrocarbons, respectively, as components of their contact pheromones.

In the subfamily Cerambycinae, *Xylotrechus colonus* was reported to utilize C25, 9Me-C25, and 3Me-C25 as the components of the contact pheromone (Ginzel et al. 2003b). In *Neoclytus acuminatus acuminatus,* 7Me-C27 is the major component of the contact pheromone, and 7Me-C25 and 9Me-C27 act as synergists (Hughes et al. 2015; Lacey et al. 2008a). *Z*9-C25 and *Z*9-C29 were reported as a component of the contact pheromone in *Megacyllene robiniae* and *M. caryae*, respectively (Ginzel et al. 2003a, 2006). 5Me17Me-C29, a female-specific compound, was identified as an essential component of the contact pheromone in *Callidiellum rufipenne* (Rutledge et al. 2009).

In Spondylidinae, the pheromone component of *Tetropium fuscum*, which induced male copulatory attempts, was revealed to be *S*-11Me-C25 (Silk et al. 2011). In *T. cinnamopterum cinnamopterum*, a blend of *Z*9-C25 and *S*-11Me-C25 exhibited sex pheromonal activity (Silk et al. 2011). In terms of the activity of enantiomers, only the (*S*)-enantiomer induced responses from male *T. fuscum*, whereas both (*R*)- and (*S*)-enantiomers are active in *P. hilaris* (Fukaya et al. 1997).

It was also reported that species in the two subfamilies, which are noted to have female-produced pheromones attracting males from a distance, possess contact pheromones. In *Mallodon dasystomus* in Prioninae, the female contact pheromone was identified as methyl-alkanes: 2Me-C26 and 2Me-C28 (Spikes et al. 2010). In *Prionus californicus*, CHCs were demonstrated to play an essential role in mate recognition (Barbour et al. 2007). In the subfamily Lepturinae, the sex- and species-specific abundance of some CHCs, and pheromonal activity of female extracts were reported for *Pidonia grallatrix* and *P. takechii* (Tanigaki et al. 2007). Males in these species are assumed to locate females from a distance by the volatile pheromone and recognize conspecific females by the contact pheromone.

Sexual dimorphism of CHC profiles was detected in many cerambycid species (e.g., Akino et al. 2001; Ginzel et al. 2003a, b, 2006; Ibeas et al. 2009; Spikes et al. 2010; Silk et al. 2011). CHCs that are more abundant in females than in males are often reported to be the pheromone components; however, this phenomenon should not be over-generalized. For example, the sex pheromone component of *P. hilaris* is monomorphic (Fig. 5.3). In *C. rufipenne*, unsaturated CHCs containing 9-pentacosyne, which are more abundant in females, did not induce mating responses from the males (Rutledge et al. 2009).

The presence of contact pheromone was also experimentally confirmed in some cerambycid species, such as *Acalolepta luxuriosa* (Kuboki et al. 1985), *Monochamus alternatus* (Kim et al. 1992), *Batocera horsfieldi* (Luo et al. 2011), laminines, and *Semanotus japonicus* (Kim et al. 1993), a cerambycine, although the chemical structures of these components have yet to be fully clarified.

5.6 Contact Pheromone of *Anoplophora malasiaca*

The white-spotted longicorn beetle, *A. malasiaca*, is widely distributed in Japan from southern Hokkaido to the Ryukyu Islands. This species is polyphagous, feeding on live woody plants such as willow, maple, alder, citrus, apple, and pear trees (Makihara 2007; Mitomi et al. 1990). *A. malasiaca* was sometimes treated as a synonym of *A. chinensis* (e.g., Lingafelter and Hoebeke 2002); however, the differences in morphological characters and mitochondrial DNA sequences support that they are two independent species (e.g., Makihara 2000, 2007; Muraji et al. 2011; Ohbayashi et al. 2009).

Extensive studies revealed uniqueness of the contact pheromone of *A. malasiaca*, which consists of 15 compounds belonging to three different chemical groups: eight hydrocarbons, four aliphatic ketones, and three lactones. All three groups of chemicals are required to induce mating attempts from males at levels similar to in the conspecific females (Fujiwara-Tsujii et al. 2019; Fukaya et al. 2000). It is the most complex pheromone blend in insects reported to date (Millar and Hanks 2017).

5.6.1 Mate-Finding Behaviors of A. malasiaca

Male and female *A. malasiaca* move between the host trees mainly by flying, and in groves by walking. In this species, the male-produced volatile pheromone components of *A. chinensis* and *A. glabripennis* were detected; however, these compounds exhibited only marginal attractancy even in the short-range (Yasui et al. 2019b). Host plant volatiles, such as sesquiterpenes in citrus trees, were found to be involved in the attraction of male *A. malasiaca* to their host plants, as explained next (Fujiwara-Tsujii et al. 2012; Yasui et al. 2007a, Yasui et al. 2011). Host plants emit large amounts of volatiles when *A. malasiaca* beetles bite the bark, twigs, or leaves of the host plants, and these volatiles were demonstrated to be absorbed by the surface wax of the body of beetles (Yasui 2009; Yasui et al. 2007b). The absorbed compounds are gradually released from the body surface and act as if a sex attractant pheromone (Yasui 2009; Yasui et al. 2007b, 2008; Yasui and Fujiwara-Tsujii 2016). Males and females are attracted to the source of the volatiles, i.e., conspecifics and wounded host plants, and land in the vicinity (≈1 m) of the odor source (Yasui et al. 2007b).

Males often encounter resting females while wandering, and sometimes meet females coming toward them (Fukaya et al. 1999). Females and males locate each other using volatile factors; a male walks toward a female in most cases (Fukaya et al. 1999). Volatiles emitted from males (females) attracted the mates from a distance of <10 cm, then the visual factor of the target acts as a synergist and increases the effects of the volatile compounds (e.g., Fukaya 2005b (movie); Fukaya et al. 2004, 2005a, b) (Fig. 5.4). When a male touches a female with his antennae, they usually immediately dash toward the female. This observation suggests that direct contact with antennae is essential for the mating of male *A. malasiaca*. If the male touched the female with his forelegs without antennal contact, he immediately grabbed her with his forelegs. The male then licked her back with his maxillary palpi, aligned his body axis with that of the female, and mounted her. He held her body with his fore- and mid-legs (mounting). Most males licked the female's back again during mounting (Fukaya et al. 1999). Then, the male bent his abdomen in an attempt to copulate. A series of precopulatory behaviors were consistently induced when a male contacted the dummy treated with the crude extract of female elytra,

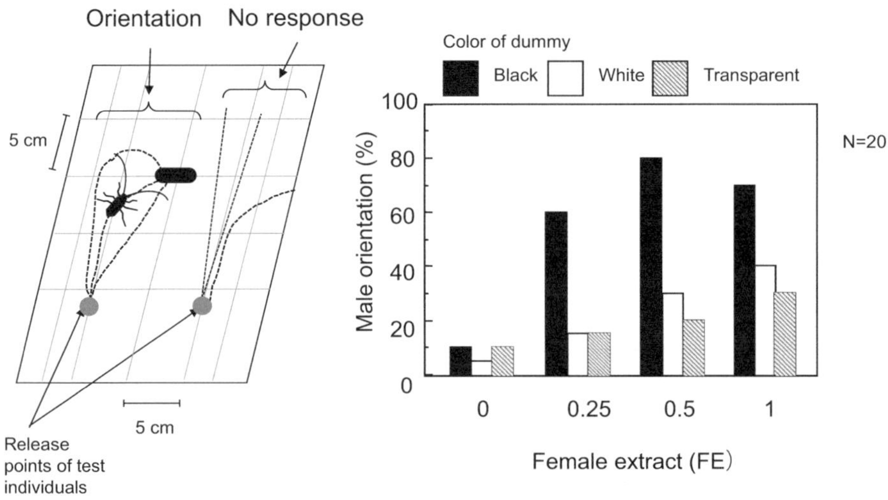

Fig. 5.4 Effects of the visual factor of the target on olfactory-mediated orientation in *Anoplophora malasiaca*. The activity of volatile attractants in female elytra extract was increased by the visual cue, the darker color of the target (Fukaya et al. 2004, 2005a). Female orientation to the dummies treated with male extract was also increased by the darker color of the dummies (Fukaya et al. 2005a). Left: evaluation of orientation response to olfactory and visual cues acting in the vicinity. A sheet of white paper was fixed on a plate and placed at 75° in an acrylic box. A dummy was fixed at the center of the sheet. Test individuals were introduced from the bottom at the left or right as illustrated. When the test individual turned toward the dummy before contacting it with its antennae, the response was evaluated as the orientation. Typical walking trails of the beetle: orientation and no response were illustrated. Right: male orientation responses to the black, white, or transparent glass dummy treated with 0.25 to 1 FE of female extract. 0 FE: solvent only. (From Fukaya et al. (2004) with the addition of data at 0.25 and 1 FE)

which contained the contact sex pheromone (Fukaya et al. 1999; Fukaya 2005a [movie: momo050921am01b]).

5.6.2 *Three Chemical Groups of Contact Sex Pheromone Components*

Three active fractions were separated from the crude extract of female elytra by silica-gel column chromatography. Each individual fraction had little pheromonal activity, but a blend of all three fractions exhibited pheromonal activity equal to that of the crude female extract (Fig. 5.5). The first, second, and third potentially active fractions contained hydrocarbons (Fukaya et al. 2000), novel aliphatic ketones (Yasui et al. 2003), and novel lactones (Fujiwara-Tsujii et al. 2019; Yasui et al. 2007a), respectively. The chemical structures of the active components are shown in Fig. 5.6. Lactone components are named "gomadalactones" after the Japanese name of *A. malasiaca*, "Gomadara-Kamikiri" (Yasui et al. 2007a).

Hydrocarbons

The hexane fraction separated from the crude extract of female *A. malasiaca* by silica-gel column chromatography contained approximately 40 hydrocarbons with C27–C37 (Fig. 5.7; Akino et al. 2001; Fukaya et al. 1999, 2000). Males exhibited a series of precopulatory behaviors toward a dummy treated with a blend of the major eight synthetic hydrocarbons (Fig. 5.6; C27, C29, 4Me-C26, 4Me-C28, 9Me-C27, 9Me-C29, 15Me-C31, and 15Me-C33) at the same ratio as those in the female elytra extracts (Fukaya et al. 2000). However, the activity was very low; 4 female equiva-

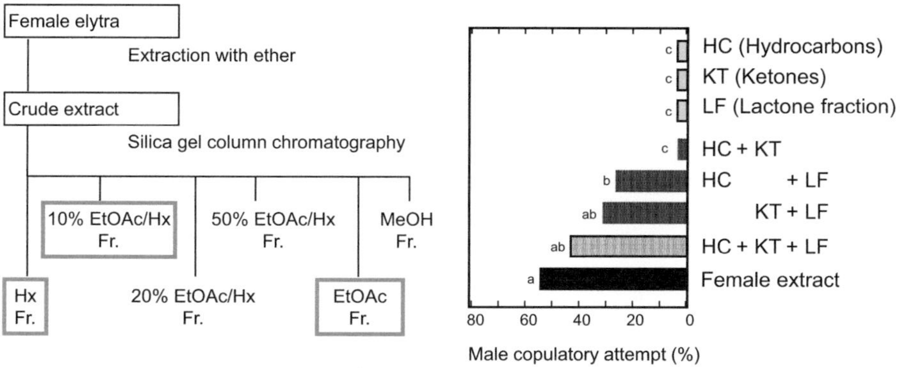

Fig. 5.5 Left: fractionation of female *A. malasiaca* elytra extracts. The boxed fractions were behaviorally active. Fr.: fraction, Hx: hexane, EtOAc: ethyl acetate. Right: behavioral response of male *A. malasiaca* to a mixture of eight synthetic hydrocarbons (HC) contained in Hx fraction (see Fig. 5.6), a mixture of four synthetic ketones (KT) contained in 10% EtOAc fraction (see Fig. 5.6), or EtOAc fraction containing lactones (LF) (N = 35). Tested at 2 FE. Columns with the same letters are not significantly different according to the paired χ^2-test or Fisher's exact test when the former test was not applicable (P = 0.05). (Modified from Yasui et al. 2003)

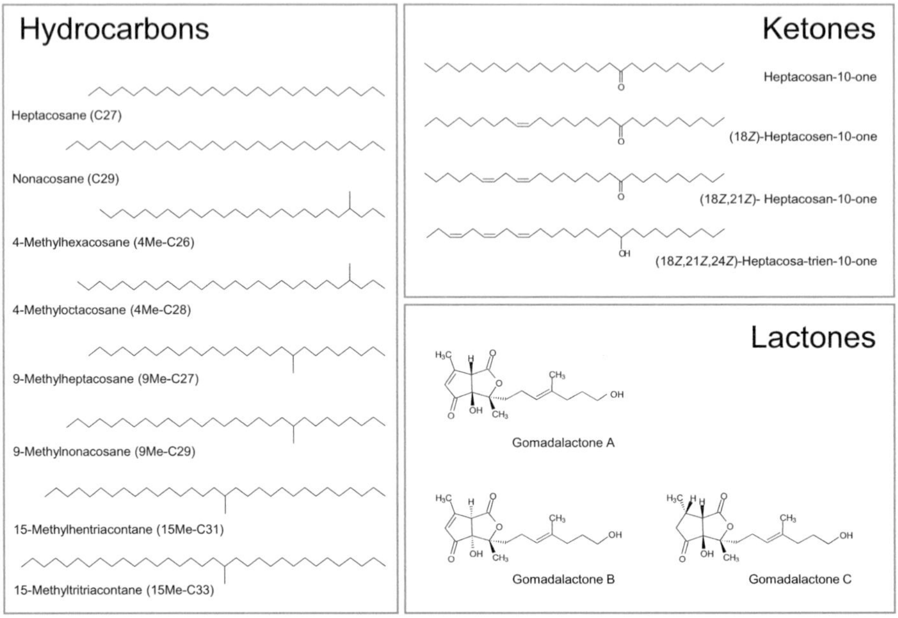

Fig. 5.6 Contact pheromone components of *A. malasiaca*. (Modified from Fukaya et al. 2000), Yasui et al. 2003, 2007a, and Fujiwara-Tsujii et al. 2019)

lents (FE) induced copulatory attempts only from 5% of the males. When the polar fraction eluted by ether was added to the hydrocarbons, however, the same dose evoked copulatory responses from >60% of the males (Fukaya et al. 2000).

Sexual difference was detected in the CHC composition; longer-chain CHCs over C27 were more abundant in females than in males (Fig. 5.7; Akino et al. 2001). When the male polar fraction was mixed with a female hydrocarbon fraction, the mating response was not induced. This suggested an essential factor that causes significant differences in male behavior is contained in the female polar fraction. However, without the hydrocarbon fraction, the female polar fraction induced only weak male mating responses. Therefore, the hydrocarbon components are considered to function synergistically with the polar compounds in the sexual recognition of *A. malasiaca.*

Ketones

The 10% EtOAc fraction of the crude extract exhibited synergistic effects on male precopulatory behavior (Yasui et al. 2003). This fraction contained five novel ketones, heptacosan-10-one, (*Z*)-18-heptacosen-10-one, (18*Z*,21*Z*)-heptacosa-18,21-dien-10-one, (18*Z*,21*Z*,24*Z*)-heptacosa-18,21,24-trien-10-one, and 12-heptacosanone. These ketones were not present in the extracts of male elytra. A mixture of these aliphatic ketones, excluding 12-heptacosanone, blended at the ratio and concentration found in the extract of female elytra exhibited greater synergistic effects than the natural fraction containing all of these ketones (Fig. 5.6; Yasui et al. 2003). The addition of 12-heptacosanone abolished the synergistic

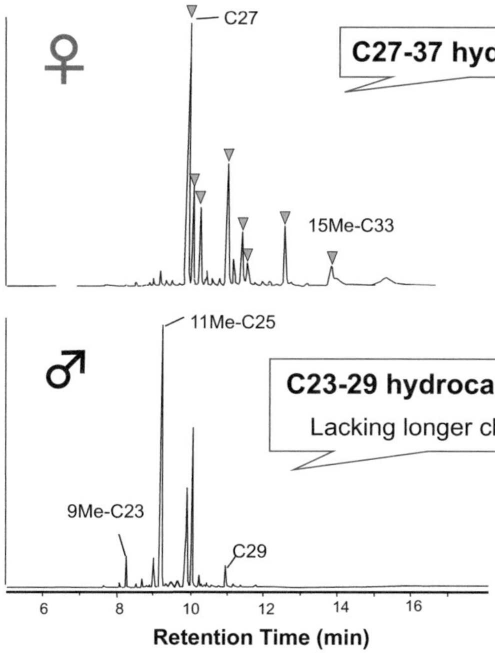

Fig. 5.7 Gas chromatogram of female and male cuticular hydrocarbons (non-polar fraction). Inverted triangles indicate peaks of CHCs involved in induction of male copulatory attempts. (Modified from Akino et al. 2001)

effects but did not suppress holding or mounting behavior of the males (Yasui et al. 2003).

Lactones

The ethyl acetate (EtOAc) fraction evoked male mating behavior in the presence of synergistic compounds of both hydrocarbons and aliphatic ketones (Fig. 5.5; Yasui et al. 2003, 2007a). Three compounds were isolated from the EtOAc fraction using HPLC. Their chemical structures were determined by spectroscopic studies, and absolute configurations were assigned by synthesizing the partial structure of the lactones (Mori 2007; Yasui et al. 2007a) as (1*S*,4*R*,5*S*) -5-hydroxy-4-[(*E*)-7-hydroxy-4-methylhept-3-enyl]-4,8-dimethyl-3-oxabicyclo[3.3.0]octan-7-en-2,6-dione, its (1*R*,4*R*,5*R*)-isomer, and (1*S*,4*R*,5*S*,8*S*)-5-hydroxy-4-[(*E*)-7-hydroxy-4-methylhept-3-enyl]-4,8-dimethyl-3-oxabi cyclo[3.3.0]-octan-2,6-dione, and named gomadalactones A, B, and C, respectively (Fig. 5.6).

Total chemical synthesis of the above three gomadalactones was recently achieved by Suzuki et al. (2017). More recently, our group successfully demonstrated that a mixture of synthetic gomadalactones A, B, and C, when combined with blends of synthetic hydrocarbons and ketones, elicits full precopulatory responses from male *A. malasiaca* beetles to the same degree as the extract of female elytra (Fig. 5.8; Fujiwara-Tsujii et al. 2019).

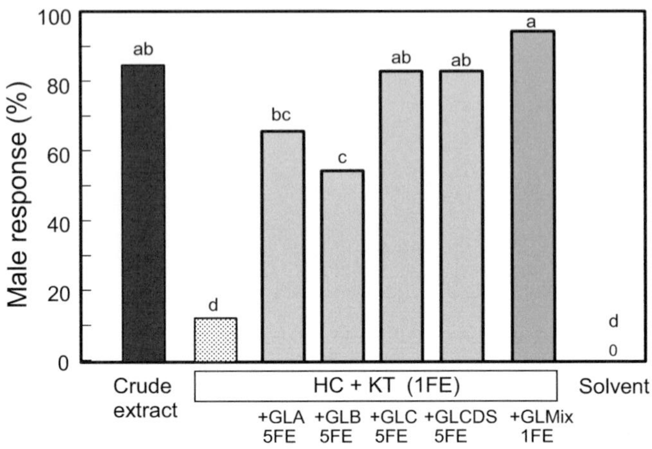

Fig. 5.8 Abdominal bending responses of male *A. malasiaca* to four synthetic gomadalactones in combination with a blend of eight synthetic hydrocarbons and four synthetic ketones at 1 FE treated on a black glass rod (N = 32). HC + KT: mixture of eight hydrocarbons and four ketones. GLA: gomadalactone A; GLB, gomadalactone B, GLC: gomadalactone C, GLCDS: gomadalactone C diastereomer; GLMix: a mixture of all four gomadalactones (1 FE). P < 0.0001 by χ^2 test (df = 7, χ^2 = 113.27). Values with the same letters did not significantly differ at $P = 0.05$ by the paired $\chi 2$ test with Bonferroni correction. (Modified from Fujiwara-Tsujii et al. 2019)

The study on the contact pheromone of *A. malasiaca* was started in 1994 by a few researchers originally working on the contact pheromone of *P. hilaris* and continued until 2019, although its participants have changed with time. In retrospect, it took a quarter of century to fully clarify the complicated composition. The identification and synthesis of gomadalactones, which have a novel basic structure, have been particularly challenging.

5.7 Suggestions for Future Studies

Cerambycid beetles have been demonstrated to use multimodal signals, volatile and contact chemicals, visual cues, vibratory cues, and presumably the "affordance" of the substrate, in mate location, mate recognition, and other behaviors (e.g., Fukaya and Honda 1996a; Fukaya et al. 2004; Fukaya et al. 2005a, b; Iwabuchi 1985; Lu et al. 2007; Takanashi et al. 2016). To discuss the evolution of sensory cues and sensing systems, and the reproductive strategy of Cerambycidae, the structures and functions of contact pheromones should be considered, as well as other infochemicals and non-chemical signals. In *Drosophila* and other flies, CHCs as contact pheromone components have been examined from behavioral, genetic, physiological, and evolutionary points of view (e.g., Blomquist et al. 1987; Buckley et al. 1997; Chung and Carroll 2015; Coyne and Oyama 1995).

In comparison with studies on volatile pheromones in cerambycid beetles, those on contact pheromones have progressed slowly. An explanation for this may be the difficulty in the practical application of contact pheromones for pest control, although they may be useful in combination with organic or microbial insecticides. In contrast, volatile attractants in cerambycids are now beginning to be utilized for monitoring of invasive pests and species to be conserved, and for control of pest species by mating disruption (Barbour et al. 2019; Collignon et al. 2019; Eyre and Haack 2017; Molander et al. 2019; Wickham et al. 2016b; Sweeney et al. 2017). Another explanation may be the difficulty in obtaining a sufficient number of beetles for laboratory bioassays; most cerambycids have a long (> 2 years in some species) life cycle and cannot be reared in the laboratory. It is also difficult to capture a sufficient number of longhorn beetles in the field.

Excluding the contact pheromone of *A. malasiaca*, which is a complex mixture of different classes of compounds, contact pheromones in Cerambycidae reported to date consist of one or a few CHCs (Table 5.1). Is the contact pheromone of *A. malasiaca* truly an exception? We hypothesize that contact pheromones of many cerambycids include polar components in addition to CHCs. In Cerambycidae, the presence of polar components in contact pheromones has been reported for *Monochamus alternatus* (Kim et al. 1992), in addition to *P. hilaris* (Fukaya et al. 1996) and *A. malasiaca*. In these studies, the importance of the polar components was confirmed by bioassays using an artificial substrate as the dummy of a female, on which separated fractions of female extracts were applied. In many studies that resulted in the identification of only CHC components, pheromonal activity was tested using the "solvent-washed" female carcass as the substrate, which was prepared by immersing the carcass in small aliquots of hexane twice for 2–5 min (Table 5.1). This method is effective for identifying CHCs that induce copulatory attempts from the males, but may overlook the effects of synergistic compounds remaining on the carcass. In our tests with some cerambycid species, the responses of males to the glass dummy treated with CHCs were significantly lower than those to the solvent-washed carcass treated with the same CHCs (unpublished observation by MF). The solvent-washed carcass of some species was reported to evoke male precopulatory behavior to some extent (Silk et al. 2011). The carcass of *P. hilaris* had to be extracted with ether using Soxhlet extractor for 12 hr. in order to completely remove its pheromonal activity (Fukaya and Honda 1992). However, the possibility that synergistic compounds, which do not exhibit pheromonal activity by themselves, are remaining on the hexane-washed female bodies cannot be excluded. This possibility should be taken into consideration when contact pheromones of cerambycids are studied in the future.

In contrast to ambiguous species specificity of volatile attractant pheromones in cerambycids, the components of which are often shared not only by congeners but also by the species in the same tribe or subfamily, contact pheromones are highly species specific (Table 5.1). Mate recognition in cerambycids may basically depend on contact sex pheromones (e.g., Ginzel and Hanks 2003; Hanks and Wang 2017; Silk et al. 2011). Identification of contact pheromones in many more cerambycids, in addi-

Table 5.1 Cerambycid species in which contact pheromones were identified. Long-range attractants and non-chemical cues are listed along with components of contact pheromones. Three species of Prioninae and Lepturinae, in which contact pheromones have not been identified, are listed for reference

Subfamily Tribe		Contact pheromone			Long-range attractants	Other cues in mate location/ recognition
		Cuticular hydrocarbons	(weakly) polar components	Bioassay substrate*		
Prioninae						
Tribe Prionini	*Mallodon dasystomus* (Say)	2Me-C26, 2Me-C28 (Spikes et al. 2010)		Hexane-washed female carcass	Female-produced PH (not identified)	
					Host plant odor (Paschen et al. 2012)	
	[Prionus californicus]	[Hexane extract containing CHCs induced male mating behavior when applied on the washed female body] (Barbour et al. 2007)		Hexane-washed female carcass	Female-produced PH: Prionic acid (Rodstein et al. 2009)	
Lepturinae						
Tribe Rhagiini	*Pidonia grallatrix* (Bates)	[Sex- and species-specific abundance of some CHCs. Female extracts had pheromonal activity] (Tanigaki et al. 2007)		Hexane-washed female carcass		
	Pidonia takechii (Kuboki)	[Sex- and species-specific abundance of some CHCs. Female extracts had pheromonal activity] (Tanigaki et al. 2007)		Hexane-washed female carcass	Floral scent volatiles: (benzyl acetate, methyl benzoate, methyl phenylacetate, linalool) (Sakakibara et al. 1998)	
Cerambycinae						

Tribe Clytini	*Megacyllene caryae* (Gahan)	Z9-C29 (Ginzel et al. 2006)		Hexane-washed female carcass	Male-produced PH: 2,3-C6-diols, terpenoids and aromatic alcohols (limonene, 2-phenylethanol, α-terpineol, nerol, neral, geranial) (Lacey et al. 2008b)	
	Megacyllene robiniae (Forster)	Z9-C25 (Ginzel et al. 2003a)		Hexane-washed female carcass		
	Neoclytus a. acuminateus (Fabricius)	7Me-C27 (major) 7Me-C25, 9Me-C25 (synergists) (Lacey et al. 2008a)		Hexane-washed female carcass	Male-produced PH: 2S,3S-C6-diol (Lacey et al. 2004)	
	Xylotrechus colonus (Fabricius)	*n*C25, 9Me-C25, 3Me-C25 (Ginzel et al. 2003b)		Hexane-washed female carcass	No reports [other species in same genera have male-produced PH, C6 - C10 alkanediol or its analogs]	
Tribe: Callidiini	*Callidiellum rufipenne* (Motschulsky)	5Me17Me-C29 (Rutledge et al. 2009)		Hexane-washed female carcass	Male-produced PH: *3R*-C6-ketol (major) 3S-C6-ketol, 2*R*- and 2*S*-C6-ketol (minor) (Zou et al. 2016)	

(continued)

Table 5.1 (continued)

Subfamily Tribe		Contact pheromone: Cuticular hydrocarbons	Contact pheromone: (weakly) polar components	Bioassay substrate*	Long-range attractants	Other cues in mate location/ recognition
					Synergist (not produced by this species): 1-(1H-pyrrol-2-yl)-1,2-propanedione (Zou et al. 2016)	
Spondylidinae						
Tribe Asemini	*Tetropium fuscum* (Fabricius)	*S*-11Me-C27** (Silk et al. 2011)		Hexane-washed female carcass	Male-produced PH: (*S*)-Fuscmol Host plant volatiles (Sweeney et al. 2004, Sweeney et al. 2010; Silk et al. 2007; Hanks and Millar 2013)	
	Tetropium cinnamopterum (Kinby)	Z9-C25 (essential), *S*-11Me-C27 (Silk et al. 2011)		Hexane-washed female carcass	Male-produced PH: (*S*)-Fuscmol Host plant volatiles (Silk et al. 2007; Sweeney et al. 2010; Hanks and Millar 2013)	
Lamiinae						
Tribe Lamiini	*Psacothea hilaris* (Pascoe)	Z8-21Me-C35 (Fukaya et al. 1996, 1997)	Evidence of existence (not identified) (Fukaya et al. 1996)	Clear gelatin capsule fixed on a paper disc (see Fig. 5.3)	Host plant volatiles (not identified) (Yokoi and Yoshii 1984, 1987)	Shapes and sizes of female (dummy) alter male behavior (Fukaya and Honda 1996a)

	Anoplophora glabripennsis (Motschulsky)	*Z*9-C23, *Z*9-C25, *Z*7-C25, *Z*7-C27, *Z*9-C27 (Zhang et al. 2003)		Plastic tube (Microcentrifuge tube)	Male-produced PH: 4-(*n*-Heptyloxy)butan-1ol 4-(*n*-Heptyloxy)butanal (Zhang et al. 2002) Host volatiles: [(−)-linalool, (*Z*)-3-hexen-1-ol, linalool oxide, *trans*-caryophyllene, pinocarveol, etc.] (Nehme et al. 2009, 2010).	Trail pheromone: 2Me-C22, Z9-C23 Z9-C25, Z7-C25 (Hoover et al. 2014)
	Anoplophora malasiaca (Thomson)	*n*C27, *n*C29, 4Me-C26, 4Me-C28, 9Me-C27, 9Me-C29, 15Me-C31, 15Me-C33 (Fukaya et al. 1999, 2000)	Gomadalactone A, B, C (essential components) Ketones (C27-10-one, Z18-C27-10-one, 18Z,21Z-C27-10-one, 18Z,21Z,24Z-C27-10-one, C27-12-one) (Yasui et al. 2003, 2007; Fujiwara-Tsujii et al. 2019)	Glass rod fixed on a paper disc (Transparent or black)	Host volatiles: active in the lab and field (sesquiterpenes) (Yasui et al. 2007b, 2008, 2011) Male-produced PH candidates: (low activity in the lab bioassay) 4-(*n*-Heptyloxy) butan-1-ol 4-(*n*-Heptyloxy)butanal (Yasui et al. 2019)	Visual cue of the conspecifics (and dummy) acts as a synergist in the vicinity (c.a. 10 cm) (Fukaya et al. 2004, 2005a, 2005b)

PH: pheromone. *Bioassay substrate: dummy of a female to which test samples were applied. Hexane–washed female carcass: the female carcass was immersed in small aliquots of hexane twice for 2–5 min each. **: The *R*-enantiomer was inactive. See the text for abbreviations of the compounds

tion to a further understanding of their phylogenetic relationships, will help clarify the evolution of mate recognition systems and the mechanisms of speciation in this family.

In addition to chemical cues, non-chemical cues, such as visual and tactile cues from the female, often play crucial roles in mate recognition (e.g., Fukaya 2005a, b; Fukaya and Honda 1995; Fukaya et al. 2004). In the bioassay of contact-sex pheromonal activity in *A. malasiaca*, the transparent glass rod used in earlier studies was replaced with a black one because the darker color of the dummy was found to synergistically act with the chemical cues (Fukaya et al. 2004, 2005a, b). All factors that possibly affect mating behavior should be isolated and examined singly or in combination with others to evaluate their effects. Invention of a new bioassay method and new techniques for chemical analysis may markedly change our current consensus on the contact pheromones of Cerambycidae.

Acknowledgments We are grateful to Prof. Y. Ishikawa for giving us the opportunity to write a chapter in this book. We thank SW, TA, TY, NFT, and all of our collaborators for working together on contact pheromones of *A. malasiaca*, and Prof. R. Iwata for his collection of literature on cerambycids. With respect and appreciation, we remember Prof. K. Mori (1935–2019), a pioneer of chiral-chemistry, for his contribution to the progress of chemical ecology.

References

Akino T, Fukaya M, Yasui H, Wakamura S (2001) Sexual dimorphism in cuticular hydrocarbons of the white-spotted longicorn beetle, *Anoplophora malasiaca* (Coleoptera: Cerambycidae). Entomol Sci 4:271–277

Akutsu K (1983) Mating behavior of the udo longicorn beetle, *Acalolepta luxuriosa* Bates (Coleoptera: Cerambycidae). Jpn J Appl Entomol Zool 27:189–196. (In Japanese with English summary)

Akutsu K, Kuboki M (1983) Mating behavior of the udo longicorn beetle, *Acalolepta luxuriosa* Bates (Coleoptera: Cerambycidae). Jpn J Appl Entomol Zool 27:247–251. (In Japanese with English summary)

Albert J, Platek M, Cizek L (2012) Vertical stratification and microhabitat selection by the great Capricorn beetle (*Cerambyx cerdo*)(Coleoptera: Cerambycidae) in open-grown, veteran oaks. Eur J Entomol 109(4):553

Allison JD, Borden JH, McIntosh RL, De Groot P, Gries R (2001) Kairomonal response by four Monochamus species (Coleoptera: Cerambycidae) to bark beetle pheromones. J Chem Ecol 27(4):633–646

Allison JD, Morewood WD, Borden JH, Hein KE, Wilson IM (2003) Differential bio-activity of Ips and Dendroctonus (Coleoptera: Scolytidae) pheromone components for *Monochamus clamator* and *M. scutellatus* (Coleoptera: Cerambycidae). Environ Entomol 32(1):23–30

Allison JD, McKenney JL, Millar JG, McElfresh JS, Mitchell RF, Hanks LM (2012) Response of the woodborers *Monochamus carolinensis* and *Monochamus titillator* (Coleoptera: Cerambycidae) to known cerambycid pheromones in the presence and absence of the host plant volatile α-pinene. Environ Entomol 41(6):1587–1596

Allison JD, McKenney JL, Miller DR, Gimmel ML (2013) Kairomonal responses of natural enemies and associates of the southern Ips (Coleoptera: Curculionidae: Scolytinae) to ipsdienol, ipsenol and cis-verbenol. J Insect Behav 26(3):321–335

Barbour JD, Lacey ES, Hanks LM (2007) Cuticular hydrocarbons mediate mate recognition in a species of longhorned beetle (Coleoptera: Cerambycidae) of the primitive subfamily Prioninae. Ann Entomol Soc Am 100(2):333–338

Barbour JD, Millar JG, Rodstein J, Ray AM, Alston DG, Rejzek M, Dutcher JD, Hanks LM (2011) Synthetic 3, 5-dimethyldodecanoic acid serves as a general attractant for multiple species of *Prionus* (Coleoptera: Cerambycidae). Ann Entomol Soc Am 104(3):588–593

Barbour JD, Alston DG, Walsh DB, Pace M, Hanks LM (2019) Mating disruption for managing *Prionus californicus* (Coleoptera: Cerambycidae) in hop and sweet cherry. J Econ Entomol 112(3):1130–1137

Bílý S, Mehl O (1989) Longhorn beetles (Coleoptera, Cerambycidae) of Fennoscandia and Denmark, vol 22. E. J. Brill, Leiden

Blomquist GJ, Nelson DR, De Renobales M (1987) Chemistry, biochemistry, and physiology of insect cuticular lipids. Arch Insect Biochem Physiol 6(4):227–265

Buckley SH, Tregenza T, Butlin RK (1997) Speciation and signal trait genetics. Trends Ecol Evol 12(8):299–301

Cardé RT (2014) Defining attraction and aggregation pheromones: teleological versus functional perspectives. J Chem Ecol 40:519–520

Chen JM, Qiao HL, Chen J, Xu CQ, Liu S, Lian ZM, Guo K (2014) Observation of antennal sensilla in *Xylotrechus grayii* (Coleoptera: Cerambycidae) with scanning electron microscopy. Microsc Res Tech 77(4):264–273

Chung H, Carroll SB (2015) Wax, sex and the origin of species: dual roles of insect cuticular hydrocarbons in adaptation and mating. BioEssays 37(7):822–830

Collignon RM, Swift IP, Zou Y, McElfresh JS, Hanks LM, Millar JG (2016) The influence of host plant volatiles on the attraction of longhorn beetles to pheromones. J Chem Ecol 42(3):215–229

Collignon RM, Halloran S, Serrano JM, McElfresh JS, Millar JG (2019) An unstable monoterpene alcohol as a pheromone component of the longhorned beetle *Paranoplium gracile* (Coleoptera: Cerambycidae). J Chem Ecol 45(4):339–347

Coyne JA, Oyama R (1995) Localization of pheromonal sexual dimorphism in *Drosophila melanogaster* and its effect on sexual isolation. Proc Natl Acad Sci 92(21):9505–9509

Crook DJ, Lance DR, Mastro VC (2014) Identification of a potential third component of the male-produced pheromone of *Anoplophora glabripennis a*nd its effect on behavior. J Chem Ecol 40(11–12):1241–1250

Curkovic T, Ferrera C (2012) Female calling and male flight orientation and searching behaviors in *Callisphyris apicicornis*: evidence for a female-produced sex attractant pheromone. Ciencia e Investigación Agraria 39(1):147–158

Dai HG, Honda H (1990) Sensilla on the antennal flagellum of the yellow spotted longicorn beetle, *Psacothea hilaris* (Pascoe)(Coleoptera: Cerambycidae). Appl Entomol Zool 25(2):273–282

Duffy EAJ (1953) A monograph of the immature stages of British and imported timber beetles (Cerambycidae). British Museum, London

Di Palma A, Pistillo M, Griffo R, Garonna AP, Germinara GS (2017) Scanning electron microscopy of the antennal sensilla and their secretion analysis in adults of *Aromia bungii* (Faldermann, 1835) (Coleoptera, Cerambycidae). Insect 10:88

Drag L, Hauck D, Bérces S, Michalcewicz J, Šerić Jelaska L, Aurenhammer S, Cizek L (2015) Genetic differentiation of populations of the threatened saproxylic beetle Rosalia longicorn, *Rosalia alpina* (Coleoptera: Cerambycidae) in Central and South-east Europe. Biol J Linn Soc 116(4):911–925

Dyer LJ, Seabrook WD (1975) Sensilla on the antennal flagellum of the sawyer beetles *Monochamus notatus* (Drury) and *Monochamus scutellatus* (Say)(Coleoptera: Cerambycidae). J Morphol 146(4):513–531

Eckelt A, Müller J, Bense U, Brustel H, Bußler H, Chittaro Y, Cizek L, Frei A, Holzer E, Kadej M, Kahlen M, Köhler F, Möller G, Mühle H, Sanchezs A, Schaffrath U, Schmidl J, Smolis A, Szallies A, Németh T, Wurst C, Thorn S, Rune T, Christensen RHB, Seibold S (2018) "Primeval forest relict beetles" of Central Europe: a set of 168 umbrella species for the protection of primeval forest remnants. J Insect Conserv 22(1):15–28

Eyre D, Haack RA (2017) Chapter 13: Invasive cerambycid pests and biosecurity measures. In: Wang Q (ed) Cerambycidae of the world: biology and pest management. CRC Press, Boca Raton, pp 563–607

Fauziah BA, Hidaka T, Tabata K (1987) The reproductive behavior of *Monochamus alternatus* HOPE:(Coleoptera: Cerambycidae). Appl Entomol Zool 22(3):272–285

Fettköther R, Dettner K, Schröder F, Meyer H, Francke W, Noldt U (1995) The male pheromone of the old house borer *Hylotrupes bajulus* (L.)(Coleoptera: Cerambycidae): identification and female response. Experientia 51(3):270–277

Fierke MK, Skabeikis DD, Millar JG, Teale SA, McElfresh JS, Hanks LM (2012) Identification of a male-produced aggregation pheromone for *Monochamus scutellatus scutellatus* and an attractant for the congener *Monochamus notatus* (Coleoptera: Cerambycidae). J Econ Entomol 105(6):2029–2034

Fujiwara-Tsujii N, Yasui H, Wakamura S, Hashimoto I, Minamishima M (2012) The white-spotted longicorn beetle, *Anoplophora malasiaca* (Coleoptera: Cerambycidae), with a blueberry as host plant, utilizes host chemicals for male orientation. Appl Entomol Zool 47(2):103–110

Fujiwara-Tsujii N, Yasui H, Yasuda T, Wakamura S, Akino T, Fukaya M, Suzuki T, Ono H (2019) Contact sex pheromone activity of synthetic Gomadalactones in male white-spotted longhorn beetle, *Anoplophora malasiaca* (Coleoptera: Cerambycidae). J Chem Ecol 45:440–446

Fukaya M (1992) Ethology and chemical ecology of the reproductive system of the yellow-spotted longicorn beetle, *Psacothea hilaris* (Pascoe) (Coleoptera: Cerambycidae). PhD Thesis, The University of Tokyo (In Japanese with English abstract)

Fukaya M (2005a) Mating behavior of the white-spotted longicorn beetle, *Anoplophora malasiaca*. 1. Male mating trial to a glass-rod treated with female pheromonal extract. [http://www.momo-p.com/index-e.html Data No.momo050921am01b] Accessed 11 July 2019

Fukaya M (2005b) Mating behavior of the white-spotted longicorn beetle, *Anoplophora malasiaca*. 2 Male mate orientation regulated by olfactory and visual cues. Movie Archive of Animal behavior [http://www.momo-p.com/index-e.html Data No. momo050921am02b] Accessed 11 July 2019

Fukaya M, Honda H (1992) Reproductive biology of the yellow-spotted longicorn beetle, *Psacothea hilaris* (Pascoe) (Coleoptera: Cerambycidae). I. Male mating behaviors and female sex pheromones. Appl Entomol Zool 27:89–97

Fukaya M, Honda H (1995) Reproductive biology of the yellow-spotted longicorn beetle, *Psacothea hilaris* (Pascoe) (Coleoptera: Cerambycidae). II. Evidence for two female pheromones with different functions. Appl Entomol Zool 30:467–460

Fukaya M, Honda H (1996a) Reproductive biology of the yellow-spotted longicorn beetle, *Psacothea hilaris* (Pascoe) (Coleoptera: Cerambycidae). IV. Effects of shape and size of female models on male mating behaviors. Appl Entomol Zool 31:51–58

Fukaya M, Honda H (1996b) Reproductive biology of the yellow-spotted longicorn beetle, *Psacothea hilaris* (Pascoe) (Coleoptera: Cerambycidae). V. Male mating response to male extract. Appl Entomol Zool 31:95–98

Fukaya M, Yasuda T, Wakamura S, Honda H (1996) Reproductive biology of the yellow-spotted longicorn beetle, *Psacothea hilaris* (Pascoe) (Coleoptera: Cerambycidae). III. Identification of the contact sex pheromone on female body. J Chem Ecol 22:259–270

Fukaya M, Wakamura S, Yasuda T, Omata T, Senda S (1997) Sex pheromonal activity of geometric and optical isomers of synthetic contact pheromone to males of the yellow-spotted longicorn beetle, *Psacothea hilaris (*Pascoe) (Coleoptera: Cerambycidae). Appl Entomol Zool 32:654–656

Fukaya M, Akino T, Yasuda T, Tatsuki S, Wakamura S (1999) Mating sequence and evidence for female sex pheromone in the white-spotted longicorn beetle, *Anoplophora malasiaca* (Thomson) (Coleoptera: Cerambycidae). Entomol Sci 2:183–187

Fukaya M, Akino T, Yasuda T, Wakamura S, Satoda S, Senda S (2000) Hydrocarbon components in contact sex pheromone of the white-spotted longicorn beetle, *Anoplophola malasiaca* (Thomson) (Coleoptera: Cerambycidae) and pheromonal activity of synthetic hydrocarbons. Entomol Sci 3:211–218

Fukaya M, Akino T, Yasuda T, Yasui H, Wakamura S (2004) Visual and olfactory cues for mate orientation behaviour in male white-spotted longicorn beetle, *Anoplophora malasiaca*. Entomol Exp Appl 111(2):111–115

Fukaya M, Akino T, Yasui H, Yasuda T, Wakamura S (2005a) Effect of size and color of female models for male mate orientation in the white spotted longicorn beetle, *Anoplophora malasiaca* (Coleoptera: Cerambycidae). Appl Entomol Zool 40:513–519

Fukaya M, Yasui H, Yasuda T, Akino T, Wakamura S (2005b) Female orientation to the male in the white-spotted longicorn beetle, *Anoplophora malasiaca* by visual and olfactory cues. Appl Entomol Zool 40:63–68

Fukaya M, Kiriyama S, Yasui H (2017) Mate-location flight of the red-necked longicorn beetle, *Aromia bungii* (Coleoptera: Cerambycidae): an invasive pest lethal to Rosaceae trees. Appl Entomol Zool 52:559–565

Fukusaki E, Satoda S, Yuasa H, Senda S, Omata T, Fukaya M, Wakamura S (1998) Synthesis of the four possible stereoisomers of 21-methyl-8-pentatriacontene, the female contact sex pheromone of the yellow-spotted longicorn beetle, *Psacothea hilaris*. J Ferment Bioeng 85(1):120–121

Galford JR (1977) Evidence for a pheromone in the locust borer (Vol. 240). USDA Forest Service Research Notes NE-240:1–3

Gatti FD, Rodrigues THA, Figueiredo LAD, Carneiro MAA (2018) Longhorn beetle (Coleoptera: Cerambycidae) assemblage and the structural heterogeneity of habitat at the Brazilian Atlantic forest. Environ Entomol 47(6):1413–1419

Ginzel MD (2010) Hydrocarbons as contact pheromones of longhorned beetles (Coleoptera: Cerambycidae). In: Insect hydrocarbons: biology, biochemistry and chemical ecology, pp 375–389

Ginzel MD, Hanks LM (2003) Contact pheromones as mate recognition cues of four species of longhorned beetles (Coleoptera: Cerambycidae). J Insect Behav 16(2):181–187

Ginzel MD, Millar JG, Hanks LM (2003a) (*Z*)-9-Pentacosene–2212; contact sex pheromone of the locust borer, *Megacyllene robiniae*. Chemoecology 13(3):135–141

Ginzel MD, Blomquist GJ, Millar JG, Hanks LM (2003b) Role of contact pheromones in mate recognition in *Xylotrechus colonus*. J Chem Ecol 29(3):533–545

Ginzel MD, Moreira JA, Ray AM, Millar JG, Hanks LM (2006) (Z)-9-Nonacosene–2014;major component of the contact sex pheromone of the beetle *Megacyllene caryae*. J Chem Ecol 32(2):435–451

Goldsmith SK (1987) The mating system and alternative reproductive behaviors of *Dendrobias mandibularis* (Coleoptera: Cerambycidae). Behav Ecol Sociobiol 20(2):111–115

Goldsmith S (2007) Density of longhorned beetles (Coleoptera: Cerambycidae) differs at different elevations in Hawaiian montane forest. Southwest Nat 52(3):364–370

Haack RA (2017) Cerambycid pests in forests and urban trees. In: Wang Q (ed) Cerambycidae of the world: biology and pest management. CRC Press, Boca Raton, pp 352–384

Halloran ST, Collignon RM, McElfresh JS, Millar JG (2018) Fuscumol and geranylacetone as pheromone components of Californian longhorn beetles (Coleoptera: Cerambycidae) in the subfamily Spondylidinae. Environ Entomol 47(5):1300–1305

Handley K, Hough-Goldstein J, Hanks LM, Millar JG, D–2019;amico V (2015) Species richness and phenology of cerambycid beetles in urban forest fragments of northern Delaware. Ann Entomol Soc Am 108(3):251–262

Hanks LM, Millar JG (2013) Field bioassays of cerambycid pheromones reveal widespread parsimony of pheromone structures, enhancement by host plant volatiles, and antagonism by components from heterospecifics. Chemoecology 23(1):21–44

Hanks LM, Millar JG (2016) Sex and aggregation-sex pheromones of cerambycid beetles: basic science and practical applications. J Chem Ecol 42(7):631–654

Hanks LM, Wang Q (2017) Reproductive biology of cerambycids. In: Cerambycidae of the world. CRC Press, Boca Raton, pp 147–174

Hanks LM, Millar JG, Mongold-Diers JA, Wong JC, Meier LR, Reagel PF, Mitchell RF (2012) Using blends of cerambycid beetle pheromones and host plant volatiles to simultaneously attract a diversity of cerambycid species. Can J For Res 42(6):1050–1059

Hansen L, Xu T, Wickham J, Chen Y, Hao D, Hanks LM, Millar JG, Teale SA (2015) Identification of a male-produced pheromone component of the citrus longhorned beetle, *Anoplophora chinensis*. PLoS One 10(8):e0134358

Heintze A (1925) Lepturinernas blombesok (Coleoptera: Cerambycidae). Entomologisk Tidskrift 46:21–34. (In Swedish)

Hoover K, Keena M, Nehme M, Wang S, Meng P, Zhang A (2014) Sex-specific trail pheromone mediates complex mate finding behavior in *Anoplophora glabripennis*. J Chem Ecol 40(2):169–180

Hoshino K, Nakaba S, Inoue H, Iwabuchi K (2015) Structure and development of male pheromone gland of longicorn beetles and its phylogenetic relationships within the tribe Clytini. J Exp Zool B Mol Dev Evol 324(1):68–76

Hughes GP, Zou Y, Millar JG, Ginzel MD (2013) (S)-fuscumol and (S)-fuscumol acetate produced by a male *Astyleiopus variegatus* (Coleoptera: Cerambycidae). Can Entomol 145(3):327–332

Hughes GP, Bello JE, Millar JG, Ginzel MD (2015) Determination of the absolute configuration of female-produced contact sex pheromone components of the longhorned beetle, *Neoclytus acuminatus acuminatus* (F). J Chem Ecol 41(11):1050–1057

Ibeas F, Gemeno CS, Díez JJ, Pajares JA (2009) Female recognition and sexual dimorphism of cuticular hydrocarbons in *Monochamus galloprovincialis* (Coleoptera: Cerambycidae). Ann Entomol Soc Am 102(2):317–325

Iwabuchi K (1985) Mating behavior of *Xylotrechus pyrrhoderus* Bates (Coleoptera: Cerambycidae) II. Female recognition by male and the existence of a female sex pheromone. Appl Entomol Zool 20(4):416–423

Iwabuchi K (1986) Mating behavior of *Xylotrechus pyrrhoderus* Bates. III. Pheromone secretion by male. Appl Entomol Zool 21:606–612

Iwabuchi K, Takahashi J, Nakagawa Y, Sakai T (1986) Behavioral responses of female grape borer *Xylotrechus pyrrhoderus* Bates (Coleoptera: Cerambycidae) to synthetic male sex pheromone components. Appl Entomol Zool 21(1):21–27

Iwabuchi K, Takahashi J, Sakai T (1987) Occurrence of 2, 3-octanediol and 2-hydroxy-3-octanone, possible male sex pheromone in *Xylotrechus chinensis* Chevrolat (Coleoptera: Cerambycidae). Appl Entomol Zool 22(1):110–111

Iwona M, Marek P, Katarzyna W, Edward B, Julia S (2018) Use of a genetically informed population viability analysis to evaluate management options for Polish populations of endangered beetle *Cerambyx cerdo* L. (1758) (Coleoptera, Cerambycidae). J Insect Conserv 22(1):69–83

Kim GH, Takabayashi J, Takahashi S, Tabata K (1992) Function of pheromones in mating behavior of the Japanese pine sawyer beetle, *Monochamus alternatus* Hope. Appl Entomol Zool 27(4):489–497

Kim GH, Takabayashi J, Takahashi S, Tabata K (1993) Function of contact pheromone in the mating behavior of the cryptomeria bark borer, *Semanotus japonicus* Lacordaire (Coleoptera: Cerambycidae). Appl Entomol Zool 28(4):525–535

Kosi AŽ, Zou Y, Hoskovec M, Vrezec A, Stritih N, Millar JG (2017) Novel, male-produced aggregation pheromone of the cerambycid beetle *Rosalia alpina*, a priority species of European conservation concern. PLoS One 12(8):e0183279

Kuboki M, Akutsu K, Sakai A, Chuman T (1985) Bioassay of the sex pheromone of the udo longicorn beetle, *Acalolepta luxuriosa* Bates (Coleoptera: Cerambycidae). Appl Entomol Zool 20(1):88–89

Kuwahara Y, Matsuyama S, Suzuki T (1987) Identification of 2, 3-octanediol, 2-hydroxy-3-octanone and 3-hydroxy-2-octanone from male *Xylotrechus chinensis* Chevrolat as possible sex pheromones (Coleoptera: Cerambycidae). Appl Entomol Zool 22(1):25–28

Lacey ES, Ginzel MD, Millar JG, Hanks LM (2004) Male-produced aggregation pheromone of the cerambycid beetle *Neoclytus acuminatus acuminatus*. J Chem Ecol 30(8):1493–1507

Lacey ES, Moreira JA, Millar JG, Ray AM, Hanks LM (2007) Male-produced aggregation pheromone of the cerambycid beetle *Neoclytus mucronatus mucronatus*. Entomol Exp Appl 122(2):171–179

Lacey ES, Ginzel MD, Millar JG, Hanks LM (2008a) 7-Methylheptacosane is a major component of the contact sex pheromone of the cerambycid beetle *Neoclytus acuminatus acuminatus*. Physiol Entomol 33(3):209–216

Lacey ES, Moreira JA, Millar JG, Hanks LM (2008b) A male-produced aggregation pheromone blend consisting of alkanediols, terpenoids, and an aromatic alcohol from the cerambycid beetle *Megacyllene caryae*. J Chem Ecol 34(3):408–417

Lingafelter SW, Hoebeke ER (2002) Revision of anoplophora (Coleoptera: Cerambycidae). Entomological Society of Washington, Washington D.C. 236 pp

Linsley EG (1959) Ecology of Cerambycidae. Annu Rev Entomol 4(1):99–138

Lopes O, Marques PC, Araújo J (2005) The role of antennae in mate recognition in *Phoracantha semipunctata* (Coleoptera: Cerambycidae). J Insect Behav 18(2):243–257

Lu W, Wang Q, Tian MY, He XZ, Zeng XL, Zhong YX (2007) Mate location and recognition in *Glenea cantor* (Fabr.)(Coleoptera: Cerambycidae: Lamiinae): roles of host plant health, female sex pheromone, and vision. Environ Entomol 36(4):864–870

Luo SL, Zhuge PP, Wang MQ (2011) Mating behavior and contact pheromones of *Batocera horsfieldi* (Hope)(Coleoptera: Cerambycidae). Entomol Sci 14(3):359–363

Macias-Samano JE, Wakarchuk D, Millar JG, Hanks LM (2012) 2-Undecyloxy-1-ethanol in combination with other semiochemicals attracts three *Monochamus* species (Coleoptera: Cerambycidae) in British Columbia, Canada. Can Entomol 144(6):764–768

Makihara H (2000) True taxonomy and distribution of *Anoplophora* cerambycid beetles in East Asia. For Pests 49:180–194

Makihara H (2007) Geneus *Anoplophora* in longicorn beetles of Japan. Tokai University Press, Tokyo, pp 583–585. (In Japanese)

Meng PS, Hoover K, Keena MA (2015) Asian longhorned beetle (Coleoptera: Cerambycidae), an introduced pest of maple and other hardwood trees in North America and Europe. J Integr Pest Manag 6(1):4

Millar JG, Hanks LM (2017) Chemical ecology of cerambycids. In: Cerambycidae of the world. CRC Press, Boca Raton, pp 175–222

Miller DR, Dodds KJ, Eglitis A, Fettig CJ, Hofstetter RW, Langor DW, Mayfield AE, Munson AS, Poland TM, Raffa KF (2013) Trap lure blend of pine volatiles and bark beetle pheromones for *Monochamus* spp.(Coleoptera: Cerambycidae) in pine forests of Canada and the United States. J Econ Entomol 106(4):1684–1692

Mitchell RF, Graham EE, Wong JC, Reagel PF, Striman BL, Hughes GP, Paschen MA, Ginzel MD, Millar JG, Hanks LM (2011) Fuscumol and fuscumol acetate are general attractants for many species of cerambycid beetles in the subfamily Lamiinae. Entomol Exp Appl 141(1):71–77

Mitchell RF, Reagel PF, Wong JC, Meier LR, Silva WD, Mongold-Diers J et al (2015) Cerambycid beetle species with similar pheromones are segregated by phenology and minor pheromone components. J Chem Ecol 41(5):431–440

Mitomi M, Kuroda E, Okamoto H (1990) Ecological study of the white-spotted longicorn beetle, *Anoplophora malasiaca* Thomson (Coleoptera: Cerambycidae). I. Investigation of adult emergence holes in citrus orchards in Kagawa Prefecture. Jpn J Appl Entomol Zool 34(1):7–13. (In Japanese with English summary)

Molander MA, Eriksson B, Winde IB, Zou Y, Millar JG, Larsson MC (2019) The aggregation-sex pheromones of the cerambycid beetles *Anaglyptus mysticus* and *Xylotrechus antilope ssp. antilope*: new model species for insect conservation through pheromone-based monitoring. Chemoecology 29(3):111–124

Monné ML, Monné MA, Wang Q (2017) General morphology, classication, and biology of cerambycidae. In: Cerambycidae of the world. CRC Press, Boca Raton, pp 15–84

Mori K (2007) Absolute configuration of gomadalactones A, B and C, the components of the contact sex pheromone of *Anoplophora malasiaca*. Tetrahedron Lett 48(32):5609–5611

Muraji M, Wakamura S, Yasui H, Arakaki N, Sadoyama Y, Ohno S, Matsuhira K (2011) Genetic variation of the white-spotted longicorn beetle *Anoplophora spp.*(Coleoptera: Cerambycidae) in Japan detected by mitochondrial DNA sequence. Appl Entomol Zool 46(3):363–373

Nehme ME, Keena MA, Zhang A, Baker TC, Hoover K (2009) Attraction of *Anoplophora glabripennis* to male-produced pheromone and plant volatiles. Environ Entomol 38(6):1745–1755

Noldt U, Fettköther R, Dettner K (1995) Structure of the sex pheromone-producing prothoracic glands of the male old house borer, *Hylotrupes bajulus* (L.)(Coleoptera: Cerambycidae). Int J Insect Morphol Embryol 24(2):223–234

Ohbayashi N, Ogawa J, Su ZH (2009) Phylogenetic analysis of the Lamiine GenusCore Anoplophora and its relatives (Coleoptera, Cerambycidae) based on the mitochondrial CO I gene. Spec Bull Jpn Soc Coleopterol, Tokyo 7:309–324

Pajares JA, Ibeas F, Diez JJ, Gallego D (2004) Attractive responses by *Monochamus galloprovincialis* (Col, Cerambycidae) to host and bark beetle semiochemicals. J Appl Entomol 128:633–638

Pajares JA, Álvarez G, Ibeas F, Gallego D, Hall DR, Farman DI (2010) Identification and field activity of a male-produced aggregation pheromone in the pine sawyer beetle, *Monochamus galloprovincialis*. J Chem Ecol 36(6):570–583

Pajares JA, Álvarez G, Hall DR, Douglas P, Centeno F, Ibarra N et al (2013) 2-(Undecyloxy)-ethanol is a major component of the male-produced aggregation pheromone of *Monochamus sutor*. Entomol Exp Appl 149(2):118–127

Paschen MA, Schiff NM, Ginzel MD (2012) Role of volatile semiochemicals in the host and mate location behavior of *Mallodon dasystomus* (Coleoptera: Cerambycidae). J Insect Behav 25(6):569–577

Ray AM, Millar JG, McElfresh JS, Swift IP, Barbour JD, Hanks LM (2009a) Male-produced aggregation pheromone of the cerambycid beetle *Rosalia funebris*. J Chem Ecol 35(1):96–103

Ray AM, Swift IP, Moreira JA, Millar JG, Hanks LM (2009b) (*R*)-3-hydroxyhexan-2-one is a major pheromone component of *Anelaphus inflaticollis* (Coleoptera: Cerambycidae). Environ Entomol 38(5):1462–1466

Ray AM, Žunič A, Alten RL, McElfresh JS, Hanks LM, Millar JG (2011) *cis*-Vaccenyl acetate, a female-produced sex pheromone component of a longhorned beetle in the subfamily Lepturinae. J Chem Ecol 37(2):173–178

Ray AM, Barbour JD, McElfresh JS, Moreira JA, Swift I, Wright IM, Graham EE, Alten RL, Millar JG, Hanks LM (2012a) 2, 3-Hexanediols as sex attractants and a female-produced sex pheromone for cerambycid beetles in the prionine genus *Tragosoma*. J Chem Ecol 38(9):1151–1158

Ray AM, Swift IP, McElfresh JS, Alten RL, Millar JG (2012b) (*R*)-Desmolactone, a female-produced sex pheromone component of the cerambycid beetle *Desmocerus californicus californicus* (subfamily Lepturinae). J Chem Ecol 38(2):157–167

Ray AM, Arnold RA, Swift I, Schapker PA, McCann S, Marshall CJ, McElfresh JS, Millar JG (2014) (*R*)-Desmolactone is a sex pheromone or sex attractant for the endangered valley elderberry longhorn beetle *Desmocerus californicus* dimorphus and several congeners (Cerambycidae: Lepturinae). PLoS One 9(12):e115498

Ray AM, Millar JG, Moreira JA, McElfresh JS, Mitchell RF, Barbour JD, Hanks LM (2015) North American species of cerambycid beetles in the genus *Neoclytu*s share a common hydroxyhexanone-hexanediol pheromone structural motif. J Econ Entomol 108(4):1860–1868

Ray AM, Francese JA, Zou Y, Watson K, Crook DJ, Millar JG (2019) Isolation and identification of a male-produced aggregation-sex pheromone for the velvet longhorned beetle, *Trichoferus campestris*. Sci Rep 9(1):4459

Rodstein J, McElfresh JS, Barbour JD, Ray AM, Hanks LM, Millar JG (2009) Identification and synthesis of a female-produced sex pheromone for the cerambycid beetle *Prionus californicus*. J Chem Ecol 35(5):590–600

Rodstein J, Millar JG, Barbour JD, McElfresh JS, Wright IM, Barbour KS, Ray AM, Hanks LM (2011) Determination of the relative and absolute configurations of the female-produced sex pheromone of the cerambycid beetle *Prionus californicus*. J Chem Ecol 37:114–124

Rutledge CE, Millar JG, Romero CM, Hanks LM (2009) Identification of an important component of the contact sex pheromone of *Callidiellum rufipenne* (Coleoptera: Cerambycidae). Environ Entomol 38(4):1267–1275

Sakai T, Nakagawa Y, Takahashi J, Iwabuchi K, Ishii K (1984) Isolation and identification of the male sex pheromone of the grape borer *Xylotrechus pyrrhoderus* Bates (Coleoptera: Cerambyci Dae). Chem Lett 13(2):263–264

Sakakibara Y, Kikuma A, Iwata R, Yamane A (1998) Performances of four chemicals with floral scents as attractants for longicorn beetles (Coleoptera: Cerambycidae) in a broadleaved forest. J For Res 3(4):221–224

Sarto i Monteys V, Torras i Tutusaus G (2018) A new alien invasive longhorn beetle, *Xylotrechus chinensis* (Cerambycidae), is infesting mulberries in Catalonia (Spain). Insects 9(2):52

Silk PJ, Sweeney J, Wu J, Price J, Gutowski JM, Kettela EG (2007) Evidence for a male-produced pheromone in *Tetropium fuscum* (F.) and *Tetropium cinnamopterum* (Kirby)(Coleoptera: Cerambycidae). Naturwissenschaften 94(8):697

Silk PJ, Sweeney J, Wu J, Sopow S, Mayo PD, Magee D (2011) Contact sex pheromones identified for two species of longhorned beetles (Coleoptera: Cerambycidae) *Tetropium fuscum* and *T. cinnamopterum* in the subfamily Spondylidinae. Environ Entomol 40(3):714–726

Smyth EG (1934) The gregarious habit of beetles. J Kansas Entomol Soc 7(3):102–108

Spikes AE, Paschen MA, Millar JG, Moreira JA, Hamel PB, Schiff NM, Ginzel MD (2010) First contact pheromone identified for a longhorned beetle (Coleoptera: Cerambycidae) in the subfamily Prioninae. J Chem Ecol 36(9):943–954

Suzuki T, Hagiwara H, Uchida H, Tsujii N, Ono H. (2017) Japan Patent JP2017–09538. [https://astamuse.com/ja/published/JP/No/2017095381.] (in Japanese). Accessed 13 Oct 2017

Sweeney JD, Silk PJ, Gutowski JM, Wu J, Lemay MA, Mayo PD, Magee DI (2010) Effect of chirality, release rate, and host volatiles on response of *Tetropium fuscum* (F.), *Tetropium cinnamopterum* Kirby, and *Tetropium castaneum* (L.) to the aggregation pheromone, fuscumol. J Chem Ecol 36(12):1309–1321

Sweeney J, De Groot P, MacDonald L, Smith S, Cocquempot C, Kenis M, Gutowski JM (2004) Host Volatile Attractants and Traps for Detection of (F.), L, and Other Longhorned Beetles (Coleoptera: Cerambycidae). Environ Entomol 33(4):844–854

Sweeney J, Silk PJ, Rhainds M, MacKay W, Hughes C, Van Rooyen K, MacKinnon W, Leclair G, Holmes S, Aromia& Kettela EG. (2017). First report of mating disruption with an aggregation pheromone: a case study with *Tetropium fuscum* (Coleoptera: Cerambycidae). J Econ Entomol, 110(3), 1078–1086

Takanashi T, Fukaya M, Nakamuta K, Skals N, Nishino H (2016) Substrate vibrations mediate behavioral responses via femoral chordotonal organs in a cerambycid beetle. Zool Lett 2(1):18

Tanigaki T, Yamaoka R, Sota T (2007) The role of cuticular hydrocarbons in mating and conspecific recognition in the closely related longicorn beetles *Pidonia grallatrix* and *P. takechii*. Zool Sci 24(1):39–46

Teale SA, Wickham JD, Zhang F, Su J, Chen Y, Xiao W, Hanks LM, Millar JG (2011) A male-produced aggregation pheromone of *Monochamus alternatus* (Coleoptera: Cerambycidae), a major vector of pine wood nematode. J Econ Entomol 104(5):1592–1598

Wang Q, Chen LY (2005) Mating behavior of a flower-visiting longhorn beetle *Zorion guttigerum* (Westwood)(Coleoptera: Cerambycidae: Cerambycinae). Naturwissenschaften 92(5):237–241

Wang Q, Zeng W, Chen L, Li J, Yin X (2002) Circadian reproductive rhythms, pair-bonding, and evidence for sex-specific pheromones in *Nadezhdiella cantori* (Coleoptera: Cerambycidae). J Insect Behav 15(4):527–539

Wickham JD, Millar JG, Hanks LM, Zou Y, Wong JC, Harrison RD, Chen Y (2015a) Pheromone of *Megopis costipennis* (Coleoptera: Cerambycidae: Prioninae). Environ Entomol 1:6

Wickham JD, Millar JG, Hanks LM, Zou Y, Wong JC, Harrison RD, Chen Y (2015b) (2*R*, 3*S*)-2, 3-octanediol, a female-produced sex pheromone of *Megopis costipennis* (Coleoptera: Cerambycidae: Prioninae). Environ Entomol 45(1):223–228

Wickham, J. D, Lu, W, Jin, T, Peng, Z, Guo, D, Millar, J. G, Hanks, L. M, & Chen, Y. (2016a). Prionic acid: an effective sex attractant for an important pest of sugarcane, *Dorysthenes granulosus* (Coleoptera: Cerambycidae: Prioninae). J Econ Entomol, 109(1), 484–486

Wickham JD, Lu W, Zhang LW, Chen Y, Zou Y, Hanks LM, Millar JG (2016b) Likely aggregation-sex pheromones of the invasive beetle *Callidiellum villosulum*, and the related Asian species *Allotraeus asiaticus*, *Semanotus bifasciatus*, and *Xylotrechus buqueti* (Coleoptera: Cerambycidae). J Econ Entomol 109(5):2243–2246

Wong JC, Mitchell RF, Striman BL, Millar JG, Hanks LM (2012) Blending synthetic pheromones of cerambycid beetles to develop trap lures that simultaneously attract multiple species. J Econ Entomol 105(3):906–915

Xu T, Yasui Y, Teale ST, Fujiwara-Tsujii N, Wickham JD, Fukaya M, Hansen L, Kiriyama S, Hao D, Nakano A, Zhang L, Watanabe T, Tokoro M, Millar JG (2017) Identification of a male-produced sex-aggregation pheromone for a highly invasive cerambycid beetle, *Aromia bungii*. Sci Rep 7(1):7330

Xu T, Yasui H, Teale SA, Fujiwara-Tsujii N, Wickham JD, Fukaya M, Hansen L, Kiriyama S, Hao D, Nakano A, Zhang L, Watanabe T, Tokoro M, Millar JG (2017) Identification of a male-produced sex-aggregation pheromone for a highly invasive cerambycid beetle, *Aromia bungii*. Sci Rep 7(1):7330

Yasui H (2009) Chemical communication in mate location and recognition in the white-spotted longicorn beetle, *Anoplophora malasiaca* (Coleoptera: Cerambycidae). Appl Entomol Zool 44(2):183–194

Yasui H, Fujiwara-Tsujii N (2016) Host plant affects the sexual attractiveness of the female white-spotted longicorn beetle, *Anoplophora malasiaca*. Sci Rep 6:29526

Yasui H, Akino T, Yasuda T, Fukaya M, Ono H, Wakamura S (2003) Ketone components in the contact sex pheromone of the white-spotted longicorn beetle, *Anoplophora malasiaca*, and pheromonal activity of synthetic ketones. Entomol Exp Appl 107:167–176

Yasui H, Akino T, Yasuda T, Fukaya M, Wakamura S, Ono H (2007a) Gomadalactones A, B, and C: novel 3-oxabicyclo [3.3.0] octane compounds in the contact sex pheromone of the white-spotted longicorn beetle, *Anoplophora malasiaca*. Tetrahedron Lett 48(13):2395–2400

Yasui H, Yasuda T, Fukaya M, Akino T, Wakamura S, Hirai Y, Kawasaki K, Ono H, Narahara M, Kousa K, Fukuda T (2007b) Host plant chemicals serve intraspecific communication in the white-spotted longicorn beetle, *Anoplophora malasiaca* (Thomson) (Coleoptera: Cerambycidae). Appl Entomol Zool 42:255–268

Yasui H, Akino T, Fukaya M, Wakamura S, Ono H (2008) Sesquiterpene hydrocarbons: kairomones with a releaser effect in the sexual communication of the white-spotted longicorn beetle, *Anoplophora malasiaca* (Thomson) (Coleoptera: Cerambycidae). Chemoecology 18:233–242

Yasui H, Fujiwara-Tsujii N, Wakamura S (2011) Volatile attractant phytochemicals for a population of white-spotted longicorn beetles *Anoplophora malasiaca* (Thomson) (Coleoptera: Cerambycidae) fed on willow differ from attractants for a population fed on citrus. Chemoecology 21:51–58

Yasui H, Fujiwara-Tsujii N, Yasuda T, Fukaya M, Kiriyama S, Nakano A, Watanabe T, Mori K (2019a) Electroantennographic responses and field attraction of an emerging invader, the red-necked longicorn beetle *Aromia bungii* (Coleoptera: Cerambycidae), to the chiral and racemic forms of its male-produced aggregation-sex pheromone. Appl Entomol Zool 54:109–114

Yasui H, Fujiwara-Tsujii N, Yasuda T (2019b) Detection of volatile pheromone candidates from the white-spotted longicorn beetle, *Anoplophora malasiaca* (Coleoptera: Cerambycidae). Appl Entomol Zool 54:203–211

Yokoi N, Yoshii T (1987) Response of each days after emergence on yellow-spotted longicorn beetle, *Psacothea hilaris* PASCOE to the attractant from mulberry stumps. J Sericultur Sci Jpn 56(6):527–528. (In Japanese)

Yokoi N, Yoshii T (1984) Attractant effect of mulberry stumps and volatile substances emitted by them on the yellow-spotted longicorn beetle, *Psacothea hilaris* PASCOE. J Sericultur Sci Jpn 53(4):363–364. (In Japanese)

Yasui H, Yasuda T, Fukaya M, Akino T, Wakamura S, Hirai Y, Kawasaki K, Ono H, Narahara M, Kousa K, Fukuda T (2007) Host plant chemicals serve intraspecific communication in the white-spotted longicorn beetle, *Anoplophora malasiaca* (Thomson) (Coleoptera: Cerambycidae). Appl Entomol Zool 42(2):255–268

Zhang A, Oliver JE, Aldrich JR, Wang BD, Mastro VC (2002) Stimulatory beetle volatiles for the Asian longhorned beetle *Anoplophora glabripennis* (Motschulsky). Z Naturforsch 57:553–558

Zhang A, Oliver JE, Chauhan K, Zhao B, Xia L, Xu Z (2003) Evidence for contact sex recognition pheromone of the Asian longhorned beetle, *Anoplophora glabripennis* (Coleoptera: Cerambycidae). Naturwissenschaften 90(9):410–413

Zou Y, Rutledge CE, Nakamuta K, Maier CT, Hanks LM, Richards AB, Lacey ES, Millar JG (2015) Identification of a pheromone component and a critical synergist for the invasive beetle *Callidiellum rufipenne* (Coleoptera: Cerambycidae). Environ Entomol 45(1):216–222

Zou Y, Rutledge CE, Nakamuta K, Maier CT, Hanks LM, Richards AB, Lacey ES, Millar JG (2016) Identification of a pheromone component and a critical synergist for the invasive beetle *Callidiellum rufipenne* (Coleoptera: Cerambycidae). Environ Entomol 45(1):216–222

Zou Y, Hansen L, Xu T, Teale SA, Hao D, Millar JG (2019) Optimizing pheromone-based lures for the invasive red-necked longhorn beetle, *Aromia bungii*. J Pest Sci 92(3):1217–1225

Part II
Biosynthesis of Sex Pheromones

Chapter 6
A Sexy Moth Model – The Molecular Basis of Sex Pheromone Biosynthesis in the Silkmoth *Bombyx mori*

J. Joe Hull and Adrien Fónagy

Abstract The reproductive behaviors of many insects are coordinated by the synthesis and release of species-specific volatiles that communicate the location of potential mates. Given their biological importance, structural elucidation of these compounds (i.e., sex pheromones) and molecular determination of the underlying biosynthetic pathways have been the focus of numerous studies. Among the various model species that have been examined, the silkmoth (*Bombyx mori*) has had an outsized impact on the research field. Indeed, it was Adolf Butenandt's pioneering publication in 1959 on chemical characterization of the silkmoth sex pheromone (E,Z)-10,12-hexadecadien-1-ol (i.e., bombykol) that ushered in a new era of chemical ecology. Since then, *B. mori* has been at the forefront of each new advancement in our understanding of the pre- and postadult eclosion processes that culminate in pheromone production – from demonstration of hormonal regulation by a neuropeptide to identification of the cognate receptors and characterization of the genes comprising the biosynthetic and regulatory pathways. In honor of the 60th anniversary of bombykol's elucidation, we provide a perspective on the spectrum of studies that have made Butenandt's "sexy" moth one of the principal models for sex pheromone biosynthesis.

Keywords *Bombyx mori* · Bombykol · Sex pheromone biosynthesis

J. J. Hull (✉)
USDA-ARS, US Arid Land Agricultural Research Center, Maricopa, AZ, USA
e-mail: joe.hull@ars.usda.gov

A. Fónagy
Plant Protection Institute, Centre for Agricultural Research of Hungarian Academy of Sciences, Budapest, Hungary
e-mail: fonagy.adrien@agrar.mta.hu

Y. Ishikawa (ed.), *Insect Sex Pheromone Research and Beyond*, Entomology Monographs, https://doi.org/10.1007/978-981-15-3082-1_6

6.1 Introduction

The reproductive behaviors of many insects are coordinated by the synchronous synthesis and release of species-specific chemical mixtures that communicate the location of potential conspecific mates. Typically, these compounds (referred to as sex pheromones) are produced as blends of volatiles with differing chemistries and varying component ratios. Over the years, sex pheromones have been the focus of numerous studies from chemical identification to elucidation of the biosynthetic pathways, olfactory detection components, and neural processes that culminate in a behavioral response. Initially viewed through the prism of pest management, pheromone-based studies have broadened to provide critical insights into the evolutionary underpinnings of insect chemical communication and speciation. Lepidopteran-based studies have arguably dominated the research field since chemical elucidation of the first sex pheromone from 5×10^5 silkmoth (*Bombyx mori*) females in 1959 (Butenandt et al. 1959). Today, the pheromone database encompasses 67 lepidopteran families representing >1500 species (El-Sayed 2018).

Elucidation of lepidopteran sex pheromone chemistries has revealed conserved pathways for most species that utilize relatively simple straight-chain C_{10}–C_{18} aliphatic compounds containing varying degrees of unsaturation and an oxygenated functional group (i.e., type I); however, straight-chain hydrocarbons and epoxide hydrocarbons (i.e., type II) have also been identified, albeit less frequently (Ando et al. 2004). In parallel with chemistry-based structure determination, research into the biochemical and molecular basis of sex pheromone biosynthetic pathways has established that fatty acid metabolism intermediates (e.g., palmitic acid/hexadecanoic acid) serve as substrates for the downstream modifications (selective β-oxidation reactions, desaturations, and diverse reductive modifications) that generate the species-specific pheromone blends (Bjostad et al. 1987). These pathways along with their regulation, evolution, and product detection have been the topic of numerous excellent reviews (Sakurai et al. 2014; Zhang et al. 2015; Groot et al. 2015; Yew and Chung 2015; Allison and Cardé 2016; Jurenka 2017). Here, however, in honor of the 60-year anniversary of the structural determination of (*E*,*Z*)-10,12-hexadecadien-1-ol (i.e., bombykol) as the principal silkmoth sex pheromone (Butenandt et al. 1959), we provide a perspective on the spectrum of studies that have made Butenandt's "sexy" moth the foremost model for understanding the pre- and posteclosion processes that culminate in moth sex pheromone production (i.e., pheromonogenesis).

6.2 Regulation of Pheromone Production

6.2.1 Circadian-Based Control

For most moths, pheromone production and release typically exhibit circadian oscillations that coincide with specific points of the day:night cycle (Groot 2014). The basis for this circadian output was the focus of early biochemical studies.

Observations that decapitation/neck ligature inhibited normal scotophase pheromone production in *Helicoverpa zea* and *Chilo suppressalis* (Raina and Klun 1984; Ohguchi et al. 1985) suggested involvement of a brain-derived hormonal factor. A subsequent study showing female brain homogenates could restore pheromone production in decapitated *H. zea* females confirmed the cephalic nature of the factor (Raina and Klun 1984). That same study provided a potential mechanism for the circadian signal as pheromonogenic activity was most pronounced in the hemolymph of scotophase female *H. zea* rather than photophase. In their model, Raina and Klun (1984) suggested that the hormonal factor accumulated during photophase and was subsequently released into the hemolymph to act on the target tissue during scotophase. Similar to other moths, *B. mori* also exhibits circadian oscillations in pheromone production, albeit with bombykol titers lowest at early photophase, peaking at midphotophase, and declining again at late photophase (Ichikawa 1998; Fujii et al. 2018). Building on the early findings of *H. zea*, Ando and coworkers (1988a) examined the relevance of the hormonal mechanism described in *H. zea* for species that exhibit photophase-based pheromone production. They found that head extracts from photophase *B. mori* females were able to stimulate pheromone production in decapitated females, suggesting that bombykol production was likewise mediated by a circadian-released neuroendocrine factor (Ando et al. 1988a). Subsequent studies performed a decade later provided further support for humoral-based circadian control of pheromone production in *B. mori* with evidence that pheromonogenic neurosecretory cells also exhibit diurnal firing activity (Ichikawa 1998; Tawata and Ichikawa 2001).

6.2.2 *Pheromone Gland as the Neuroendocrine Target Tissue*

The site of sex pheromone production and release for most female moths is typically a bulbous, extrudable gland (i.e., pheromone gland, PG) located at the intersegmental membrane of the eighth and ninth segments (Bjostad et al. 1987). Early histological and ultrastructural analyses in a number of species, including pioneering work with *B. mori* PGs (Hayashi and Ito 1933; Steinbrecht 1964; Waku and Sumimoto 1969), revealed the glands are well suited for hydrocarbon-based pheromone biosynthesis and secretion. PGs are composed of hypertrophied secretory cells and modified epidermal cells with an apical brush border, a well-developed smooth ER network, and varying numbers of electron-lucent lipid vesicles (Ma and Ramaswamy 2003). In *B. mori*, the PG is a symmetrical pair of ventrolateral sacs termed *sacculi laterales* (Percy-Cunningham and MacDonald 1987; Fónagy et al. 2001) composed of two distinct layers – an outer cuticular layer and an inner layer consisting of 9000–10,000 homogenous epidermal cells arranged in a monolayer (Fónagy et al. 2000). This inner layer is characterized by the presence of cytoplasmic lipid droplets that accumulate 1–2 days prior to adult eclosion and which fluctuate in size and number in accordance with bombykol production and photoperiod (Fónagy et al. 2001). In contrast, the cuticle overlaying the gland consists of a lamellate endocuticle, protein epicuticle, a thin electron-dense cuticulin layer, and an outer epicuticle (Steinbrecht 1964; Waku and Sumimoto 1969). Using a microsome-based bioassay consisting of fractions generated from each of the *B. mori* PG layers, Fónagy and coworkers (2000) confirmed that bombykol production was restricted to the inner epidermal cell layer.

Although the role of the PG in sex pheromone production and release had been well established, it was less clear if it was also the target of the brain-derived pheromonogenic factor as suggested for some species (Soroker and Rafaeli 1989; Rafaeli et al. 1990). Or, as suggested by other studies, the brain-derived factor acted on a non-PG target tissue(s) that either released a secondary stimulus to trigger pheromone production or provided pheromone precursor substrates (Tang et al. 1989; Teal et al. 1989). Multiple in vitro studies using isolated *B. mori* PGs (Arima et al. 1991; Fónagy et al. 1992b, c; Ozawa et al. 1993; Matsumoto et al. 1995a, Ozawa et al. 1995; Fónagy et al. 2001), however, revealed the glands to be the principal target of neuroendocrine control and confirmed that all of the cellular machinery necessary for bombykol biosynthesis and release are specific to the PG.

6.2.3 HPLC-Based Purification of the Neuroendocrine Factor

Neuroendocrine-based regulation of sex pheromone synthesis was initially proposed by Riddiford and Williams based on *corpora cardiaca* excision studies using saturniid moths (Riddiford and Williams 1971). Later observations that decapitation/neck ligature likewise inhibited sex pheromone production (Raina and Klun 1984; Ohguchi et al. 1985; Ando et al. 1988a) and that multiple moth species had a brain-derived pheromonotropic factor (Raina and Klun 1984; Raina et al. 1987; Ando et al. 1988a) provided further support for neuroendocrine regulation. Using high-performance size exclusion chromatography, a *B. mori* factor was shown to be both susceptible to protease treatment and peptidergic in nature (Ando et al. 1988a). Using a 10-step isolation procedure, a pheromonotropic peptide was purified to homogeneity from 75,000 adult male *B. mori* heads (Nagasawa et al. 1988). N-terminal sequencing of the purified peptide yielded a 10-amino acid (aa) sequence (Leu-Ser-Glu-Asp-Met-Pro-Ala-Thr-Pro-Ala-) with 100% identity to a previously isolated neuropeptide, melanization and reddish coloration hormone (MRCH)-I (Matsumoto et al. 1986). Sufficient quantities of the pheromonotropic peptide were subsequently purified from 6 x 10^5 adult male *B. mori* heads for complete sequence determination of the factor as a 33-aa, C-terminal amidated peptide termed pheromone biosynthesis activating neuropeptide (PBAN) (Kitamura et al. 1989). A second, later eluting pheromonotropic fraction yielded *B. mori* PBAN-II, which is differentiated from the first *B. mori* PBAN by the inclusion of a single N-terminal Arg residue (Kitamura et al. 1990). A similar 33-aa, C-terminal amidated PBAN was independently purified from 5000 adult *H. zea* male/female brain-suboesophageal ganglion (SOG) complexes (Raina et al. 1989). High sequence identity (79%) between the *B. mori* and *H. zea* PBANs and pheromonotropic cross-reactivity in species with differing sex pheromone chemistries confirmed that neuroendocrine regulation of pheromone production was evolutionarily conserved (Raina et al. 1989; Fónagy et al. 1992b).

6.2.3.1 Structure–Function Analysis of PBAN

Initial structure–function analyses of *B. mori* PBAN (BmPBAN or BommoPBAN) highlighted the importance of the C terminus, rather than the N terminus, in mediating pheromonotropic activity. Loss of the C-terminal Leu33 abolished activity, whereas deletion of residues 1–23 had no effect on activity (Kitamura et al. 1989; Kitamura et al. 1990). Comparison of amidated, hydroxylated, and methyl ester versions of the peptide underscored the critical importance of the C-terminal amide (Kitamura et al. 1989; Kitamura et al. 1990; Kuniyoshi et al. 1991a; Nagasawa et al. 1994). Further studies determined that the minimal sequence necessary for activity (albeit reduced) resided in the pentapeptide (Phe-Ser-Pro-Arg-Leu-NH_2); higher activity, however, was retained in the hexapeptide (Tyr-Phe-Ser-Pro-Arg-Leu-NH_2) (Kuniyoshi et al. 1991a). Replacing the first three residues (Tyr, Phe, and Ser) of the hexapeptide had inconsequential effects on pheromonotropic activity, whereas substitution of Pro, Arg, and Leu had severe effects (Kuniyoshi et al. 1991a). A more recent substitution study revealed that the Arg functional group is essential for receptor binding (Kawai et al. 2012).

Kitamura and coworkers (1989) reported that purified BommoPBAN co-eluted with the oxidized version of the synthetic peptide, suggesting that the natural peptide was in the fully oxidized state. Furthermore, the oxidized peptide was reported to have significantly higher activity (Kitamura et al. 1989; Kitamura et al. 1990). To explore the basis for this modification, Nagasawa and coworkers generated a series of oxidation products examining the importance of each of the three Met residues (Met5, Met14, and Met22). Oxidation of single residues enhanced activity over the nonoxidized form, but the most pronounced effects were only apparent when all three residues were in their oxidized state (Nagasawa et al. 1994). Additional peptide modifications designed to generate highly active analogs of the penta- and hexapeptides were also examined for potential use in developing PBAN mimetics. Addition of acetyl, benzoyl, ethyl, benzyl, pyroglutamyl, or D-alanyl groups to the N-terminal position enhanced activity relative to nonmodified peptides. The observed increase in activity though was the result of enhanced hemolymph stability rather than receptor activation (Kuniyoshi et al. 1991b; Kuniyoshi et al. 1992b). Attempts at peptide cyclization, which restricts the available conformational space of a linear peptide, via derivatization of the Arg and Leu sites had limited activity, whereas cyclization outside of the core sequence [e.g., cyclo(-Asn-Thr-Ser-Phe-Thr-Pro-Arg-Leu)] yielded an analog of equal potency as the linear hexapeptide alone (Nagasawa et al. 1994). More recent efforts to develop pheromonotropic agonists/antagonists based on the BommoPBAN decapeptide sequence (i.e., residues 24–33) determined that a linear synthetic derivative with a Tyr-Asn substitution (Ser-Lys-Thr-Arg-Asn-Phe-Ser-Pro-Leu-NH_2) of the carbon 6 aa upstream of the terminal residue (i.e., C6 position) was a partial agonist (Kawai et al. 2010). Cyclic peptides [cyclo(Thr-Cys-Asn-Phe-Ser-Pro-Arg-Leu) and cyclo(Thr-Cys-Tyr-Phe-Ser-Pro-Arg-Leu)] based on this substitution likewise yielded partial agonists and demonstrated that the C6 position and the restricted conformational space were important for receptor recognition (Kawai et al. 2011). These findings provided a biological context for an earlier NMR study (Okada et al. 2009) that showed the bioactive decapeptide (Ser-Arg-Thr-Arg-Tyr-Phe-Ser-Pro-Arg-Leu-NH_2) and the inactive free acid form assume different conformations with the amide exhibiting a type I β-turn similar to that reported for the *H. zea* peptide (Nachman et al. 1993). Given their potential for next-generation pest management tactics, PBAN peptidomimetics, many based on the initial *B. mori* structure–function studies, have been the focus of multiple research groups with developments the topic of numerous reviews (Altstein et al. 2007; Nachman 2014).

6.2.4 Molecular Cloning of the BommoPBAN cDNA

Determination of the BommoPBAN aa sequence facilitated molecular elucidation of the encoding mRNA transcript and provided a means for localizing PBAN expression. Using dissimilar degenerate methods with cDNAs generated from adult female brain–SOG complexes or pooled SOG from day 0 pupae, two groups with differing aims separately identified a transcript encoding a 192-aa peptide precursor that included the PBAN sequence (Kawano et al. 1992; Sato et al. 1993). Although alternative splicing is a common mechanism for enhancing neuropeptide diversity, subsequent molecular studies failed to find any evidence for alternative splice variants of the *B. mori* transcript (Sato et al. 1993; Sato et al. 1994). Post-translational proteolytic processing of the prepropeptide was predicted to yield the PBAN sequence along with four additional amidated peptides including diapause hormone (DH), a previously purified peptide critical for induction of embryonic diapause (Imai et al. 1991) that was the target of the pupae-based study. The DH active core (Phe-Gly-Pro-Arg-Leu-NH_2) is similar to the critical PBAN pentapeptide motif as is the C terminus of the three other peptides (Phe-Thr-Pro-Arg-Leu-NH_2; Phe-Ile-Pro-Arg-Leu-NH_2; Phe-Ser-Pro-Arg-Leu-NH_2) termed suboesophageal ganglion neuropeptides (SGNPs) α, β, and γ (Kawano et al. 1992; Sato et al. 1993). Structurally, the prepropeptide consists of the initial signal peptide (23 aa) followed by the DH sequence (residues 24–47), then α-SGNP (residues 97–103), β-SGNP (residues 106–122), PBAN (residues 125–158), and γ-SGNP (residues 161–168) with each of the peptides flanked by mono- or dibasic cleavage sites and amidating Gly signals. Processing of the PBAN peptide via the dibasic Arg125-Arg126 site yields PBAN-I, whereas alternative processing at Arg125 alone, which follows the "rule" for single Arg processing sites with a basic aa (Arg122) in the −4 position (Veenstra 2000), would generate PBAN-II (Kitamura et al. 1990). HPLC-based purification of *B. mori* SOG extractions confirmed that the respective SGNPs were generated following proteolytic processing of the PBAN prepropeptide (Sato et al. 1993). Consistent with the early structure–function analyses that highlighted the importance of the PBAN C-terminal pentapeptide (Kuniyoshi et al. 1991a), both DH and the SGNPs are pheromonotropic and activate heterologously expressed PBAN receptor at low to submicromolar concentrations (Sato et al. 1993; Watanabe et al. 2007). PBAN encoding transcripts with similar multipeptide processing sites have since been cloned from a number of moths (see Jurenka and Nusawardani 2011; Jurenka 2015).

6.2.4.1 Expression Profiling of the PBAN Transcript

Given the two main products of the DH-PBAN transcript regulate different physiological processes (i.e., pheromone production and diapause induction), the timing and location of DH-PBAN expression can have profound biological effects. Initial expression profiling of the PBAN transcript (commonly referred to as DH-PBAN) based on Northern blots showed a strong hybridization signal in the SOG of day 0 pupae as well as brain–SOG complexes of day 4 pharate adult females (Sato et al. 1993; Sato et al. 1994). A somewhat reduced signal was also apparent in brain–SOG complexes of day 4 pharate adult males. More extensive profiling revealed robust hybridization signals in day 8 eggs as well as SOGs from day 5 fifth instar larvae, day 3 pupae, and day 6 pharate adults with a weaker signal in day 3 fourth instar larvae and newly eclosed adults (Xu et al. 1995a). PCR-based analyses using primers designed to anneal to the 5′-ends of the DH (sense primer) and PBAN (antisense primer) coding sequences revealed that DH-PBAN expression was limited to the late pupal stage of nondiapausing silkworms, whereas peaks of expression were observed throughout embryonic, larval, and pupal development of diapause silkworms (Xu et al. 1995a). The observed expression in late-stage pharate adults is most consistent with the role of PBAN in pheromone production, whereas expression across developmental stages is more reflective of the role DH has in diapause induction. Similarly, given the critical role environmental conditions have on the induction of embryonic diapause in *B. mori* (Watanabe 1924), reports of temperature-linked DH-PBAN expression (Xu et al. 1995a) would be associated with the DH-driven activity of the transcript rather than pheromone production. Dopamine-induced expression of the DH-PBAN transcript is also associated with DH activity as dopamine and/or DOPA exposure switches the diapause programming state (Noguchi and Hayakawa 2001).

6.2.4.2 Localization of the PBAN Transcript/Peptide

In situ hybridization using a probe designed to the DH coding sequence localized the *B. mori* DH-PBAN transcript to 12 neurosecretory cells grouped into three clusters along the ventral surface of the mandibular, maxillary, and labial neuromeres of the SOG (Sato et al. 1994). Similar cell groupings corresponding to four mandibular cells (also termed anterior or SMd cells), six maxillary cells (medial or SMx cells), and two labial cells (posterior or SLb cells) were likewise identified immunohistochemically using antibodies against DH, PBAN, and/or peptides with the shared FXPRLamide (Phe-xxx-Pro-Arg-Leu-NH_2, where x = any aa) pentapeptide C terminus (Ichikawa et al. 1995; Ichikawa et al. 1996; Sato et al. 1998; Morita et al. 2003; Shiomi et al. 2007; Hagino et al. 2010). Similar sets of PBAN-immunoreactive neurosecretory cells have been identified in other Lepidoptera, including *H. zea*, *Mamestra brassicae*, *Ostrinia nubilalis*, *Manduca sexta*, and *Pseudaletia separata* (Kingan et al. 1992; Tips et al. 1993; Ma and Roelofs 1995a; Davis et al. 1996; Raina et al. 2003). Based on the SOG immunoreactivity, translation of the DH-PBAN transcript occurs throughout *B. mori* larval, pupal, and adult development as well as embryonically with earliest detection at the embryonic blastokinesis stage (Morita et al. 2003). Surprisingly, targeted ablation of the differing immunoreactive cells in *B. mori* resulted in cell-specific phenotypes with SLb ablation affecting ovariole accumulation of 3-hydroxykynurenine (a marker of diapause-destined eggs) and SMd/SMx removal resulting in reduced sex pheromone production (Ichikawa et al. 1996). This apparent functional differentiation of the DH-PBAN immunoreactive neurosecretory cells is supported by reports of diapause-linked differences in the immunoreactive staining and firing activity patterns of SLb cells during pupal-adult development (Sato et al. 1998; Ichikawa and Suenobu 2003; Hagino et al. 2010). Similar differences were not observed in the firing of SMd/SMx cells, and their activity during a diapause sensitive state had no effect on diapause induction (Ichikawa and Kamimoto 2003), further supporting functional differentiation. Reports of weak PBAN-specific immunoreactivity observed in SLb cells relative to SMd/SMx cells coupled with the ablation studies support anatomical and functional differentiation of these cells and suggest that they are specialized for diapause determination (Hagino et al. 2010). The staining profile of neurite projections from DH-PBAN immunoreactive cells suggests that axonal transport from the SMd/SMx and SLb cells also differs (Ichikawa et al. 1995). SMd/SMx-derived axons that terminate in the *corpus cardiacum* (a neurohemal organ that functions in neuropeptide storage for hemolymph release) pass through the maxillary nerve and *nervus corporis cardiaci-ventralis* (also referred to as NCC-V), whereas those from SLb cells utilize circumesophageal connectives and NCC-3. Intriguingly, the respective DH-PBAN peptidergic activities also exhibit differences in neurosecretory control with SLb-derived DH activity controlled by the brain (Matsutani and Sonobe 1987) and likely involve γ-aminobutyric acid (Shimizu et al. 1989; Hasegawa and Shimizu 1990; Ichikawa et al. 1997). In contrast, PBAN is regulated by a circadian pacemaker (Ando et al. 1988c; Ichikawa 1998; Tawata and Ichikawa 2001).

6.2.5 *Identification of the PBAN Gene*

By using radiolabeled probes corresponding to the DH and PBAN coding regions to screen a genomic library, Xu and coworkers (1995b) determined that the *B. mori* DH-PBAN gene was organized across 5 introns and 6 exons. The exons are arranged such that the signal peptide and the first four aa of DH are on exon 1, the remaining

20 DH aa on exon 2, an uncharacterized peptidergic sequence on exon 3, complete sequences for the α and β SGNPs and first 15 aa of PBAN on exon 4, the remaining 19 PBAN aa and γ SGNP on exon 5, and the stop codon on exon 6. Splicing of the five introns follows the "GT-AG" rule (Mount 1982) and consists of a 0, 2, 1, 2, 1 phase pattern (0 = codon that does not include sites from 3′ donor, 1 = codon that has two sites from 3′ donor, and 2 = codon that has one site from 3′ donor). The genomic organization of the DH-PBAN gene appears to be evolutionarily conserved in Lepidoptera with identical exon coding regions reported in other moth species (Zhang et al. 2005; Jing et al. 2007; Chang and Ramasamy 2014; Fodor et al. 2017; Senthilkumar and Srinivasan 2019); however, the size of the intronic regions varies with larger introns typically occurring in *B. mori*.

6.2.5.1 PBAN Promoter Region

An initial scan of the *B. mori* DH-PBAN gene revealed a number of potential promoter sites, including a canonical TATA box (position −46) and a potentially modified CAAT box (position −90) (Xu et al. 1995b). Subsequent in vitro analyses, however, demonstrated transcriptional activity of a proximal promoter (position −75 to −65) upstream of the TATA box corresponding to a POU-binding domain, and expression of the corresponding transcription factor (POU-M1/M2) coincided with DH-PBAN expression (Zhang et al. 2004). That study also identified an enhancer element 3.5–5 kb upstream of the DH-PBAN transcription site. In vivo studies also identified a *cis*-acting element upstream of the DH-PBAN start site (Shiomi et al. 2003), refinement of which localized the regulatory element to a Pitx-binding core sequence that resides in a different region (position −1117 to −1088) of the DH-PBAN gene than the POU motif (Shiomi et al. 2007). As with POU-M1/M2, the expression of the BmPitx transcription factor paralleled that of the DH-PBAN transcript and co-localized in the 12 DH-PBAN-producing neurosecretory cells of the SOG. Viral-based overexpression of the transcription factor enhanced DH-PBAN expression in pupal SOGs, whereas knockdown reduced expression (Shiomi et al. 2007). The identification of different DH-PBAN promoter regions may reflect limitations of the disparate assays used (POU-M1/M2 – in vitro; Pitx – in vivo) and/or *B. mori* strain-specific sequence variations (POU-M1/M2 – Jingsong x Haoyue; Pitx – Tokai x Asahi). However, reports of transcriptional activation involving physical cooperativity between Pitx homologs and POU domain factors (Tremblay and Drouin 1999; Quentien et al. 2002) may indicate that DH-PBAN transcription is similarly regulated. This proposed mechanism does not appear to be evolutionarily conserved across Lepidoptera as the POU-M1/M2 transcription factor was unable to drive expression from a conserved region of the *Helicoverpa armigera* DH-PBAN gene (Zhang et al. 2005). In that species, an E-box element (CAGCTG) rather than the POU domain is critical for transcriptional activation (Hong et al. 2006). Two ecdysone responsive element-like regions (position −753 to −743 and position −709 to −609) have also been identified in the 5′-region of the *B. mori* DH-PBAN gene (Xu et al. 1995b). Given the role ecdysteroids have in lepidopteran reproduction, the response elements may link PBAN transcription with reproductive competence; however, their role in pheromone production remains to be ascertained.

6.2.6 *PBAN-Related Family of Peptides*

The C-terminal FXPRLamide pentapeptide motif that characterizes BommoPBAN is critical for pheromone production (Nagasawa et al. 1988; Kitamura et al. 1989) as well as different physiological processes in the silkworm, including induction of embryonic diapause (Imai et al. 1991) and ecdysteroidogenesis in larval prothoracic glands (Watanabe et al. 2007). The peptide was also shown to exhibit pheromonotropic activity in other species (Matsumoto et al. 1990; Fónagy et al. 1992b; Fónagy et al. 1992d) in addition to having myotropic (Kuniyoshi et al. 1992a; Fónagy et al. 1992d) and cuticular melanization (Matsumoto et al. 1990) activities. Reciprocal cross-species pheromonotropic activity in *B. mori* has also been reported for FXPRLamide peptides from other species (Matsumoto et al. 1992a; Kuniyoshi et al. 1992a; Fónagy et al. 1992a). In the years since the initial purification and identification of BommoPBAN, the C-terminal pentapeptide has become a defining feature of the FXPRLamide family of pleiotropic neuropeptides (i.e., pyrokinins, PBANs, myotropins, DH, and SGNPs), which are expressed throughout the Insecta via the *capa* and *DH-PBAN/hugin* genes (Jurenka 2015; Yaginuma and Niimi 2015; Schoofs et al. 2017). In addition to pheromonotropic activity in moths, these peptides also regulate an astounding array of functions, including the induction of cuticular melanization in moth larvae (Matsumoto et al. 1992b; Altstein et al. 1996), the termination of pupal diapause in heliothine moths (Xu and Denlinger 2003; Zhang and Denlinger 2012), the induction of embryonic diapause and seasonal polyphenism in moths (Imai et al. 1991; Uehara et al. 2011), prothoracic gland ecdysteroidogenesis (Watanabe et al. 2007), visceral muscle contraction in cockroaches (Holman et al. 1986; Schoofs et al. 1993), acceleration of puparium formation in flies (Žďárek et al. 1998; Žďárek et al. 2004), pheromone synthesis in male *H. armigera* (Bober and Rafaeli 2010; Zhao et al. 2018), and trail pheromone biosynthesis in *Solenopsis invicta* (Choi and Vander Meer 2012).

6.3 The PBAN Receptor

6.3.1 *Molecular Cloning of the PBAN Receptor*

Although the molecular identities of BommoPBAN and its encoding gene had both been well elucidated by the mid 1990s, similar identification of the corresponding receptor proved more elusive. Based on the hypothesis that receptor/ligand coevolution would yield closely aligned receptor families, *Drosophila melanogaster* receptors phylogenetically related to the mammalian neuromedin U receptor (NmUR), which is activated by a peptide with a C-terminal FRPRNamide sequence, were assayed for activation by peptides with similar C-terminal cores (Park et al. 2002). Among the receptors assayed in that study, the Drosophila receptors CG8795 (AF522190) and CG8784 (AF522189) were activated by FXPRLamides. A homology-based cloning approach based on the two Drosophila receptors facilitated amplification of a *B. mori* PG-derived transcript encoding a 413-aa protein with significant sequence similarity to class A rhodopsin-like G protein-coupled receptors (GPCRs) and which was dose-dependently activated by *B. mori* PBAN (Hull et al. 2004). Based on the transcript expression profile, ligand activation, and similarity with the NmUR family, the *B. mori* protein was identified as a PBAN receptor (PBANR). Subsequent RNA interference (RNAi)-mediated knockdown confirmed the role of the receptor in bombykol production (Ohnishi et al. 2006). Similar cloning strategies enabled amplification of receptors from *H. zea* PGs (Choi et al. 2003) and *B. mori* ovaries (Homma et al. 2006) that were activated by synthetic PBAN and DH, respectively. Surprisingly, sequence identity between the *B. mori* PBANR and DH receptor (41%) is less than between the two PBANRs (83%), suggesting differences in efficacy between DH and PBAN (Sato et al. 1993; Homma et al. 2006; Watanabe et al. 2007) may be linked to receptor-based ligand discrimination. PBANRs have since been amplified and/or identified from a number of moth species (Zheng et al. 2007; Rafaeli et al. 2007; Kim et al. 2008; Cheng et al. 2010; Lee et al. 2012a; Nusawardani et al. 2013; Wu et al. 2015; Ding and Löfstedt 2015; Fodor et al. 2018; Jiang et al. 2018; Cha et al. 2018).

Initial comparative analyses of PBANRs were confounded by a 67-aa C-terminal extension critical for ligand-induced receptor internalization in BommoPBANR (Hull et al. 2004; Hull et al. 2005). The absence of this C-terminal extension suggested that the "shorter" PBANRs, such as the *H. zea* and *H. armigera* PBANRs, utilized a different regulatory mechanism and lead to speculation that the varied C-terminal lengths reflected species-specific differences in the cellular signaling pathways activated. Equally perplexing were results from previous photoaffinity-based PBAN-binding studies that had identified a ~ 50-kDa membrane protein in isolated *H. armigera* intersegmental membrane preparations (Rafaeli and Gileadi 1997; Rafaeli et al. 2003) that was closer in size to BommoPBANR (45.9 kDa) than either of the cloned heliothine PBANRs at that time (*H. armigera* PBANR – 38.7 kDa or *H. zea* PBANR – 38.6 kDa). Subsequent modification of PBANR amplification conditions resolved the controversy as transcripts encoding multiple variants (PBANR-As, −A, -B, and -C) that differed only in the length of their respective C-terminal ends were amplified from PG cDNAs in *B. mori* and a number of other species (Kim et al. 2008; Lee et al. 2012a; Nusawardani et al. 2013; Fodor et al. 2018). Although the BommoPBANR-C variant was the first identified, the nomenclature was amended to better reflect that used in other moths. BommoPBANR-As is a 306-aa receptor with an incomplete seventh transmembrane (TM) domain that does not traffic to the cell surface and is instead retained in the endoplasmic reticulum and/or Golgi complex. BommoPBANR-A, which more closely resembles the initial *H. zea* PBANR, is a 345-aa receptor with a 20-aa C-terminal truncation relative to the C variant. BommoPBANR-B is the largest (475-aa) of the *B. mori* receptors and, like the C variant, undergoes ligand-induced internalization (Lee et al. 2012a). Overall, transcripts for the "shorter" A variants, which correspond to the initial *H. zea* and *H. armigera* PBANRs, were found to be less abundant, had lower PBAN efficacies, and exhibited different internalization kinetics (Lee et al. 2012a; Lee et al. 2012b). Preferential amplification of the shorter variants reported by the earlier studies was likely methodological based given that the high GC content (55–80%) of the extended C-terminal ends can adversely affect PCR amplification efficiencies (McDowell et al. 1998).

6.3.2 Identification of the PBANR Gene

The modular nature of the cloned *PBANR* variants (i.e., differences restricted to the C terminus) is consistent with known GPCR alternative splicing events (Markovic and Challiss 2009; Maggio et al. 2016). When aligned to the *B. mori* genome (Shimomura et al. 2009), the *BommoPBANR* gene localized to a > 50 kb segment of chromosome 12 on scaffold Bm_scaf84 (Lee et al. 2012a). The gene is composed of six exons and five introns with exon 1 consisting of the 5′-untranslated region, exons 2–4 encoding the N terminus through TM7, and exons 5–6 encoding the C terminus and stop codon. Based on sequence analyses, BommoPBANR-As and BommoPBANR-A appear to be generated from the introduction of premature stop codons following retention of introns 3 or 4, respectively. BommoPBANR-C results from a five-nucleotide (nt) insertion at the 3′-end of exon 5 that shifts the codon usage for residues 404–413 and introduces a stop codon that generates a C-terminal tail truncated 62-aa compared to BommoPBANR-B, which is generated from conventional splicing of exons 2–6 (Lee et al. 2012a).

6.3.3 Expression of Multiple PBANR Variants

Although the biological significance underlying the concomitant expression of multiple PBANR variants in PGs remains to be fully elucidated, similar variants in other systems have been reported to impact ligand specificity/potency, receptor trafficking, endocytotic regulation, and spatial/temporal regulated expression (Markovic and Challiss 2009; Maggio et al. 2016; Donaldson and Beazley-Long 2016). We have speculated previously that the multiplicity of transcripts may provide a mechanism for fine-tuning cellular responsiveness to the PBAN signal. In one theoretical model, expression of the shorter, less active PBANR-A at the cell surface could function as a ligand sink that competes with PBANR-C for ligand binding. The net result would be a reduction in peptide available to activate the GPCR-mediated cellular response, thus reducing overall sensitivity to the extracellular signal. Alternatively, heterodimerization of the longer variants (PBANR-B and PBANR-C) with the shorter variants (PBANR-As and PBANR-A) could impact normal cell surface trafficking, ligand specificity, and/or cellular signaling. In support of this, "short" (i.e., truncated) GPCRs exert dominant negative effects on their full-length isoforms when the two are co-expressed (Seck et al. 2005; Zmijewski and Slominski 2009; Chow et al. 2012). Rhodopsin-like GPCR dimerization, however, remains a controversial topic with support both for (Fotiadis et al. 2006; Ferré et al. 2014) and against (Chabre and le Maire 2005; Whorton et al. 2008) the event. In addition to altered ligand binding/receptor activation, receptor variants have also been reported to exhibit distinct spatial and temporal expression profiles (Markovic and Challiss 2009). Consequently, multiple PBANR transcripts may reflect a spatiotemporal dependence of functionality. This hypothesis is especially attractive given the pleiotropic complexity of the *DH-PBAN* gene and evidence for BommoPBANR activation by the various FXPRLamide peptides (Watanabe et al. 2007).

6.3.4 PBANR Structure–Function Relationships

GPCRs function at the cell surface to turn an extracellular signal into a cellular response. Mechanistically, ligand (e.g., PBAN) binding triggers conformational changes in the receptor that promote the activation of a complex signaling network that culminate in the associated cellular response and feedback regulation (i.e., desensitization and internalization) of the receptor (Cong et al. 2017; Wang et al. 2018). Although elucidation of the specific GPCR structural motifs that mediate these processes is an area of active research among vertebrate receptors, it is not as well developed for insect GPCRs. Of the insect GPCRs that have been characterized, structure–function studies of BommoPBANR are among the most extensive and have provided insights into PBAN functionality in terms of ligand-mediated internalization, which rapidly proceeds via a clathrin-mediated pathway that requires C-terminal interactions and protein kinase C phosphorylation (Hull et al. 2004; Hull et al. 2005; Hull et al. 2011).

6.3.4.1 G-Protein Coupling

Typically, propagation and termination of a peptide ligand signal requires GPCR coupling and activation of specific classes of heterotrimeric (α, β, and γ) G proteins. Interactions between these proteins and the GPCR promote release of G protein-bound GDP (guanosine 5′-diphosphate), which results in recruitment of GTP and subsequent dissociation of the α and β/γ subunits from the receptor, freeing them to activate downstream effector proteins. Hydrolysis of GTP to GDP leads to reassembly of the heterotrimeric complex, effectively resetting the system (Duc et al. 2015). Receptor-G protein coupling has been reported to involve ionic interactions between basic residues in intracellular loop 3 (ICL3) and anionic residues in the C terminus of the G protein (Yang et al. 2002; Kleinau et al. 2010). A dibasic site (R263 and R264) near TM6 of BommoPBANR that is highly conserved among other PBANRs is critical for feedback regulation of the receptor post-activation. Site-directed mutagenesis of the double Arg site with either neutral (Ala) or anionic (Glu) residues significantly reduced internalization of the ligand-bound receptor (Hull et al. 2011). The disruption in internalization is consistent with perturbed PBANR signaling and provided the first evidence for this region in PBANR-G protein coupling.

6.3.4.2 C-Terminal Motifs Critical to Ligand-Induced Internalization

A number of C-terminal motifs critical for vertebrate GPCR desensitization and endocytosis have been identified (Ferguson 2001; Kristiansen 2004; Pandey 2009; Calebiro and Godbole 2018), two of which are present in the longer BommoPBANRs: NPxxY (residues 325–329) and YxxΦ (residues 360–363). The NPxxY sequence (N = Asn, P = Pro, x = any aa, Y = Tyr) has been reported to function in the internalization of some (Barak et al. 1995; Gripentrog et al. 2000; He et al. 2001; Bouley et al. 2003) but not all vertebrate GPCRs (Slice et al. 1994). The YxxΦ internalization motif (Y = Tyr, x = any aa, and Φ = aa with a bulky hydrophobic side chain) typically located 10–40 aa downstream of TM7 has been likewise implicated in receptor internalization (Collawn et al. 1990; Paing et al. 2004; Pandey 2009). In this tetrapeptide sequence, the Tyr residue appears to be the most critical signal (Pandey 2009). C-terminal truncations of BommoPBANR-C localized the internalization motif to a 10-aa region spanning residues 357–367 that encompass the YxxΦ motif (Hull et al. 2005). Impaired internalization following Ala-substitution of the Y and Φ residues (Tyr383 and Leu 386) confirmed the importance of the signal. The YxxΦ motif, YSAL, is highly conserved in lepidopteran PBANRs and a number of related FXPRLamide receptors (i.e., pyrokinin 2 receptor) but has diverged somewhat in DHRs (YTAM/V) and is not readily apparent in pyrokinin 1 receptors. This variance suggests that regulation of those receptors either utilizes a different internalization signal or proceeds via a nonendocytotic pathway. Whether or not this sequence is sufficient in and of itself to promote internalization has yet to be experimentally determined.

6.3.4.3 Phosphorylation-Dependent Internalization of BommoPBANR

Ligand-induced receptor internalization is typically triggered following phosphorylation of sites in the ICLs or C terminus. For BommoPBANR, protein kinase C (PKC) appears to fulfill this function as endocytosis was blocked by the general kinase inhibitor staurosporine (Hull et al. 2005), mutations to consensus PKC sites (S333 and S366) in the C terminus of BommoPBANR-C (Hull et al. 2011), and RNAi knockdown of endogenous PKC in cultured insect cells expressing the receptor (Hull et al. 2011). The consensus PKC sites are highly conserved in other

PBANRs, which may indicate that feedback regulation of this class of receptors is evolutionarily conserved.

6.3.4.4 Effects of the N Terminus

For some GPCRs, structural determinants in the N terminus, such as N-linked glycosylation, can impact efficient cell surface trafficking and functional activation (Duvernay et al. 2005; Dong et al. 2007). Glutamine substitution of two consensus N-glycosylation sites (N19 and N22) in the *H. zea* PBANR N terminus significantly reduced receptor activation (Choi et al. 2007). Deletion of the first 27 residues from the BommoPBANR-C N terminus, which contains the corresponding N-glycosylation sites (N18 and N21), had no noticeable effects on receptor trafficking or ligand-induced internalization (Hull et al. 2011). While the variation in responses between the studies may simply be methodological based, it might also reflect functional differentiation of the two receptors given the low (37%) sequence identity across the two N termini.

6.3.4.5 Mapping the PBANR Ligand-Binding Pocket

To provide insights into the structural determinants underlying PBAN–PBANR interactions, an in silico model of BommoPBANR-C was constructed using spatial coordinates derived from crystal structures for the human β2 and A_{2A} adrenergic receptors, both of which, like PBANR, are class A GPCRs (Kawai et al. 2014). Based on alignment with known ligand recognition sites in the two human receptors and conserved residues across FXPRLamide receptors, 27 aa interspersed across the TM domains and extracellular loops (ECLs) were predicted to comprise a portion of the PBAN-binding pocket. The functional importance of the individual residues in terms of cell surface trafficking, ligand binding, and receptor activation was evaluated via sequential Ala-substations (Kawai et al. 2014). Substitution of four residues (Ser207, Phe211, Phe212, and His284) affected cell surface expression and, given their interhelical localization, are predicted to contribute to receptor stabilization. Eleven residues (Glu95, Glu120, Asn124, Val195, Phe276, Trp280, Phe283, Arg287, Tyr307, Thr311, and Phe319) influenced ligand-binding and receptor activation, three residues (Phe209, Phe303, GLy315) influenced ligand-binding alone, and a single residue (Tyr318) influenced receptor activation alone. This latter substitution was the only change that resulted in a receptor with normal ligand binding but impaired signaling abilities, which suggests that the Tyr side chain may play a significant role in the PBAN-induced conformational change in PBANR that results in receptor activation. The effects of Ala-substitutions on TM residues Phe212, Phe276, Trp280, Phe283, and Phe319, which are highly conserved in class A GPCRs, are likewise consistent with a proposed receptor conformational switch (Holst et al. 2010; Trzaskowski et al. 2012; Hulme 2013; Tehan et al. 2014).

Extension of the proposed in silico BommoPBANR model to include molecular docking simulations of a pentapeptide FXPRLamide analog identified a number of the same TM bundle localized receptor–ligand interaction sites (Kawai et al. 2014). Many of these sites are evolutionarily conserved across insect FXPRLamide receptors and human NmURs with Glu residues in TM2 and TM3 (i.e., E95 and E120) predicted to play a role in FXPRLamide binding as conservation of those sites in other class A GPCRs is more limited. Incorporation of a 10-aa PBAN analog into the molecular docking analyses identified two additional residues (V195 in ECL2 and F303 in ECL3) as contact points that were not predicted for the smaller 5-aa analog (Kawai et al. 2014). Interactions between these ECL-localized contact points and noncritical portions of the peptide could potentially stabilize ligand binding and/or serve as a selectivity filter for differentiating PBAN and DH, as suggested by the reported differences in peptide efficacies (Sato et al. 1993; Homma et al. 2006; Watanabe et al. 2007). Indeed, among the functionally relevant ligand contact points in BommoPBANR, only V195 (Glu in DHR) and F303 (Pro in DHR) are not conserved in *B. mori* DHR.

A similar approach that incorporated spatial coordinates from the bovine rhodopsin crystal structure to map potential ligand-binding sites in the *H. zea* PBANR predicted an inner-binding pocket surface of 20 contact points (Stern et al. 2007). Of those potential ligand-binding points, only three were also identified in the *B. mori* study (Ile113, Lys196, and Tyr307) and only Ala-substitution of Tyr307 perturbed receptor functionality via altered ligand binding and receptor activation (Kawai et al. 2014). The discrepancy in potential contact sites likely reflects methodological variances, which include the use of an early GPCR structural template (i.e., rhodopsin) that is now recognized to be less optimal for modeling peptide receptors (Sabio et al. 2008; Mobarec et al. 2009; Congreve et al. 2011). In contrast, all of the residues identified based on evolutionary trace analysis of PBANR-related sequences (Jurenka and Nusawardani 2011), which sought to identify a conserved TM-bounded pocket, are represented in the *B. mori* model.

Elucidation of the structure–function relationships underlying BommoPBANR functionality has expanded our understanding of the PBAN mode of action and provided insights into the molecular determinants that discriminate diverging peptidergic signals (i.e., PBAN vs DH). These insights into the endocrinological/peptidergic control of critical insect physiological functions can be used to further develop peptidomimetic agonists and/or antagonists of various insect GPCRs identified as potential targets for next-generation pest management strategies (Scherkenbeck and Zdobinsky 2009; Van Hiel et al. 2010; Bai and Palli 2013; Audsley and Down 2015). Indeed, PBAN antagonists have been the focus of much research over the years (Altstein 2001; Altstein et al. 2007; Altstein and Hariton 2009; Nachman 2009; Hariton et al. 2010; Nachman 2014).

6.4 PBAN Signal Transduction

6.4.1 Molecular Events Pre-Ca^{2+} Influx

6.4.1.1 Activation of the Secondary Messenger Cascade

A driving focus for numerous studies over the years was elucidating the signal transduction cascade activated in response to PBAN binding. The initial steps in most GPCR-linked cascades require receptor-mediated dissociation of an associated G protein complex with subsequent activation of the downstream effector proteins involved in generating the secondary messenger molecules that drive the signal transduction cascades. The predominant Gα subunits involved in receptor signaling include the following: Gαs – stimulate cAMP production, Gαi/o – inhibit cAMP production, and Gαq – stimulate Ca^{2+} influx. Although four Gα subunits (2 – Gαs, 1 – Gαo, and 1 – Gαq) have been identified to date in *B. mori* PGs, RNAi-mediated knockdown demonstrated that only the Gαq subunit functions in PBAN signal transduction (Hull et al. 2007a; Hull et al. 2010).

Determination that Gαq plays a role in PBAN signaling provided the initial molecular basis for early studies that reported pharmacological manipulation of intracellular Ca^{2+} levels via chelators (EGTA), ionophores (ionomycin, A23187, thapsigargin), or Ca^{2+} channel blockers (lanthanum, SKF-96365, 2-APB) could affect pheromone production (Fónagy et al. 1992b; Fónagy et al. 1992c; Matsumoto et al. 1995a). Fluorescent Ca^{2+} imaging techniques provided more direct evidence for PBAN-mediated Ca^{2+} influx in *B. mori* PGs (Hull et al. 2007a). The role of Ca^{2+} in PBAN signal transduction appears to be invariant as the pheromonotropic effects of the second messenger have been reported in a number of species (Jurenka et al. 1991a; Fónagy et al. 1992b; Rafaeli 1994; Jurenka et al. 1994; Soroker and Rafaeli 1995; Ma and Roelofs 1995b; Matsumoto et al. 1995b; Zhao et al. 2002; Choi and Jurenka 2004; Choi and Jurenka 2006). In contrast, the utilization of cAMP (cyclic adenosine-3′, 5′-monophosphate) as a co-messenger is species dependent. Pharmacological compounds (cAMP analogs, IBMX, and forskolin) that affect intracellular cAMP levels are pheromonotropic in a number of species (Rafaeli and Soroker 1989; Jurenka et al. 1991a; Jurenka et al. 1994; Soroker and Rafaeli 1995; Jurenka 1996), and cAMP levels are elevated following PBAN stimulation (Rafaeli and Soroker 1989; Rafaeli 1994; Soroker and Rafaeli 1995; Rafaeli and Gileadi 1996). In *B. mori*, however, the compounds do not exhibit pheromonotropic effects (Fónagy et al. 1992c; Hull et al. 2007a), and there is no PBAN-mediated increase in cAMP levels (Hull et al. 2007a). In species that utilize the second messenger, PBAN regulates the activity of acetyl-CoA carboxylase (ACC), an enzyme in fatty acid biosynthesis (Tang et al. 1989; Jurenka et al. 1991b; Tsfadia et al. 2008; Du et al. 2017a). Conversely, in species that utilize cAMP-independent pathways, PBAN regulates later steps in biosynthesis, which in *B. mori* includes both lipase (release of pheromone precursors from storage droplets) and fatty acyl reductase (FAR) activities, the final step in bombykol biosynthesis (Ozawa A et al. 1993; Ozawa and Matsumoto 1996; Fónagy et al. 2000; Ohnishi et al. 2011a; Du et al. 2012a).

6.4.1.2 Role of Phospholipase C

The influx of extracellular Ca^{2+} is tightly regulated by various cation channels including receptor-activated Ca^{2+} channels, which open in response to receptor-mediated phospholipase C (PLC) activation (Berridge et al. 2000). PLC hydrolysis of PIP_2 (phosphatidylinositol-4,5-bisphosphate) yields two products, the phosphoinositide IP_3 (inositol 1, 4, 5-triphosphate) and diacylglycerol (DAG), which act on two different types of Ca^{2+} channels. IP_3 works through endoplasmic reticulum Ca^{2+} stores to trigger an influx of extracellular Ca^{2+} via store-operated channels (SOC), whereas DAG and/or its metabolites act directly on a different subset of channels (Shuttleworth et al. 2004; Parekh 2006; Hardie 2007). Pharmacological manipulation of the two channel types implicated the involvement of SOC activity in *B. mori* pheromone production (Hull et al. 2007b). Building on that study, a PBAN-mediated increase in *B. mori* PG phosphoinositides coupled with reduced pheromonotropic effects of PBAN in the presence of PLC inhibitors confirmed a role for the enzyme downstream of PBANR activation (Hull et al. 2010). To date, three PLC transcripts (PLCβ1, PLCβ4, and PLCγ) and an IP_3 receptor (IP_3R) have been amplified from *B. mori* PGs with RNAi-mediated knockdown revealing PBAN-linked functionalities for the IP_3R as well as PLCβ1 and PLCγ (Hull et al. 2010). While the specific mechanistic roles the two PLCs have in propagating the PBAN signal remain to be fully elucidated, findings from other systems suggest that PLCβ1 likely functions in PIP_2 hydrolysis and that specific domains in PLCγ stabilize protein–protein interactions essential to formation of the SOC complex (Patterson et al. 2005).

6.4.1.3 Role of Ca^{2+} Channels

SOC complexes typically consist of stromal interaction molecule 1 (STIM1) functioning as the sensor of stored Ca^{2+} levels and Orai1 as the pore-forming unit at the plasma membrane (Derler et al. 2016; Nwokonko et al. 2017). Depletion of ER Ca^{2+} levels in response to an extracellular signal triggers redistribution of STIM1 to areas near the cell surface where interactions with Orai1 promote conformational changes in the pore sufficient for Ca^{2+} influx. Consistent with a role in propagating the PBAN signal, RNAi-mediated knockdown of the two *B. mori* homologs (BmSTIM1 and BmOrai1) impacted pheromone production without affecting non-pheromonotropic enzyme activities (Hull et al. 2009). In vitro analyses conducted in parallel showed that BmSTIM1 and BmOrai1 interactions were stoichiometrically dependent and required a cluster of basic residues in BmSTIM1. Expression of only the shorter of two BmOrai1 splice variants in the PG, in contrast to the expression of both variants, suggests that its role in the PBAN signaling cascade may be mechanistically distinct than in other tissues (Derler et al. 2016).

6.4.2 Post-Ca^{2+} Influx Signaling

6.4.2.1 Role of Calmodulin

Intracellular Ca^{2+} functions as a highly versatile signaling molecule that controls the regulation of diverse cellular processes (Berridge et al. 2003). As discussed above, PBAN-mediated Ca^{2+} influx is critical to not only pheromone biosynthesis in *B. mori* but also all other moth species. The predominant post-influx mediator of Ca^{2+} signaling is calmodulin, a multifunctional Ca^{2+}-binding protein that regulates the activity of a host of downstream processes (Sharma and Parameswaran 2018; Urrutia et al. 2019). Although transcriptional and pharmacological evidence support a role for calmodulin in the *B. mori* PBAN-signaling pathway, in vivo functionality remains to be demonstrated. Calmodulin has been purified from *B. mori* PGs (Iwanaga et al. 1998), and digital gene expression analyses of the tissue revealed significant upregulation of the encoding transcript within 72 hr of adult eclosion (Du et al. 2012a). Furthermore, pharmacological agents (W-7 and trifluoperazine) that inhibit calmodulin activity blocked in vitro pheromone production (Matsumoto et al. 1995a). Similar pheromonostatic effects of calmodulin inhibition have been reported in other species (Matsumoto et al. 1995b; Rafaeli and Gileadi 1996); however, contradictory results with other inhibitory compounds (Soroker and Rafaeli 1995) suggest that inferences regarding the presumed role of calmodulin will require more direct in vivo demonstration of functionality.

6.4.2.2 Phosphorylation Cascade

GPCR-mediated signaling pathways typically proceed via a phosphorylation cascade involving diverse kinase (phosphorylation) and phosphatase (dephosphorylation) steps, many of which are regulated by Ca^{2+}-bound calmodulin (Sharma and Parameswaran 2018). Although early studies assessing the pharmacological effects of kinase inhibition (H-89, PKA; staurosporine, PKC; KN-62, calmodulin-dependent kinase II, CaMKII) or activation (phorbol 12-myrstate 13-acetate, PKC) found no effect on in vitro pheromone production in *B. mori* (Matsumoto et al. 1995a; Ozawa et al. 1995), subsequent demonstration of PKC function in PBANR feedback regulation (Hull et al. 2011) implicated a PBAN-linked phosphorylation cascade. PG-derived immunoblots probed with anti-phosphoamino acid antibodies provided direct demonstration of kinase activity with multiple proteins shown to undergo rapid PBAN-mediated phosphorylation (Ohnishi et al. 2011a). RNAi-targeted knockdown of three kinases (PKA, PKC, and CaMKII), two of which are expressed in the PG (PKC and CaMKII), revealed CaMKII functions in *B. mori* sex pheromone production by regulating the lipolytic release of stored pheromone precursors (Ohnishi et al. 2011a). Phosphoproteomic analysis of *H. armigera* PGs revealed similar PBAN-induced phosphorylation of multiple proteins (Du et al. 2017a); however, in that species, PKA activity is critical for pheromone biosynthesis in both males and females (Du et al. 2017b; Du et al. 2017a). Unlike *B. mori*, the pheromone biosynthetic point regulated by PBAN in heliothines is ACC, a fatty acid biosynthesis enzyme that catalyzes carboxylation of acetyl-CoA to generate the malonyl-CoA used in fatty acid chain elongation. Based on their phosphoproteomic analyses and in vivo knockdown, Du and coworkers suggest that PKA indirectly regulates PBAN-mediated ACC activity by inhibiting the kinase (AMP-activated protein kinase) that maintains ACC in the inactive state.

In contrast to the kinase studies, pharmacological inhibition of phosphatase activity had pronounced in vitro pheromonostatic effects in *B. mori* (Matsumoto et al. 1995a; Ozawa and Matsumoto 1996; Fónagy et al. 1999). Inhibitor profiling implicated calcineurin (also called protein phosphatase 2b), a serine/threonine phosphatase activated by Ca^{2+}-bound calmodulin, in the PBAN-signaling cascade (Fónagy et al. 1999). In support of this role, a calcineurin homolog with 85% sequence identity to the *D. melanogaster* protein was amplified from a *B. mori* PG-specific cDNA library (Yoshiga et al. 2002). Although the encoding transcript is expressed in multiple tissues, it undergoes significant upregulation around the time of adult eclosion in concert with other transcripts in the *B. mori* sex pheromone biosynthetic pathway (Yoshiga et al. 2002). Determination of the rate-limiting steps in *B. mori* suggests that calcineurin or calcineurin-like phosphatase activity regulates the terminal fatty acyl reduction (Ozawa et al. 1993, 1995; Ozawa and Matsumoto 1996). Direct demonstration of calcineurin on FAR activity in *B. mori*, however, remains to be experimentally demonstrated. Knockdown studies have recently demonstrated the importance of calcineurin for sex pheromone biosynthesis in both male and female *H. armigera* (Du et al. 2017a; Zhao et al. 2018). In females, the phosphatase appears to directly regulate ACC activity, which is inactive when phosphorylated (Du et al. 2017a). In males, the control point remains to be elucidated; however, a FAR has been reported to be critical for the biosynthesis of some of the pheromonal components (Du et al. 2017b). It is intriguing to speculate that PBAN-mediated activation of calcineurin as the principal control point for FAR activity may be evolutionarily conserved.

6.5 Pheromone Gland Lipid Droplet

6.5.1 Lipid Droplet Dynamics

Unlike many other moth species, bioactive sex pheromone (i.e., bombykol) can be extracted from the PGs of newly eclosed *B. mori* females (Ando et al. 1988c). The pheromone-producing cells that comprise the inner layer of the PG are morphologically distinguished by the presence of large cytosolic lipid droplets that begin to accumulate 2 days prior to adult eclosion (Fónagy et al. 2000; Fónagy et al. 2001; Yokoyama et al. 2003). The size and number of lipid droplets fluctuate in relation to the circadian clock as well as pheromonotropic stimuli (Fónagy et al. 2000; Fónagy et al. 2001). Under pheromonostatic conditions, the lipid droplets are large and limited in number; conversely, smaller lipid droplets predominate in response to a pheromonotropic stimulus. Structural and chemical characterization revealed that the lipid droplets are largely composed of triacylglyerols (TAGs) that sequester five long-chain fatty acyls: two unsaturated C16 fatty acyl bombykol precursors (i.e., Δ10,12-hexadecadienoate and Δ11-hexadecenoate) at the *sn*-1/*sn*-3 positions and some combination of conventional diet-derived oleate, linoleate, and linolenate C18 fatty acyls (Matsumoto et al. 2002). Knockdown of the *B. mori* PBANR confirmed that lipolytic release of the stored precursors for entry into the pheromone biosynthetic pathway is regulated by the PBAN signal (Ohnishi et al. 2006). The utilization of lipid droplets for storage of pheromone precursor presumably predates silkworm domestication (Fujii et al. 2018) and may be a conserved storage mechanism for rapid pheromone production/release in some moth species (Fang et al. 1996; Foster 2001; Foster 2005).

6.5.2 *Lipid Droplet Formation*

Although our understanding of the mechanisms driving lipid droplet biogenesis in moth PGs remains to be as fully developed as that of vertebrate systems (see Guo et al. 2009; Pol et al. 2014; Gao and Goodman 2015), significant advances in elucidating aspects of the molecular framework have been made. Using an initial *B. mori* PG expressed-sequence tag database (Yoshiga et al. 2000) as well as digital gene expression (Du et al. 2012a) and iTRAQ-based quantitative proteomic (Du et al. 2015) analyses, a number of transcripts were found to be specifically/predominantly PG expressed and significantly upregulated at the time of adult eclosion, suggesting potential roles in pheromone biosynthesis. Among the transcripts identified to date are a number of genes critical for TAG synthesis, and by extension, cytoplasmic lipid droplet formation.

6.5.2.1 Fatty Acid Transport Protein

Fatty acid transport proteins and/or lipid transport proteins typically facilitate extracellular import of long-chain fatty acids and diacylglycerols (DAGs) across the plasma membrane (Shapiro et al. 1988) with subsequent ATP-dependent esterification to the corresponding acyl-CoA derivatives (Stahl 2004). Although the *B. mori* sex pheromone is de novo synthesized from acetyl-CoA via fatty acid biosynthesis, targeted knockdown of the *B. mori* FATP (BmFATP) affected pheromone production via impaired lipid droplet accumulation (Ohnishi et al. 2009). Chemical analysis revealed disrupted stoichiometric integration of fatty acyl components. Typically, a subset of the TAGs that comprise the PG lipid droplets are composed of two unsaturated C18 fatty acyls (i.e., oleate, linoleate, linolenate) and one of the C16 bombykol precursors – Δ10,12-hexadecadienoate or Δ11-hexadecenoate (Matsumoto et al. 2002). In BmFATP knockdown PGs, these TAGs were significantly reduced. The disrupted availability of C18 fatty acyls for incorporation into these TAGs impeded normal lipid droplet formation and, as a result, less pheromone precursor was available for flux into the pheromone biosynthetic pathway. Thus, BmFATP functions in pheromone production by importing the extracellular diet-derived C18 fatty acids critical for synthesizing a subset of lipid droplet TAGs that store sex pheromone precursors.

6.5.2.2 Triacylglyceride Synthesis Enzymes

Two major pathways contribute to TAG biosynthesis: the glycerol phosphate pathway, which proceeds from acylation of glycerol-3-phosphate, and the monoacylglycerol pathway, which utilizes diet-derived monoacylglycerol to generate TAGs (Takeuchi and Reue 2009; Hussain 2014). The two pathways converge at the final reaction point that converts DAG into the final TAG molecule via diacylglycerol acyltransferase (DGAT). The rate-limiting steps in the two pathways are the acylation of glycerol 3-phosphate by glycerol-3-phosphate acyltransferase (GPAT) and conversion of monoacylglycerol to diacylglycerol by acyl-CoA:monoacylglycerol acyltransferase (MGAT). Analysis of *B. mori* PG transcripts at various points in pupal-adult development revealed that the expression of GPAT and DGAT is consistent with a role in pheromone biosynthesis (Du et al. 2012a). Further analysis of the PG DGAT indicated that it was most similar to DGAT2 family members, which are characterized by two TM domains rather than the nine TMs of DGAT1. Although both DGAT1 and DGAT2 catalyze similar reactions, there was no evidence in the digital gene expression analyses for *B. mori* DGAT1 upregulation in the post-eclosion PG (Du et al. 2012a). RNAi-mediated knockdown of DGAT2 and GPAT both affected sex pheromone production with GPAT also reported to impact lipid droplet formation (Du et al. 2012b; Du et al. 2015). Similar to DGAT1, there is currently no transcriptional support for MGAT functionality in generating the pheromone-associated TAGs (Du et al. 2012a), suggesting TAG formation in the PG is largely dependent on the glycerol phosphate pathway.

6.5.2.3 Acyl-CoA-Binding Proteins

Similar disruption of *B. mori* TAG synthesis and PG lipid droplet formation was reported following RNAi-mediated knockdown of two transcripts encoding acyl-CoA-binding proteins (ACBPs) – mgACBP and pgACBP (Ohnishi et al. 2006). Although ACBPs typically bind and protect long-chain (C14-C22) acyl-CoA esters from hydrolysis (Gossett et al. 1996), mgACBP and pgACBP are thought to impact

TAG-dependent lipid droplet synthesis, and by extension downstream pheromone production, differently (Ohnishi et al. 2006). Unlike pgACBP, which is predominantly expressed in the PG, mgACBP is also expressed in the larval midgut, suggesting a potential dietary role (Matsumoto et al. 2001). Given the different expression patterns of the two transcripts, the current model for PG TAG synthesis suggests that mgACBP donates acyl-CoAs from the C18 fatty acyls imported by BmFATP, whereas pgACBP provides the de novo synthesized pheromone precursor fatty acyl-CoAs.

6.5.2.4 Acyl Carrier Protein

Since bombykol is de novo synthesized from acetyl-CoA through palmitate (16: acyl) (Yamaoka et al. 1984; Ando et al. 1988b; Arima et al. 1991), it is not surprising that disruption of steps in the fatty acid biosynthetic pathway also affects lipid droplet formation in the PG. An acyl carrier protein (ACP) homolog was among a group of transcripts identified in a PG EST database (Yoshiga et al. 2000) with an expression profile consistent with a pheromonogenic role (Ohnishi et al. 2011b). ACPs function early on in fatty acid biosynthesis by binding and presenting acyl chain intermediates to other enzymes in the pathway (Byers and Gong 2007; Chan and Vogel 2010). Targeted knockdown of the ACP transcript impaired lipid droplet production and specifically affected the synthesis of TAGs storing the C16 pheromone precursor products Δ10,12-hexadecadienoate and Δ11-hexadecenoate (Ohnishi et al. 2011b).

6.5.3 *Lipolysis of the Lipid Droplet*

6.5.3.1 Lipid Storage Droplet Protein

Studies in other organisms have revealed a complex network of protein interactions and phosphorylation events underlying lipid droplet dynamics (Barbosa et al. 2015; D'Andrea 2016; Sztalryd and Brasaemle 2017). For *B. mori*, the role of PBAN on lipid droplet dynamics (i.e., changes in size and number in response to pheromonotropic stimuli) has been well documented (Fónagy et al. 2000; Fónagy et al. 2001; Matsumoto et al. 2002; Yokoyama et al. 2003; Fónagy et al. 2005; Ohnishi et al. 2006) as has the importance of PBAN-mediated phosphorylation (Ohnishi et al. 2011a). Among the group of proteins phosphorylated in response to PBAN stimulation is a member of the perilipin family of lipid droplet proteins, *B. mori* lipid storage droplet protein-1 (BmLsd1). BmLsd1 has high sequence identity with a similar protein in *Manduca sexta* implicated in adipokinetic hormone-mediated lipolysis of fat body lipid droplets (Patel et al. 2005). Similar to that protein, BmLsd1 localizes to lipid droplets and is critical for lipolytic release of pheromone precursors stored in the PG lipid droplets (Ohnishi et al. 2011a). BmLsd1, however, does not appear to be the lipolytic enzyme. Rather, it is thought that the protein functions analogously to mammalian perilipins with PBAN-activated CaMKII phosphorylation of Ser/Thr residues in BmLds1, promoting a conformational change that either exposes the lipid droplet surface to associated lipases or allows for lipase binding in conjunction with other sequestered co-activators such as the CGI-58 protein that has been described in adipocytes (Sztalryd and Brasaemle 2017). Although homologs of CGI-58 are present in the *B. mori* genome (accession #s XP_004927228.1, XP_004927229.1, XP_012546725.1), there is currently no transcriptional or biochemical evidence for their role in PBAN-mediated lipolysis.

6.5.3.2 Lipases

Expression analyses by differing groups have identified seven lipase-like genes that are upregulated in the PG within 72 hr of adult eclosion (Ohnishi et al. 2011b; Du et al. 2012a; Zhang et al. 2013), a time period that is consistent with a role in pheromone production. RNAi-mediated knockdown, however, revealed that only four of the lipases have a role in pheromone production. Among the lipases identified include homologs of triacylglycerol lipase (NRPG0023/BGIBMGA005695), *D. melanogaster* lipase 3 (NRPG1187), *Aedes aegypti* lipase (NRPG1885), and a pancreatic lipase-like gene (BmPLLG/BGIBMGA011864). Although stimulated lipolysis typically proceeds through a phosphorylation cascade the culminates in lipase activation (D'Andrea 2016), the role of PBAN-mediated phosphorylation on lipase activity remains to be determined.

6.6 Bombykol Biosynthetic Pathway Enzymes

Unlike other Type I pheromones, the bombykol biosynthetic pathway is relatively simple in that the palmitic acid backbone does not undergo chain-shortening reactions or further modification of the terminal hydroxyl group. Rather, the bioactive pheromone is generated by stepwise conversion of fatty acid biosynthesis-derived palmitate via two desaturation steps and a terminal fatty-acyl reduction step (Ando et al. 1988b; Arima et al. 1991). The first desaturation step is a general Z11 desaturase reaction common in the pheromone biosynthetic pathways of numerous moth species (Roelofs et al. 2002). In contrast, the second desaturation step, which generates a conjugated diene system through 1,4-elimination of two allylic hydrogens at the double bond in the Z11-monoene C16 intermediate, is less common. Neither of the desaturation steps, however, is controlled by PBAN; rather, the peptide regulates the terminal fatty acyl reduction reaction (Ozawa A et al. 1993; Ozawa et al. 1995; Ozawa and Matsumoto 1996). As with other components of the pheromone pathway, transcripts encoding the respective enzymes are predominantly expressed in the PG and are upregulated at adult eclosion. Molecular characterization of the enzymes via a yeast expression system revealed the desaturase (Desat1 also referred to as Bmpgdesat1) catalyzes both desaturation steps and that the reductase (pgFAR) exhibits strong substrate specificity for the immediate precursor, Δ10,12-hexadecadienoate (Moto et al. 2003; Moto et al. 2004). RNAi-mediated knockdown of the respective transcripts confirmed the role of the two enzymes in bombykol biosynthesis (Ohnishi et al. 2006). Since the initial identification of the enzymes, additional multifunctional desaturases critical for moth pheromone production have been reported (Serra et al. 2006b; Serra et al. 2006a; Matoušková et al. 2007; Serra et al. 2007) as have a number of FARs selective for pheromone precursors (Antony et al. 2009; Liénard et al. 2010; Hagström et al. 2012; Lassance et al. 2013).

6.7 Summary

Based on the aggregate of the studies highlighted above, a model for the cellular and molecular processes that govern bombykol biosynthesis has emerged (Fig. 6.1). Prior to adult eclosion, DH-PBAN expression is upregulated in a subset of neurosecretory cells and post-translationally processed PBAN is transported along axons that pass through the maxillary nerve to the *corpus cardiacum*. Concomitantly in the PG, an array of pheromonogenic genes are upregulated and active TAG biosynthesis leads to the accumulation and enlargement of cytoplasmic lipid droplets (upper panel, Fig. 6.1). Soon after eclosion, stimulation of a central circadian pacemaker triggers release of stored PBAN into the hemolymph with subsequent binding of PBAN to cell surface localized PBANRs in the PG. The ensuing PBANR conformational change results in dissociation of the heterotrimeric G protein complex with subsequent Gαq activation of PLCβ1-mediated hydrolysis of PIP2 into DAG and IP_3. The soluble IP_3 diffuses through the cytosol to activate IP_3 receptors in the endoplasmic reticulum membrane resulting in the subsequent release of stored Ca^{2+}. The drop in luminal Ca^{2+} levels promotes translocation of STIM1 to the plasma membrane where it triggers an influx of extracellular Ca^{2+} through Orai1 channels, presumably via interactions with a scaffolding complex that includes PLCγ (lower left panel, Fig. 6.1). The concomitant rise in intracellular Ca^{2+} allows for the formation of Ca^{2+}-calmodulin complexes, at which point the PBAN pathway exhibits species-dependent divergence. In *B. mori,* and presumably species in which PBAN regulates a step late in pheromonogenesis, the Ca^{2+}-calmodulin complexes activate both calcineurin and CamKII. Calcineurin, in turn, activates the FAR that catalyzes the terminal step in pheromone biosynthesis (i.e., reduction of Δ10,12-hexadecadienoate to 10,12-hexadecadien-1-ol), while CamKII-dependent phosphorylation of BmLdsp-1 promotes lipolytic release of stored pheromone precursors from the cytoplasmic lipid droplets (lower right panel, Fig. 6.1).

Building on the *B. mori* framework, other groups have shown in species that utilize cAMP as a secondary messenger, and calcineurin-mediated dephosphorylation promotes ACC (i.e., the rate-limiting step in fatty acid biosynthesis) activity. In concert with this action, elevation of cAMP levels in response to Ca^{2+}-calmodulin activation of an adenylate cyclase leads to a PKA-initiated cascade that inhibits the

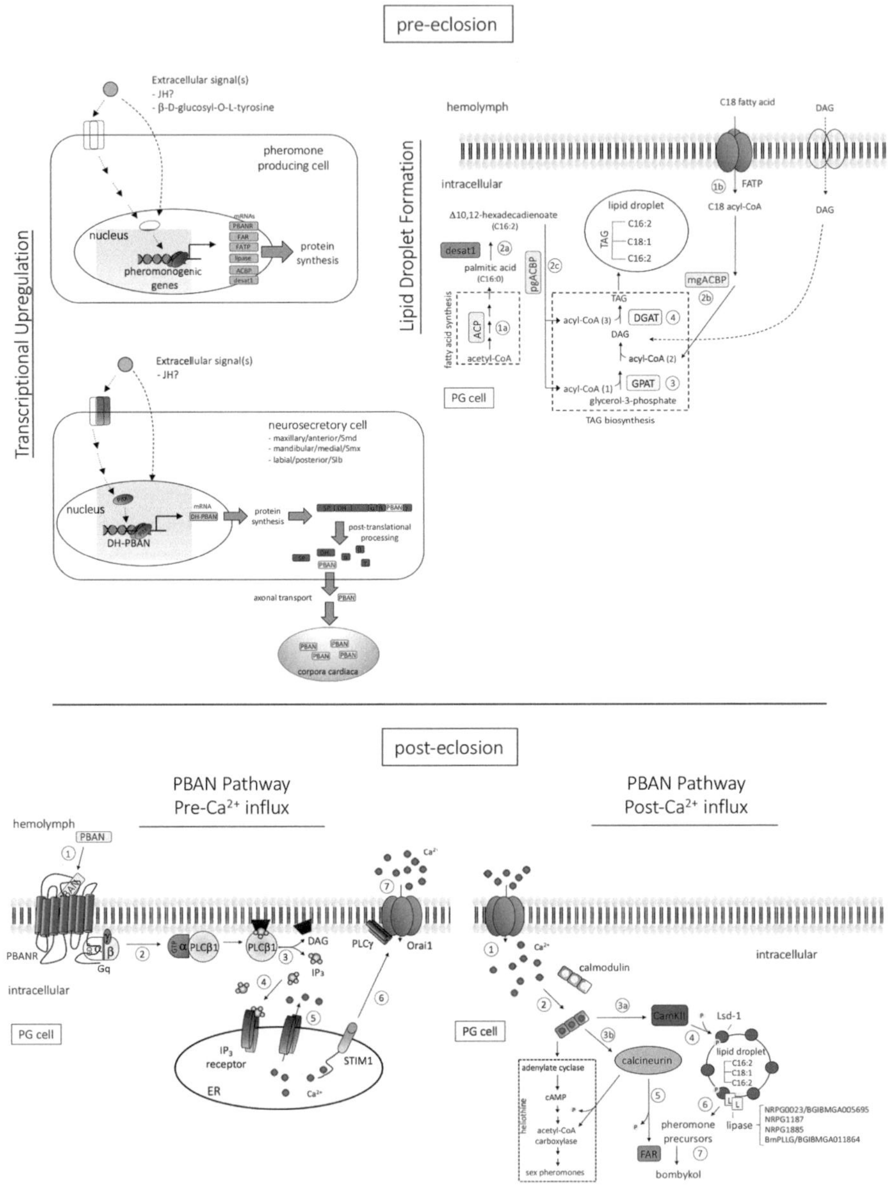

Fig. 6.1 Pre- and posteclosion cellular and molecular events driving bombykol production. Upper panel: Pre-eclosion events. (Left) Transcriptional upregulation of pheromonogenic genes in the developing PG and upregulation of DH-PBAN transcription in a subset of neurosecretory cells with axonal transport of processed PBAN peptide to the *corpora cardiaca* for subsequent circadian-controlled release into the hemolymph. (Right) Lipid droplet formation in the cytoplasm of developing PG cells. Fatty acid synthesis-derived palmitic acid is converted to pheromone precursor and incorporated into TAG via the glycerol-3-phosphate pathway along with diet-derived C18 acyl-CoAs and imported DAGs. Lower panel: Post-eclosion events. (Left) Signal transduction steps downstream of PBAN/PBANR binding prior to the influx of extracellular Ca^{2+}. (Right) PBAN pathway steps post-Ca^{2+} influx involve a calmodulin-dependent cascade that culminates in lipolytic release of pheromone precursors and their subsequent FAR-dependent modification to the final bioactive product – bombykol

phosphorylation event that maintains ACC in an inactive state (see Du et al. 2017a; Jurenka 2017).

6.8 Conclusion

The six decades of research since Butenandt's pioneering study have witnessed an explosion in both identification of pheromonal compounds and molecular elucidation of the associated biosynthetic pathways. Among the Lepidoptera, *B. mori* has been at the forefront of each new advancement in our understanding of the processes underlying moth pheromonogenesis – from purification of the regulatory peptide (PBAN) to identification of the cognate receptor to characterization of the genes comprising the biosynthetic and regulatory pathways. Going forward, advances in gene editing and transgenesis techniques hold great promise for *B. mori* to continue providing critical discoveries in relation to pheromonogenesis (see Moto and Matsumoto 2012; Shiomi et al. 2015).

To date, each new insight into the processes that comprise pheromonogenesis has highlighted the complexity of the system and provided new puzzles for us to unravel. Some of the questions raised by the current paradigm of pheromonogenesis that we find the most intriguing, and for which *B. mori* is well positioned to address, include:

1. What is the pre-eclosion signal that initiates transcription of pheromonogenic genes? Early studies linked β-D-glucosyl-*O*-L-tyrosine with transcription of pgACBP (Ohnishi et al. 2005), whereas more recent studies suggested a role for juvenile hormone signaling in priming the PG for pheromonogenesis (Zhang et al. 2014). Additional studies will be needed to determine if a lone signal drives transcription or if multiple signals are involved.
2. What is the molecular basis for regulation of the pleiotropic FxPRLamide peptide/receptor system?
3. What mechanism drives SLb-specific enrichment of DH relative to PBAN?
4. How is ligand selectivity of PBANRs/DHRs achieved?
5. What biological role do the concomitantly expressed PBANR variants play in PBAN signaling?
6. How is alternative splicing of PBANRs regulated?
7. What is the evolutionary significance of the different PBAN-mediated control points (fatty acid biosynthesis vs terminal fatty acyl reduction), and how did this divergence arise?

Acknowledgments We are grateful to Dr. József Fodor for critical reading of the text and insightful comments. We also thank Dr. Shogo Matsumoto (retired; RIKEN Advanced Science Institute) for his support of the Japan Society for the Promotion of Science which played a pivotal role in our respective careers. In addition, we thank the many members of the former Molecular Entomology Laboratory at the RIKEN Advanced Science Institute and the numerous colleagues and peers who have contributed to advancing our understanding of the cellular processes that govern biosynthesis of the pheromone that first piqued Butenstadt's interest 60 years ago. Mention of trade names or commercial products in this article is solely for the purpose of providing specific information and does not imply recommendation or endorsement by the United States Department of Agriculture (USDA). USDA is an equal opportunity provider and employer.

References

Allison JD, Cardé RT (2016) Pheromone communication in moths: evolution, behavior, and application. University of California Press

Altstein M (2001) Insect neuropeptide antagonists. Biopolymers 60:460–473

Altstein M, Hariton A (2009) Rational design of insect control agents: the PK/PBAN family as a study case. In: Biorational control of arthropod pests. Springer, Dordrecht, pp 49–81

Altstein M, Gazit Y, Aziz OB et al (1996) Induction of cuticular melanization in *Spodoptera littoralis* larvae by PBAN/MRCH: development of a quantitative bioassay and structure function analysis. Arch Insect Biochem Physiol 31:355–370

Altstein M, Ben-Aziz O, Zeltser I et al (2007) Inhibition of PK/PBAN-mediated functions in insects: discovery of selective and non-selective inhibitors. Peptides 28:574–584

Ando T, Arima R, Uchiyama M et al (1988a) Pheromone biosynthesis activating neuropeptide hormone in heads of the silkworm moth. Agric Biol Chem 52:881–883

Ando T, Hase T, Arima R, Uchiyama M (1988b) Biosynthetic pathway of bombykol, the sex pheromone of the female silkworm moth. Agric Biol Chem 52:473–478

Ando T, Hase T, Funayoshi A et al (1988c) Sex pheromone biosynthesis from^{14}C-Hexadecanoic acid in the silkworm moth. Agric Biol Chem 52:141–147

Ando T, Inomata SI, Yamamoto M (2004) Lepidopteran sex pheromones. In: Schulz S (ed) The chemistry of pheromones and other semiochemicals I. Springer-Verlag, Berlin Heidelberg, pp 51–96

Antony B, Fujii T, Moto K et al (2009) Pheromone-gland-specific fatty-acyl reductase in the adzuki bean borer, *Ostrinia scapulalis* (Lepidoptera: Crambidae). Insect Biochem Mol Biol 39:90–95

Arima R, Takahara K, Kadoshima T et al (1991) Hormonal regulation of pheromone biosynthesis in the silkworm moth, *Bombyx mori* (Lepidoptera: Bombycidae). Appl Entomol Zool 26:137–147

Audsley N, Down RE (2015) G protein coupled receptors as targets for next generation pesticides. Insect Biochem Mol Biol 67:1–32

Bai H, Palli SR (2013) G protein-coupled receptors as target sites for insecticide discovery. In: Advanced technologies for managing insect pests. Springer, Dordrecht, pp 57–82

Barak LS, Ménard L, Ferguson SS et al (1995) The conserved seven-transmembrane sequence NP(X)2,3Y of the G-protein-coupled receptor superfamily regulates multiple properties of the beta 2-adrenergic receptor. Biochemistry 34:15407–15414

Barbosa AD, Savage DB, Siniossoglou S (2015) Lipid droplet–organelle interactions: emerging roles in lipid metabolism. Curr Opin Cell Biol 35:91–97

Berridge MJ, Lipp P, Bootman MD (2000) The versatility and universality of calcium signalling. Nat Rev Mol Cell Biol 1:11–21

Berridge MJ, Bootman MD, Roderick HL (2003) Calcium signalling: dynamics, homeostasis and remodelling. Nat Rev Mol Cell Biol 4:517–529

Bjostad LB, Wolf WA, Roelofs WL (1987) Pheromone biosynthesis in lepidopterans: desaturation and chain shortening. In: Pheromone biochemistry. Academic Press, Orlando, pp 77–120

Bober R, Rafaeli A (2010) Gene-silencing reveals the functional significance of pheromone biosynthesis activating neuropeptide receptor (PBAN-R) in a male moth. Proc Natl Acad Sci 107:16858–16862

Bouley R, Sun T-X, Chenard M et al (2003) Functional role of the NPxxY motif in internalization of the type 2 vasopressin receptor in LLC-PK1 cells. Am J Physiol Cell Physiol 285:C750–C762

Butenandt A, Beckmann R, Stamm D, Hecker ET (1959) Uber den sexual-lockstoff des seidenspinners *Bombyx mori* -reindarstellung und konstitution. Zeitschrift Fur Naturforschung Part B-Chemie Biochemie Biophysik Biologie Und Verwandten Gebiete 14:283–284

Byers DM, Gong H (2007) Acyl carrier protein: structure–function relationships in a conserved multifunctional protein family. Biochem Cell Biol 85:649–662

Calebiro D, Godbole A (2018) Internalization of G-protein-coupled receptors: implication in receptor function, physiology and diseases. Best Pract Res Clin Endocrinol Metab 32:83–91

Cha WH, Jung JK, Lee D-W (2018) Identification of G protein-coupled receptors in the pheromone gland of *Maruca vitrata* by transcriptomic analysis. J Asia-Pacific Entomol 21:1203–1210

Chabre M, le Maire M (2005) Monomeric G-protein-coupled receptor as a functional unit. Biochemistry 44:9395–9403

Chan DI, Vogel HJ (2010) Current understanding of fatty acid biosynthesis and the acyl carrier protein. Biochem J 430:1–19

Chang J-C, Ramasamy S (2014) Identification and expression analysis of diapause hormone and pheromone biosynthesis activating neuropeptide (DH-PBAN) in the legume pod borer, *Maruca vitrata* Fabricius. PLoS One 9:e84916–e84911

Cheng Y, Luo L, Jiang X et al (2010) Expression of pheromone biosynthesis activating neuropeptide and its receptor (PBANR) mRNA in adult female *Spodoptera exigua* (Lepidoptera: Noctuidae). Arch Insect Biochem Physiol 75:13–27

Choi M-Y, Jurenka RA (2004) PBAN stimulation of pheromone biosynthesis by inducing calcium influx in pheromone glands of *Helicoverpa zea*. J Insect Physiol 50:555–560

Choi M-Y, Jurenka RA (2006) Role of extracellular Ca^{2+} and calcium channel activated by a G protein-coupled receptor regulating pheromone production in *Helicoverpa zea* (Lepidoptera: Noctuidae). Ann Entomol Soc Am 99:905–909

Choi M-Y, Vander Meer RK (2012) Ant trail pheromone biosynthesis is triggered by a neuropeptide hormone. PLoS One 7:e50400

Choi M-Y, Fuerst E-J, Rafaeli A, Jurenka RA (2003) Identification of a G protein-coupled receptor for pheromone biosynthesis activating neuropeptide from pheromone glands of the moth *Helicoverpa zea*. Proc Natl Acad Sci U S A 100:9721–9726

Choi M-Y, Fuerst E-J, Rafaeli A, Jurenka R (2007) Role of extracellular domains in PBAN/pyrokinin GPCRs from insects using chimera receptors. Insect Biochem Mol Biol 37:296–306

Chow KBS, Sun J, Chu KM et al (2012) The truncated ghrelin receptor polypeptide (GHS-R1b) is localized in the endoplasmic reticulum where it forms heterodimers with ghrelin receptors (GHS-R1a) to attenuate their cell surface expression. Mol Cell Endocrinol 348:247–254

Collawn JF, Stangel M, Kuhn LA et al (1990) Transferrin receptor internalization sequence YXRF implicates a tight turn as the structural recognition motif for endocytosis. Cell 63:1061–1072

Cong X, Topin J, Golebiowski J (2017) Class A GPCRs: structure, function, modeling and structure-based ligand design. Curr Pharm Des 23:4390–4409

Congreve M, Langmead CJ, Mason JS, Marshall FH (2011) Progress in structure-based drug design for G protein-coupled receptors. J Med Chem 54:4283–4311

D'Andrea S (2016) Lipid droplet mobilization: the different ways to loosen the purse strings. Biochimie 120:17–27

Davis NT, Homberg U, Teal PEA et al (1996) Neuroanatomy and immunocytochemistry of the median neuroendocrine cells of the subesophageal ganglion of the tobacco hawkmoth, *Manduca sexta*: immunoreactivities to PBAN and other neuropeptides. Microsc Res Tech 35:201–229

Derler I, Jardín I, Romanin C (2016) Molecular mechanisms of STIM/Orai communication. Am J Physiol Cell Physiol 310:C643–C662

Ding B-J, Löfstedt C (2015) Analysis of the *Agrotis segetum* pheromone gland transcriptome in the light of sex pheromone biosynthesis. BMC Genomics:1–21

Donaldson LF, Beazley-Long N (2016) Alternative RNA splicing: contribution to pain and potential therapeutic strategy. Drug Discov Today 21:1787–1798

Dong C, Filipeanu CM, Duvernay MT, Wu G (2007) Regulation of G protein-coupled receptor export trafficking. Biochim Biophys Acta 1768:853–870

Du M, Yin X, Zhang S et al (2012a) Identification of lipases involved in PBAN stimulated pheromone production in *Bombyx mori* using the DGE and RNAi approaches. PLoS One 7:e31045
Du M, Zhang S, Zhu B et al (2012b) Identification of a diacylglycerol acyltransferase 2 gene involved in pheromone biosynthesis activating neuropeptide stimulated pheromone production in *Bombyx mori*. J Insect Physiol 58:699–703
Du M, Liu X, Liu X et al (2015) Glycerol-3-phosphate O-acyltransferase is required for PBAN-induced sex pheromone biosynthesis in *Bombyx mori*. Sci Rep 5:8110
Du M, Liu X, Ma N et al (2017a) Calcineurin-mediated dephosphorylation of acetyl-coA carboxylase is required for pheromone biosynthesis activating neuropeptide (PBAN)-induced sex pheromone biosynthesis in *Helicoverpa armigera*. Mol Cell Proteomics 16:2138–2152
Du M, Zhao W, Jurenka R et al (2017b) Transcriptome analysis of *Helicoverpa armigera* male hairpencils: alcohol biosynthesis and requirement for mating success. Insect Biochem Mol Biol 87:154–164
Duc NM, Kim HR, Chung KY (2015) Structural mechanism of G protein activation by G protein-coupled receptor. Eur J Pharmacol 763:214–222
Duvernay MT, Filipeanu CM, Wu G (2005) The regulatory mechanisms of export trafficking of G protein-coupled receptors. Cell Signal 17:1457–1465
El-Sayed AM (2018) The Pherobase: database of Pheromones and Semiochemicals www.pherobase.com
Fang N, Teal PE, Tumlinson JH (1996) Effects of decapitation and PBAN injection on amounts of triacylglycerols in the sex pheromone gland of *Manduca sexta* (L.). Arch Insect Biochem Physiol 32:249–260
Ferguson S (2001) Evolving concepts in G protein-coupled receptor endocytosis: the role in receptor desensitization and signaling. Pharmacol Rev 53:1–24
Ferré S, Casadó V, Devi LA et al (2014) G protein-coupled receptor oligomerization revisited: functional and pharmacological perspectives. Pharmacol Rev 66:413–434
Fodor J, Köblös G, Kákai et al (2017) Molecular cloning, mRNA expression and biological activity of the pheromone biosynthesis activating neuropeptide (PBAN) from the European corn borer, *Ostrinia nubilalis*. Insect Mol Biol 31:355–317
Fodor J, Hull JJ, Köblös G et al (2018) Identification and functional characterization of the pheromone biosynthesis activating neuropeptide receptor isoforms from *Mamestra brassicae*. Gen Comp Endocrinol 258:60–69
Fónagy A, Matsumoto S, Schoofs L et al (1992a) *In vivo* and *in vitro* pheromonotropic activity of two locustatachykinin peptides in *Bombyx mori*. Biosci Biotechnol Biochem 56:1692–1693
Fónagy A, Matsumoto S, Uchiumi K et al (1992b) Action of pheromone biosynthesis activating neuropeptide on pheromone glands of *Bombyx mori* and *Spodoptera litura*. J Pest Sci 17:47–54
Fónagy A, Matsumoto S, Uchiumi K, Mitsui T (1992c) Role of calcium ion and cyclic nucleotides in pheromone production in *Bombyx mori*. J Pest Sci 17:115–121
Fónagy A, Schoofs L, Matsumoto S et al (1992d) Functional cross-reactivities of some locustamyotropins and Bombyx pheromone biosynthesis activating neuropeptide. J Insect Physiol 38:651–657
Fónagy A, Yokoyama N, Ozawa R et al (1999) Involvement of calcineurin in the signal transduction of PBAN in the silkworm, *Bombyx mori* (Lepidoptera). Comp Biochem Physiol B 124:51–60
Fónagy A, Yokoyama N, Okano K et al (2000) Pheromone-producing cells in the silkmoth, *Bombyx mori*: identification and their morphological changes in response to pheromonotropic stimuli. J Insect Physiol 46:735–744
Fónagy A, Yokoyama N, Matsumoto S (2001) Physiological status and change of cytoplasmic lipid droplets in the pheromone-producing cells of the silkmoth, *Bombyx mori* (Lepidoptera, Bombycidae). Arthropod Struct Dev 30:113–123
Fónagy A, Ohnishi A, Esumi Y et al (2005) Further studies of lipid droplets in the bombykol-producing pheromone gland of *Bombyx mori*. Ann N Y Acad Sci 1040:310–314

Foster SP (2001) Fatty acyl pheromone analogue-containing lipids and their roles in sex pheromone biosynthesis in the lightbrown apple moth, *Epipyhas postvittana* (Walker). J Insect Physiol 47:433–443

Foster SP (2005) Lipid analysis of the sex pheromone gland of the moth *Heliothis virescens*. Arch Insect Biochem Physiol 59:80–90

Fotiadis D, Jastrzebska B, Philippsen A et al (2006) Structure of the rhodopsin dimer: a working model for G-protein-coupled receptors. Curr Opin Struct Biol 16:252–259

Fujii T, Sakurai T, Ito K et al (2018) Lipid droplets in the pheromone gland of wild silkmoth *Bombyx mandarina*. J Insect Biotechnol Sericol 87:2029–2034

Gao Q, Goodman JM (2015) The lipid droplet—a well-connected organelle. Front Cell Dev Biol 3:49

Gossett RE, Frolov AA, Roths JB et al (1996) Acyl-CoA binding proteins: multiplicity and function. Lipids 31:895–918

Gripentrog JM (2000) A single amino acid substitution (N297A) in the conserved NPXXY sequence of the human N-formyl peptide receptor results in inhibition of desensitization and endocytosis, and a dose-dependent shift in p42/44 mitogen-activated protein kinase activation and chemotaxis. Biochem J 352:399–407

Groot AT (2014) Circadian rhythms of sexual activities in moths: a review. Front Ecol Evol 2:43

Groot AT, Dekker T, Heckel DG (2015) The genetic basis of pheromone evolution in moths. Annu Rev Entomol 61:99–117

Guo Y, Cordes KR, Farese RV, Walther TC (2009) Lipid droplets at a glance. J Cell Sci 122:749–752

Hagino A, Kitagawa N, Imai K et al (2010) Immunoreactive intensity of FXPRL amide neuropeptides in response to environmental conditions in the silkworm, *Bombyx mori*. Cell Tissue Res 342:459–469

Hagström AK, Liénard MA, Groot AT et al (2012) Semi-selective fatty acyl reductases from four heliothine moths influence the specific pheromone composition. PLoS One 7:e37230

Hardie RC (2007) TRP channels and lipids: from Drosophila to mammalian physiology. J Physiol Lond 578:9–24

Hariton A, Ben-Aziz O, Davidovitch M, Altstein M (2010) Bioavailability of backbone cyclic PK/PBAN neuropeptide antagonists--inhibition of sex pheromone biosynthesis elicited by the natural mechanism in *Heliothis peltigera* females. FEBS J 277:1035–1044

Hasegawa K, Shimizu I (1990) GABAergic control of the release of diapause hormone from the suboesophageal ganglion of the silkworm, *Bombyx mori*. J Insect Physiol 36:909–915

Hayashi T, Ito U (1933) Histological observations of the alluring gland of the female silkmoth. J Seric Sci 4:308–314

He R, Browning DD, Ye RD (2001) Differential roles of the NPXXY motif in formyl peptide receptor signaling. J Immunol 166:4099–4105

Holman GM, Cook BJ, Nachman RJ (1986) Primary structure and synthesis of a blocked myotropic neuropeptide isolated from the cockroach, *Leucophaea maderae*. Comp Biochem Physiol Part C: Comp Pharmacol 85:219–224

Holst B, Nygaard R, Valentin-Hansen L et al (2010) A conserved aromatic lock for the tryptophan rotameric switch in TM-VI of seven-transmembrane receptors. J Biol Chem 285:3973–3985

Homma T, Watanabe K, Tsurumaru S et al (2006) G protein-coupled receptor for diapause hormone, an inducer of Bombyx embryonic diapause. Biochem Biophys Res Commun 344:386–393

Hong B, Zhang Z-F, Tang S-M et al (2006) Protein–DNA interactions in the promoter region of the gene encoding diapause hormone and pheromone biosynthesis activating neuropeptide of the cotton bollworm, *Helicoverpa armigera*. Biochim Biophys Acta Gene Struct Expr 1759:177–185

Hull JJ, Ohnishi A, Moto K et al (2004) Cloning and characterization of the pheromone biosynthesis activating neuropeptide receptor from the silkmoth, *Bombyx mori*. Significance of the carboxyl terminus in receptor internalization. J Biol Chem 279:51500–51507

Hull JJ, Ohnishi A, Matsumoto S (2005) Regulatory mechanisms underlying pheromone biosynthesis activating neuropeptide (PBAN)-induced internalization of the *Bombyx mori* PBAN receptor. Biochem Biophys Res Commun 334:69–78
Hull JJ, Kajigaya R, Imai K, Matsumoto S (2007a) The *Bombyx mori* sex pheromone biosynthetic pathway is not mediated by cAMP. J Insect Physiol 53:782–793
Hull JJ, Kajigaya R, Imai K, Matsumoto S (2007b) Sex pheromone production in the silkworm, *Bombyx mori*, is mediated by store-operated Ca2+ channels. Biosci Biotechnol Biochem 71:1993–2001
Hull JJ, Lee JM, Kajigaya R, Matsumoto S (2009) *Bombyx mori* homologs of STIM1 and Orai1 are essential components of the signal transduction cascade that regulates sex pheromone production. J Biol Chem 284:31200–31213
Hull JJ, Lee JM, Matsumoto S (2010) Gqalpha-linked phospholipase Cbeta1 and phospholipase Cgamma are essential components of the pheromone biosynthesis activating neuropeptide (PBAN) signal transduction cascade. Insect Mol Biol 19:553–566
Hull JJ, Lee JM, Matsumoto S (2011) Identification of specific sites in the third intracellular loop and carboxyl terminus of the *Bombyx mori* pheromone biosynthesis activating neuropeptide receptor crucial for ligand-induced internalization. Insect Mol Biol 20:801–811
Hulme EC (2013) GPCR activation: a mutagenic spotlight on crystal structures. Trends Pharmacol Sci 34:67–84
Hussain MM (2014) Intestinal lipid absorption and lipoprotein formation. Curr Opin Lipidol 25:200–206
Ichikawa T (1998) Activity patterns of neurosecretory cells releasing pheromonotropic neuropeptides in the moth *Bombyx mori*. Proc Natl Acad Sci U S A 95:4055–4060
Ichikawa T, Kamimoto S (2003) Firing activities of neurosecretory cells producing diapause hormone and its related peptides in the female silkmoth, *Bombyx mori*. II. Mandibular and maxillary cells. Zool Sci 20:979–983
Ichikawa T, Suenobu A (2003) Firing activity of "diapause hormone" producing cells in the male silkmoth, *Bombyx mori*. Zool Sci 20:957–962
Ichikawa T, Hasegawa K, Shimizu I et al (1995) Structure of neurosecretory cells with immunoreactive diapause hormone and pheromone biosynthesis activating neuropeptide in the silkworm, *Bombyx mori*. Zool Sci 12:703–712
Ichikawa T, Shiota T, Shimizu I, Kataoka H (1996) Functional differentiation of neurosecretory cells with immunoreactive diapause hormone and pheromone biosynthesis activating neuropeptide of the moth, *Bombyx mori*. Zool Sci 13:21–25
Ichikawa T, Aoki S, Shimizu I (1997) Neuroendocrine control of diapause hormone secretion in the silkworm, *Bombyx mori*. J Insect Physiol 43:1101–1109
Imai K, Konno T, Nakazawa Y et al (1991) Isolation and structure of diapause hormone of the silkworm, *Bombyx mori*. Proc Jpn Acad Ser B Phys Biol Sci 67:98–101
Iwanaga M, Dohmae N, Francke A et al (1998) Isolation and characterization of calmodulin in the pheromone gland of the silkworm, *Bombyx mori*. Comp Biochem Physiol B Biochem Mol Biol 120:761–767
Jiang L, Zhang F, Hou Y et al (2018) Isolation and functional characterization of the pheromone biosynthesis activating neuropeptide receptor of Chinese oak silkworm, *Antheraea pernyi*. Int J Biol Macromol 117:42–50
Jing T-Z, Wang Z-Y, Qi F-H, Liu K-Y (2007) Molecular characterization of diapause hormone and pheromone biosynthesis activating neuropeptide from the black-back prominent moth, *Clostera anastomosis* (L.) (Lepidoptera, Notodontidae). Insect Biochem Mol Biol 37:1262–1271
Jurenka R (1996) Signal transduction in the stimulation of sex pheromone biosynthesis in moths. Arch Insect Biochem Physiol 33:245–258
Jurenka R (2015) The PRXamide neuropeptide signalling system: conserved in animals. In: Jurenka R (ed) Advances in insect physiology. Academic Press, Oxford, pp 123–170
Jurenka R (2017) Regulation of pheromone biosynthesis in moths. Curr Opin Insect Sci 24:29–35

Jurenka R, Nusawardani T (2011) The pyrokinin/ pheromone biosynthesis-activating neuropeptide (PBAN) family of peptides and their receptors in Insecta: evolutionary trace indicates potential receptor ligand-binding domains. Insect Mol Biol 20:323–334

Jurenka RA, Jacquin E, Roelofs WL (1991a) Stimulation of pheromone biosynthesis in the moth *Helicoverpa zea*: action of a brain hormone on pheromone glands involves Ca2+ and cAMP as second messengers. Proc Natl Acad Sci U S A 88:8621–8625

Jurenka RA, Jacquin E, Roelofs WL (1991b) Control of the pheromone biosynthetic pathway in *Helicoverpa zea* by the pheromone biosynthesis activating neuropeptide. Arch Insect Biochem Physiol 17:81–91

Jurenka RA, Fabriás G, DeVoe L, Roelofs WL (1994) Action of PBAN and related peptides on pheromone biosynthesis in isolated pheromone glands of the redbanded leafroller moth, *Argyrotaenia velutinana*. Comp Biochem Physiol Pharmacol Toxicol Endocrinol 108:153–160

Kawai T, Hull JJ, Matsumoto S et al (2010) Studies on the structure-activity relationship of pheromone biosynthesis-activating neuropeptide (PBAN). In: Okamoto K (ed) Peptide science 2009: proceedings of the 46th Japanese peptide symposium. The Japanese Peptide Society Press, Osaka, pp 231–234

Kawai T, Nagata K, Okada K et al (2011) Structure-activity relationship studies of the pheromone biosynthesis-activating neuropeptide of the silkworm, *Bombyx mori*. In: Fujii N, Kiso Y (eds) Peptide SCIENCE 2010: proceedings of the fifth international peptide symposium. The Japanese Peptide Society Press, Osaka, p 126

Kawai T, Lee JM, Nagata K et al (2012) The arginine residue within the C-terminal active core of *Bombyx mori* pheromone biosynthesis-activating neuropeptide is essential for receptor binding and activation. Front Endocrinol 3:42

Kawai T, Katayama Y, Guo L et al (2014) Identification of functionally important residues of the silkmoth pheromone biosynthesis-activating neuropeptide receptor, an insect ortholog of the vertebrate neuromedin U receptor. J Biol Chem 289:19150–19163

Kawano T, Kataoka H, Nagasawa H, Isogai A (1992) cDNA cloning and sequence determination of the pheromone biosynthesis activating neuropeptide of the silkworm, *Bombyx mori*. Biochem Biophys Res Commun 189:221–226

Kim Y-J, Nachman R, Aimanova K et al (2008) The pheromone biosynthesis activating neuropeptide (PBAN) receptor of *Heliothis virescens*: identification, functional expression, and structure-activity relationships of ligand analogs. Peptides 29:268–275

Kingan TG, Blackburn MB, Raina AK (1992) The distribution of pheromone-biosynthesis-activating neuropeptide (PBAN) immunoreactivity in the central nervous system of the corn earworm moth, *Helicoverpa zea*. Cell Tissue Res 270:229–240

Kitamura A, Nagasawa H, Kataoka H et al (1989) Amino acid sequence of pheromone-biosynthesis-activating neuropeptide (PBAN) of the silkworm, *Bombyx mori*. Biochem Biophys Res Commun 163:520–526

Kitamura A, Nagasawa H, Kataoka H et al (1990) Amino acid sequence of pheromone biosynthesis activating neuropeptide-II (PBAN-II) of the silkmoth, *Bombyx mori*. Agric Biol Chem 54:2495–2497

Kleinau G, Jaeschke H, Worth CL et al (2010) Principles and determinants of G-protein coupling by the rhodopsin-like thyrotropin receptor. PLoS One 5:e9745

Kristiansen K (2004) Molecular mechanisms of ligand binding, signaling, and regulation within the superfamily of G-protein-coupled receptors: molecular modeling and mutagenesis approaches to receptor structure and function. Pharmacol Ther 103:21–80

Kuniyoshi H, Kitamura A, Nagasawa H et al (1991a) Structure-activity relationship of pheromone biosynthesis activating neuropeptide (PBAN) from the silkmoth, *Bombyx mori*. Pept Chem 1990:251–254

Kuniyoshi H, Nagasawa H, Ando T, Suzuki A (1991b) Structure-activity relationship and metabolism of pheromone biosynthesis activating neuropeptide (PBAN) from the silkmoth *Bombyx mori*. Pept Chem:221–226

Kuniyoshi H, Nagasawa H, Ando T et al (1992a) Cross-activity between pheromone biosynthesis activating neuropeptide (PBAN) and myotropic pyrokinin insect peptides. Biosci Biotechnol Biochem 56:167–168

Kuniyoshi H, Nagasawa H, Ando T, Suzuki A (1992b) N-terminal modified analogs of C-terminal fragments of PBAN with pheromonotropic activity. Insect Biochem Mol Biol 22:399–403

Lassance J-M, Liénard MA, Antony B et al (2013) Functional consequences of sequence variation in the pheromone biosynthetic gene pgFAR for Ostrinia moths. Proc Natl Acad Sci U S A 110:3967–3972

Lee JM, Hull JJ, Kawai T et al (2012a) Re-evaluation of the PBAN receptor molecule: characterization of PBANR variants expressed in the pheromone glands of moths. Front Endocrinol 3:6

Lee JM, Hull JJ, Kawai T et al (2012b) Establishment of Sf9 transformants constitutively expressing PBAN receptor variants: application to functional evaluation. Front Endocrinol 3:56

Liénard MA, Hagström AK, Lassance J-M, Löfstedt C (2010) Evolution of multicomponent pheromone signals in small ermine moths involves a single fatty-acyl reductase gene. Proc Natl Acad Sci U S A 107:10955–10960

Ma PW, Ramaswamy SB (2003) Biology and ultrastructure of sex pheromone-producing tissue. In: Insect pheromone biochemistry and molecular biology. Academic Press, Boston, pp 19–51

Ma PWK, Roelofs WL (1995a) Sites of synthesis and release of PBAN-like factor in the female European corn borer, *Ostrinia nubilalis*. J Insect Physiol 41:339–350

Ma PWK, Roelofs WL (1995b) Calcium involvement in the stimulation of sex pheromone production by PBAN in the European corn borer, *Ostrinia nubilalis* (Lepidoptera: Pyralidae). Insect Biochem Mol Biol 25:467–473

Maggio R, Fasciani I, Rossi M et al (2016) Variants of G protein-coupled receptors: a reappraisal of their role in receptor regulation. Biochem Soc Trans 44:589–594

Markovic D, Challiss RAJ (2009) Alternative splicing of G protein-coupled receptors: physiology and pathophysiology. Cell Mol Life Sci 66:3337–3352

Matoušková P, Pichová I, Svatos A (2007) Functional characterization of a desaturase from the tobacco hornworm moth (*Manduca sexta*) with bifunctional Z11-and 10, 12-desaturase activity. Insect Biochem Mol Biol 37:601–610

Matsumoto S, Isogai A, Suzuki A (1986) Isolation and amino terminal sequence of melanization and reddish coloration hormone (MRCH) from the silkworm, *Bombyx mori*. Insect Biochem 16:775–779

Matsumoto S, Kitamura A, Nagasawa H et al (1990) Functional diversity of a neurohormone produced by the suboesophageal ganglion: molecular identity of melanization and reddish colouration hormone and pheromone biosynthesis activating neuropeptide. J Insect Physiol 36:427–432

Matsumoto S, Fónagy A, Kurihara M, Uchiumi K (1992a) Isolation and primary structure of a novel pheromonotropic neuropeptide structurally related to leucopyrokinin from the armyworm larvae, *Pseudaletia separata*. Biochem Biophys Res Commun 182:534–539

Matsumoto S, Yamashita O, Fónagy A (1992b) Functional diversity of a pheromonotropic neuropeptide: induction of cuticular melanization and embryonic diapause in lepidopteran insects by Pseudaletia pheromonotropin. J Insect Physiol 38:847–851

Matsumoto S, Ozawa R, Nagamine T et al (1995a) Intracellular transduction in the regulation of pheromone biosynthesis of the silkworm, *Bombyx mori*: suggested involvement of calmodulin and phosphoprotein phosphatase. Biosci Biotechnol Biochem 59:560–562

Matsumoto S, Ozawa R, Uchiumi K et al (1995b) Intracellular signal transduction of PBAN action in the common cutworm, *Spodoptera litura*: effects of pharmacological agents on sex pheromone production in vitro. Insect Biochem Mol Biol 25:1055–1059

Matsumoto S, Yoshiga T, Yokoyama N et al (2001) Characterization of acyl-CoA-binding protein (ACBP) in the pheromone gland of the silkworm, *Bombyx mori*. Insect Biochem Mol Biol 31:603–609

Matsumoto S, Fónagy A, Yamamoto M (2002) Chemical characterization of cytoplasmic lipid droplets in the pheromone-producing cells of the silkmoth, *Bombyx mori*. Insect Biochem Mol Biol 32:1447–1455

Matsutani K, Sonobe H (1987) Control of diapause-factor secretion from the suboesophageal ganglion in the silkworm, *Bombyx mori*: the roles of the protocerebrum and tritocerebrum. J Insect Physiol 33:279–285
McDowell DG, Burns NA, Parkes HC (1998) Localised sequence regions possessing high melting temperatures prevent the amplification of a DNA mimic in competitive PCR. Nucleic Acids Res 26:3340–3347
Mobarec JC, Sanchez R, Filizola M (2009) Modern homology modeling of G-protein coupled receptors: which structural template to use? J Med Chem 52:5207–5216
Morita A, Niimi T, Yamashita O (2003) Physiological differentiation of DH-PBAN-producing neurosecretory cells in the silkworm embryo. J Insect Physiol 49:1093–1102
Moto K-I, Matsumoto S (2012) Construction of an in vivo system for functional analysis of the genes involved in sex pheromone production in the silkmoth, *Bombyx mori*. Front Endocrinol 3:30
Moto K, Yoshiga T, Yamamoto M et al (2003) Pheromone gland-specific fatty-acyl reductase of the silkmoth, *Bombyx mori*. Proc Natl Acad Sci U S A 100:9156–9161
Moto K, Suzuki MG, Hull JJ et al (2004) Involvement of a bifunctional fatty-acyl desaturase in the biosynthesis of the silkmoth, *Bombyx mori*, sex pheromone. Proc Natl Acad Sci U S A 101:8631–8636
Mount SM (1982) A catalogue of splice junction sequences. Nucleic Acids Res 10:459–472
Nachman RJ (2009) Agonists/antagonists of the insect kinin and pyrokinin/PBAN neuropeptide classes as tools for rational pest control. In: Biorational control of arthropod pests: application and resistance management. Springer, Dordrecht, pp 21–48
Nachman RJ (2014) Peptidomics applied: a new strategy for development of selective antagonists/agonists of insect pyrokinin (FXPRLamide) family using a novel conformational-mimetic motif. EuPA Open Proteom 3:138–142
Nachman RJ, Kuniyoshi H, Roberts VA et al (1993) Active conformation of the pyrokinin/PBAN neuropeptide family for pheromone biosynthesis in the silkworm. Biochem Biophys Res Commun 193:661–666
Nagasawa H, Kitamura A, Inoue T et al (1988) Isolation of pheromone biosynthesis activating neuropeptide of the silkworm, *Bombyx mori*. Agric Biol Chem 52:2985–2987
Nagasawa H, Kuniyoshi H, Arima R et al (1994) Structure and activity of Bombyx PBAN. Arch Insect Biochem Physiol 25:261–270
Noguchi H, Hayakawa Y (2001) Dopamine is a key factor for the induction of egg diapause of the silkworm, *Bombyx mori*. Eur J Biochem 268:774–780
Nusawardani T, Kroemer JA, Choi M-Y, Jurenka RA (2013) Identification and characterization of the pyrokinin/pheromone biosynthesis activating neuropeptide family of G protein-coupled receptors from *Ostrinia nubilalis*. Insect Mol Biol 22:331–340
Nwokonko RM, Cai X, Loktionova NA et al (2017) The STIM-Orai pathway: conformational coupling between STIM and Orai in the activation of store-operated Ca2+ entry. In: Store-operated Ca^{2+} entry (SOCE) pathways. Springer, Cham, pp 83–98
Ohguchi Y, Tatsuki S, Usui K, Arai K (1985) Hormone-like substance present in the cephalic organs of the female moth, *Chilo suppressalis* (Walker)(Lepidoptera:Pyralidae) and controlling sex pheromone production. Jpn J Appl Entomol Zool 29:265–269
Ohnishi A, Koshino H, Takahashi S et al (2005) Isolation and characterization of a humoral factor that stimulates transcription of the acyl-CoA-binding protein in the pheromone gland of the silkmoth, *Bombyx mori*. J Biol Chem 280:4111–4116
Ohnishi A, Hull JJ, Matsumoto S (2006) Targeted disruption of genes in the *Bombyx mori* sex pheromone biosynthetic pathway. Proc Natl Acad Sci U S A 103:4398–4403
Ohnishi A, Hashimoto K, Imai K, Matsumoto S (2009) Functional characterization of the *Bombyx mori* fatty acid transport protein (BmFATP) within the silkmoth pheromone gland. J Biol Chem 284:5128–5136
Ohnishi A, Hull JJ, Kaji M et al (2011a) Hormone signaling linked to silkmoth sex pheromone biosynthesis involves Ca2+/calmodulin-dependent protein kinase II-mediated phosphorylation

of the insect PAT family protein *Bombyx mori* lipid storage droplet protein-1 (BmLsd1). J Biol Chem 286:24101–24112
Ohnishi A, Kaji M, Hashimoto K, Matsumoto S (2011b) Screening for the genes involved in bombykol biosynthesis: identification and functional characterization of *Bombyx mori* acyl carrier protein. Front Endocrinol 2:92
Okada A, Kawai T, Sugisaka A, Ohtsuka J, Hull JJ, Moto K, Matsumoto S, Nagasawa H, Nagata K, Tanokura M (2009) Structural analysis of the active and inactive fragments of pheromone biosynthesis-activating neuropeptide (PBAN) from the silkmoth Bombyx mori. In: Peptide science: proceedings of the Japanese peptide symposium, pp 535–538
Ozawa R, Matsumoto S (1996) Intracellular signal transduction of PBAN action in the silkworm, *Bombyx mori*: involvement of acyl CoA reductase. Insect Biochem Mol Biol 26:259–265
Ozawa AR, Ando T, Nagasawa H et al (1993) Reduction of the acyl group: the critical step in bombykol biosynthesis that is regulated in vitro by the neuropeptide hormone in the pheromone gland of *Bombyx mori*. Biosci Biotechnol Biochem 57:2144–2147
Ozawa R, Matsumoto S, Kim GH et al (1995) Intracellular signal transduction of PBAN action in lepidopteran insects: inhibition of sex pheromone production by compactin, an HMG CoA reductase inhibitor. Regul Pept 57:319–327
Paing MM, Temple BRS, Trejo J (2004) A tyrosine-based sorting signal regulates intracellular trafficking of protease-activated receptor-1: multiple regulatory mechanisms for agonist-induced G protein-coupled receptor internalization. J Biol Chem 279:21938–21947
Pandey KN (2009) Functional roles of short sequence motifs in the endocytosis of membrane receptors. Front Biosci 14:5339–5360
Parekh AB (2006) On the activation mechanism of store-operated calcium channels. Pflugers Arch 453:303–311
Park Y, Kim Y-J, Adams ME (2002) Identification of G protein-coupled receptors for Drosophila PRXamide peptides, CCAP, corazonin, and AKH supports a theory of ligand-receptor coevolution. Proc Natl Acad Sci U S A 99:11423–11428
Patel RT, Soulages JL, Hariharasundaram B, Arrese EL (2005) Activation of the lipid droplet controls the rate of lipolysis of triglycerides in the insect fat body. J Biol Chem 280:22624–22631
Patterson R, Vanrossum D, Nikolaidis N et al (2005) Phospholipase C-γ: diverse roles in receptor-mediated calcium signaling. Trends Biochem Sci 30:688–697
Percy-Cunningham JE, MacDonald JA (1987) Biology and ultrastructure of sex pheromone-producing glands. In: Blomquist GJ, Prestwich GD (eds) Pheromone biochemistry. Academic Press, Orlando, pp 27–75
Pol A, Gross SP, Parton RG (2014) Biogenesis of the multifunctional lipid droplet: lipids, proteins, and sites. J Cell Biol 204:635–646
Quentien MH, Manfroid I, Moncet D et al (2002) Pitx factors are involved in basal and hormone-regulated activity of the human prolactin promoter. J Biol Chem (46):44408–44416
Rafaeli A (1994) Pheromonotropic stimulation of moth pheromone gland cultures in vitro. Arch Insect Biochem Physiol 25:287–299
Rafaeli A, Gileadi C (1996) Down regulation of pheromone biosynthesis: cellular mechanisms of pheromonostatic responses. Insect Biochem Mol Biol 26:797–807
Rafaeli A, Gileadi C (1997) Neuroendocrine control of pheromone production in moths. Invertebr Neurosci 3:223–229
Rafaeli A, Soroker V (1989) Cyclic AMP mediation of the hormonal stimulation of ^{14}C-acetate incorporation by *Heliothis armigera* pheromone glands in vitro. Mol Cell Endocrinol 65:43–48
Rafaeli A, Soroker V, Kamensky B, Raina AK (1990) Action of pheromone biosynthesis activating neuropeptide on in vitro pheromone glands of *Heliothis armigera* females. J Insect Physiol 36:641–646
Rafaeli A, Zakharova T, Lapsker Z, Jurenka RA (2003) The identification of an age- and female-specific putative PBAN membrane-receptor protein in pheromone glands of *Helicoverpa armigera*: possible up-regulation by juvenile hormone. Insect Biochem Mol Biol 33:371–380

Rafaeli A, Bober R, Becker L et al (2007) Spatial distribution and differential expression of the PBAN receptor in tissues of adult Helicoverpa spp. (Lepidoptera: Noctuidae). Insect Mol Biol 16:287–293

Raina AK, Klun JA (1984) Brain factor control of sex pheromone production in the female corn earworm moth. Science 225:531–533

Raina AK, Jaffe H, Klun JA et al (1987) Characteristics of a neurohormone that controls sex pheromone production in *Heliothis zea*. J Insect Physiol 33:809–814

Raina A, Jaffe H, Kempe T et al (1989) Identification of a neuropeptide hormone that regulates sex pheromone production in female moths. Science 244:796–798

Raina AK, Kingan TG, Kochansky JP (2003) A pheromonotropic peptide of *Helicoverpa zea*, with melanizing activity, interaction with PBAN, and distribution of immunoreactivity. Arch Insect Biochem Physiol 53:147–157

Riddiford LM, Williams CM (1971) Role of the corpora cardiaca in the behavior of saturniid moths. I. Release of sex pheromone. Biol Bull 140:1–7

Roelofs WL, Liu W, Hao G et al (2002) Evolution of moth sex pheromones via ancestral genes. Proc Natl Acad Sci U S A 99:13621–13626

Sabio M, Jones K, Topiol S (2008) Use of the X-ray structure of the beta2-adrenergic receptor for drug discovery. Part 2: identification of active compounds. Bioorg Med Chem Lett 18:5391–5395

Sakurai T, Namiki S, Kanzaki R (2014) Molecular and neural mechanisms of sex pheromone reception and processing in the silkmoth *Bombyx mori*. Front Physiol 5:125

Sato Y, Oguchi M, Menjo N et al (1993) Precursor polyprotein for multiple neuropeptides secreted from the suboesophageal ganglion of the silkworm *Bombyx mori*: characterization of the cDNA encoding the diapause hormone precursor and identification of additional peptides. Proc Natl Acad Sci U S A 90:3251–3255

Sato Y, Ikeda M, Yamashita O (1994) Neurosecretory cells expressing the gene for common precursor for diapause hormone and pheromone biosynthesis-activating neuropeptide in the suboesophageal ganglion of the silkworm, *Bombyx mori*. Gen Comp Endocrinol 96:27–36

Sato Y, Shiomi K, Saito H et al (1998) Phe-X-Pro-Arg-Leu-NH_2 peptide producing cells in the central nervous system of the silkworm, *Bombyx mori*. J Insect Physiol 44:333–342

Scherkenbeck J, Zdobinsky T (2009) Insect neuropeptides: structures, chemical modifications and potential for insect control. Bioorganic Med Chem 17:4071–4084

Schoofs L, Vanden Broeck J, De Loof A (1993) The myotropic peptides of *Locusta migratoria*: structures, distribution, functions and receptors. Insect Biochem Mol Biol 23:859–881

Schoofs L, De Loof A, Van Hiel MB (2017) Neuropeptides as regulators of behavior in insects. Annu Rev Entomol 62:35–52

Seck T, Pellegrini M, Florea AM et al (2005) The delta e13 isoform of the calcitonin receptor forms a six-transmembrane domain receptor with dominant-negative effects on receptor surface expression and signaling. Mol Endocrinol 19:2132–2144

Senthilkumar R, Srinivasan R (2019) Sex-specific spatial and temporal gene expressions of pheromone biosynthesis activating neuropeptide (PBAN) and binding proteins (PBP/OBP) in *Spoladea recurvalis*. Sci Rep 9:3515

Serra M, Gauthier LT, Fabrias G, Buist PH (2006a) Delta11 desaturases of *Trichoplusia ni* and *Spodoptera littoralis* exhibit dual catalytic behaviour. Insect Biochem Mol Biol 36:822–825

Serra M, Piña B, Bujons J et al (2006b) Biosynthesis of 10, 12-dienoic fatty acids by a bifunctional Δ11desaturase in *Spodoptera littoralis*. Insect Biochem Mol Biol 36:634–641

Serra M, Piña B, Abad JL et al (2007) A multifunctional desaturase involved in the biosynthesis of the processionary moth sex pheromone. Proc Natl Acad Sci U S A 104:16444–16449

Shapiro JP, Law JH, Wells MA (1988) Lipid transport in insects. Annu Rev Entomol 33:297–318

Sharma RK, Parameswaran S (2018) Calmodulin-binding proteins: a journey of 40 years. Cell Calcium 75:89–100

Shimizu I, Matsui T, Hasegawa K (1989) Possible involvement of GABAergic neurons in regulation of diapause hormone secretion in the silkworm, *Bombyx mori*. Zool Sci 6:809–812

Shimomura M, Minami H, Suetsugu Y et al (2009) KAIKObase: an integrated silkworm genome database and data mining tool. BMC Genomics 10:486
Shiomi K, Kajiura Z, Nakagaki M, Yamashita O (2003) Baculovirus-mediated efficient gene transfer into the central nervous system of the silkworm, *Bombyx mori*. J Insect Biotech Sericology 72:149–155
Shiomi K, Fujiwara Y, Yasukochi Y et al (2007) The Pitx homeobox gene in *Bombyx mori*: regulation of DH-PBAN neuropeptide hormone gene expression. Mol Cell Neurosci 34:209–218
Shiomi K, Takasu Y, Kunii M et al (2015) Disruption of diapause induction by TALEN-based gene mutagenesis in relation to a unique neuropeptide signaling pathway in Bombyx. Sci Rep:1–10
Shuttleworth TJ, Thompson JL, Mignen O (2004) ARC channels: a novel pathway for receptor-activated calcium entry. Physiology (Bethesda) 19:355–361
Slice LW, Wong HC, Sternini C et al (1994) The conserved NPXnY motif present in the gastrin-releasing peptide receptor is not a general sequestration sequence. J Biol Chem 269:21755–21761
Soroker V, Rafaeli A (1989) In vitro hormonal stimulation of [14C]acetate incorporation by *Heliothis armigera* pheromone glands. Insect Biochemistry 19:1–5
Soroker V, Rafaeli A (1995) Multi-signal transduction of the pheromonotropic response by pheromone gland incubations of *Helicoverpa armigera*. Insect Biochem Mol Biol 25:1–9
Stahl A (2004) A current review of fatty acid transport proteins (SLC27). Pflugers Arch 447:722–727
Steinbrecht RA (1964) Feinstruktur und Histochemie der Sexualduftdrüse des Seidenspinners *Bombyx mori* L. Z Zellforsch Mikrosk Anat 64:227–261
Stern PS, Yu L, Choi M-Y et al (2007) Molecular modeling of the binding of pheromone biosynthesis activating neuropeptide to its receptor. J Insect Physiol 53:803–818
Sztalryd C, Brasaemle DL (2017) The perilipin family of lipid droplet proteins: gatekeepers of intracellular lipolysis. Biochim Biophyis Acta - Mol Cell Biol Lipids 1862:1221–1232
Takeuchi K, Reue K (2009) Biochemistry, physiology, and genetics of GPAT, AGPAT, and lipin enzymes in triglyceride synthesis. Am J Physiol Endocrinol Metab 296:E1195–E1209
Tang JD, Charlton RE, Jurenka RA et al (1989) Regulation of pheromone biosynthesis by a brain hormone in two moth species. Proc Natl Acad Sci U S A 86:1806–1810
Tawata M, Ichikawa T (2001) Circadian firing activities of neurosecretory cells releasing pheromonotropic neuropeptides in the silkmoth, *Bombyx mori*. Zool Sci 18:645–649
Teal PEA, Tumlinson JH, Oberlander H (1989) Neural regulation of sex pheromone biosynthesis in Heliothis moths. Proc Natl Acad Sci U S A 86:2488–2492
Tehan BG, Bortolato A, Blaney FE et al (2014) Unifying family A GPCR theories of activation. Pharmacol Ther 143:51–60
Tips A, Schoofs L, Paemen L et al (1993) Co-localization of locustamyotropin-like and pheromone biosynthesis activating neuropeptide-like immunoreactivity in the central-nervous-system of five insect species. Comp Biochem Physiol A Comp Physiol 106:195–207
Tremblay JJ, Drouin J (1999) Egr-1 is a downstream effector of GnRH and synergizes by direct interaction with Ptx1 and SF-1 to enhance luteinizing hormone beta gene transcription. Mol Cell Biol 19:2567–2576
Trzaskowski B, Latek D, Yuan S et al (2012) Action of molecular switches in GPCRs--theoretical and experimental studies. Curr Med Chem 19:1090–1109
Tsfadia O, Azrielli A, Falach L et al (2008) Pheromone biosynthetic pathways: PBAN-regulated rate-limiting steps and differential expression of desaturase genes in moth species. Insect Biochem Mol Biol 38:552–567
Uehara H, Senoh Y, Yoneda K et al (2011) An FXPRLamide neuropeptide induces seasonal reproductive polyphenism underlying a life-history tradeoff in the tussock moth. PLoS One 6:e24213
Urrutia J, Aguado A, Muguruza-Montero A et al (2019) The crossroad of ion channels and calmodulin in disease. Int J Mol Sci 20:400–422
Van Hiel MB, Van Loy T, Poels J et al (2010) Neuropeptide receptors as possible targets for development of insect pest control agents. Adv Exp Med Biol 692:211–226

Veenstra JA (2000) Mono- and dibasic proteolytic cleavage sites in insect neuroendocrine peptide precursors. Arch Insect Biochem Physiol 43:49–63

Waku Y, Sumimoto K-I (1969) Ultrastructure and secretory mechanism of the alluring gland cell in the silkworm, *Bombyx mori* L. (Lepidoptera: Bombycidae). Appl Entomol Zool 4:135–146

Wang W, Qiao Y, Li Z (2018) New insights into modes of GPCR activation. Trends Pharmacol Sci 39:367–386

Watanabe K (1924) Studies on the voltinism of the silkworm, *Bombyx mori*. Bull Sericult Exp Sta (Tokyo) 6:411–455

Watanabe K, Hull JJ, Niimi T et al (2007) FXPRL-amide peptides induce ecdysteroidogenesis through a G-protein coupled receptor expressed in the prothoracic gland of *Bombyx mori*. Mol Cell Endocrinol 273:51–58

Whorton MR, Jastrzebska B, PS-H P et al (2008) Efficient coupling of transducin to monomeric rhodopsin in a phospholipid bilayer. J Biol Chem 283:4387–4394

Wu SF, Yu HY, Jiang TT et al (2015) Superfamily of genes encoding G protein-coupled receptors in the diamondback moth *Plutella xylostella* (Lepidoptera: Plutellidae). Insect Mol Biol 24:442–453

Xu W-H, Denlinger DL (2003) Molecular characterization of prothoracicotropic hormone and diapause hormone in *Heliothis virescens* during diapause, and a new role for diapause hormone. Insect Mol Biol 12:509–516

Xu W-H, Sato Y, Ikeda M, Yamashita O (1995a) Stage-dependent and temperature-controlled expression of the gene encoding the precursor protein of diapause hormone and pheromone biosynthesis activating neuropeptide in the silkworm, *Bombyx mori*. J Biol Chem 270:3804–3808

Xu W-H, Sato Y, Ikeda M, Yamashita O (1995b) Molecular characterization of the gene encoding the precursor protein of diapause hormone and pheromone biosynthesis activating neuropeptide (DH-PBAN) of the silkworm, *Bombyx mori* and its distribution in some insects. Biochim Biophys Acta Gene Struct Expr 1261:83–89

Yaginuma T, Niimi T (2015) FXPRLamide peptide family. In: handbook of hormones: comparative endocrinology for basic and. Clin Res:395

Yamaoka R, Taniguchi Y, Hayashiya K (1984) Bombykol biosynthesis from deuterium-labeled (*Z*)-11-hexadecenoic acid. Experientia 40:80–81

Yang M, Wang W, Zhong M et al (2002) Lysine 270 in the third intracellular domain of the oxytocin receptor is an important determinant for G alpha(q) coupling specificity. Mol Endocrinol 16:814–823

Yew JY, Chung H (2015) Insect pheromones: an overview of function, form, and discovery. Prog Lipid Res 59:88–105

Yokoyama N, Francke A, Tatsuki S et al (2003) Ultrastructural studies on the pheromone-producing cells in the silkmoth, *Bombyx mori*: formation of cytoplasmic lipid droplets before adult eclosion. Acta Biol Hung 54:299–311

Yoshiga T, Okano K, Mita K et al (2000) cDNA cloning of acyl-CoA desaturase homologs in the silkworm, *Bombyx mori*. Gene 246:339–345

Yoshiga T, Yokoyama N, Imai N et al (2002) cDNA cloning of calcineurin heterosubunits from the pheromone gland of the silkmoth, *Bombyx mori*. Insect Biochem Mol Biol 32:477–486

Ždárek J, Nachman RJ, Hayes TK (1998) Structure-activity relationships of insect neuropeptides of the pyrokinin/PBAN family and their selective action on pupariation in fleshfly (*Neobelleria bullata*) larvae (Diptera:Sarcophagidae). Eur J Entomol 95:9–16

Ždárek J, Verleyen P, Mares M et al (2004) Comparison of the effects of pyrokinins and related peptides identified from arthropods on pupariation behaviour in flesh fly (*Sarcophaga bullata*) larvae (Diptera: Sarcophagidae). J Insect Physiol 50:233–239

Zhang Q, Denlinger DL (2012) Dynamics of diapause hormone and prothoracicotropic hormone transcript expression at diapause termination in pupae of the corn earworm, *Helicoverpa zea*. Peptides 34:120–126

Zhang T-Y, Kang L, Zhang Z-F, Xu W-H (2004) Identification of a POU factor involved in regulating the neuron-specific expression of the gene encoding diapause hormone and pheromone biosynthesis-activating neuropeptide in *Bombyx mori*. Biochem J 380:255–263

Zhang T-Y, Sun J-S, Liu W-Y et al (2005) Structural characterization and transcriptional regulation of the gene encoding diapause hormone and pheromone biosynthesis activating neuropeptide in the cotton bollworm, *Helicoverpa armigera*. Biochim Biophys Acta 1728:44–52
Zhang S, Li X, Bin Z et al (2013) Molecular identification of a pancreatic lipase-like gene involved in sex pheromone biosynthesis of *Bombyx mori*. Insect Sci 21:459–468
Zhang S, Liu X, Zhu B et al (2014) Identification of differentially expressed genes in the pheromone glands of mated and virgin *Bombyx mori* by digital gene expression profiling. PLoS One 9:e111003
Zhang J, Walker WB, Wang G (2015) Pheromone reception in moths: from molecules to behaviors. Prog Mol Biol Trans Sci 130:109–128
Zhao C-H, Li Q, Gao W (2002) Stimulation of sex pheromone production by PBAN-like substance in the pine caterpillar moth, *Dendrolimus punctatus* (Lepidoptera: Lasiocampidae). Arch Insect Biochem Physiol 49:137–148
Zhao W, Li L, Zhang Y et al (2018) Calcineurin is required for male sex pheromone biosynthesis and female acceptance. Insect Mol Biol 72:173–110
Zheng L, Lytle C, Njauw C-N et al (2007) Cloning and characterization of the pheromone biosynthesis activating neuropeptide receptor gene in *Spodoptera littoralis* larvae. Gene 393:20–30
Zmijewski MA, Slominski AT (2009) CRF1 receptor splicing in epidermal keratinocytes: potential biological role and environmental regulations. J Cell Physiol 218:593–602

Chapter 7
Molecular Bases for the Biosynthesis of Species-Specific Sex Pheromones in the Genus *Ostrinia* (Lepidoptera: Crambidae)

Yukio Ishikawa and Takeshi Fujii

Abstract The genus *Ostrinia* (Lepidoptera; Crambidae) consists of 21 species worldwide, 9 of which inhabit Japan. These closely related species, which utilize different host plants, provide an excellent model for studies on the evolution of sex pheromone communication systems as well as speciation. In this chapter, we review the progress of research on the sex pheromone communication system of *Ostrinia* inhabiting Japan. We focus on the molecular bases for the biosynthesis of sex pheromone components, differential production of which is responsible for the divergence of sex pheromones.

Keywords *Ostrinia* · Biosynthesis · Sex pheromone · Desaturase · Reductase · Limited β-oxidation · Acetyltransferase

7.1 Introduction

The genus *Ostrinia* (Lepidoptera: Crambidae) consists of 21 species worldwide, 9 of which inhabit Japan (Mutuura and Munroe 1970; Ohno et al. 2003, 2006). This genus has long attracted the attention of entomologists because it includes two serious pests of maize (*Zea mays*), the European corn borer (ECB; *O. nubilalis*), which is distributed in Europe, Russia, and North America, and the Asian corn borer (ACB; *O. furnacalis*), which is distributed in Asia. Although ECB and ACB are known as pests of maize, which belongs to Poaceae, both species are highly polyphagous, feeding on numerous plants belonging to a wide range of plant families such as

Y. Ishikawa (✉) · T. Fujii
Graduate School of Agricultural and Life Sciences, The University of Tokyo, Tokyo, Japan
e-mail: ayucky@mail.ecc.u-tokyo.ac.jp

Y. Ishikawa (ed.), *Insect Sex Pheromone Research and Beyond*, Entomology Monographs, https://doi.org/10.1007/978-981-15-3082-1_7

Asteraceae, Polygonaceae, Moraceae, Fabaceae, and Solanaceae (Ishikawa et al. 1999b). Of note, *Ostrinia* species other than ECB and ACB do not feed on maize but on plants mainly belonging to Polygonaceae and Asteraceae (Mutuura and Munroe 1970; Ishikawa et al. 1999b), suggesting ECB and ACB to be exceptional *Ostrinia* species in that they can feed on maize. Regarding the unpalatability of maize, it is attributable, at least in part, to a toxic allelochemical in the plant, 2,4-dihydroxy-7-methoxy-1,4-benzoxazin-3-one (DIMBOA); animals that feed on maize must cope with this compound (Kojima et al. 2010; Phuong et al. 2016, 2018). In this regard, as ECB and ACB are distantly related within the genus (Kim et al. 1999), they must have independently adapted to maize during the course of evolution. Studies on the changes in the use of host plants in *Ostrinia*, the use of maize in particular, may provide insights into the relationship between the shift of host plants and speciation in the genus.

Speciation of lepidopteran insects is often associated with the changes in sex pheromone communication systems, reflecting their essential role in premating reproductive isolation (Groot et al. 2016). In the sex pheromone communication system of moths, the signal detection in the receiver, usually males, is tuned to the highly canalized signal and, thus, a slight change in the signal may result in the failure of its recognition. Therefore, the sex pheromone communication system is considered to be under strong stabilizing selection. On the other hand, the presence of closely related species, which originated from the common ancestor, suggests that divergence of the pheromone communication system has certainly occurred despite stabilizing selection. The process of this divergence remains largely enigmatic. The *Ostrinia* species, a group of closely related species, provides an excellent model for studies on the evolution of sex pheromone communication systems as well as speciation. For this reason, the sex pheromone communication systems of *Ostrinia* have been intensively studied, and accordingly, a cornucopia of knowledge is available (Ishikawa et al. 1999b; Lassance 2010, 2016; Tabata and Ishikawa 2016). In this chapter, we review the progress of research on the sex pheromone communication system of *Ostrinia* inhabiting Japan. We focus on the molecular bases for the biosynthesis of sex pheromone components, differential production of which is responsible for the divergence of sex pheromones.

7.2 *Ostrinia* spp. in Japan

The nine *Ostrinia* species inhabiting Japan are shown in Fig. 7.1, along with their representative host plant families and sex pheromone components. *Ostrinia* spp. in Japan are largely classified into two groups, Group II (three species) and Group III (six species), based on the morphology of male genitalia (Mutuura and Munroe 1970). The single species belonging to Group I, *O. penitalis*, is not distributed in Japan. Group II is further divided into two subgroups, one containing a single species, *O. palustralis*, and the other containing two similar species, *O. latipennis* and *O. ovalipennis*. Group III includes six species, which are so similar to each other

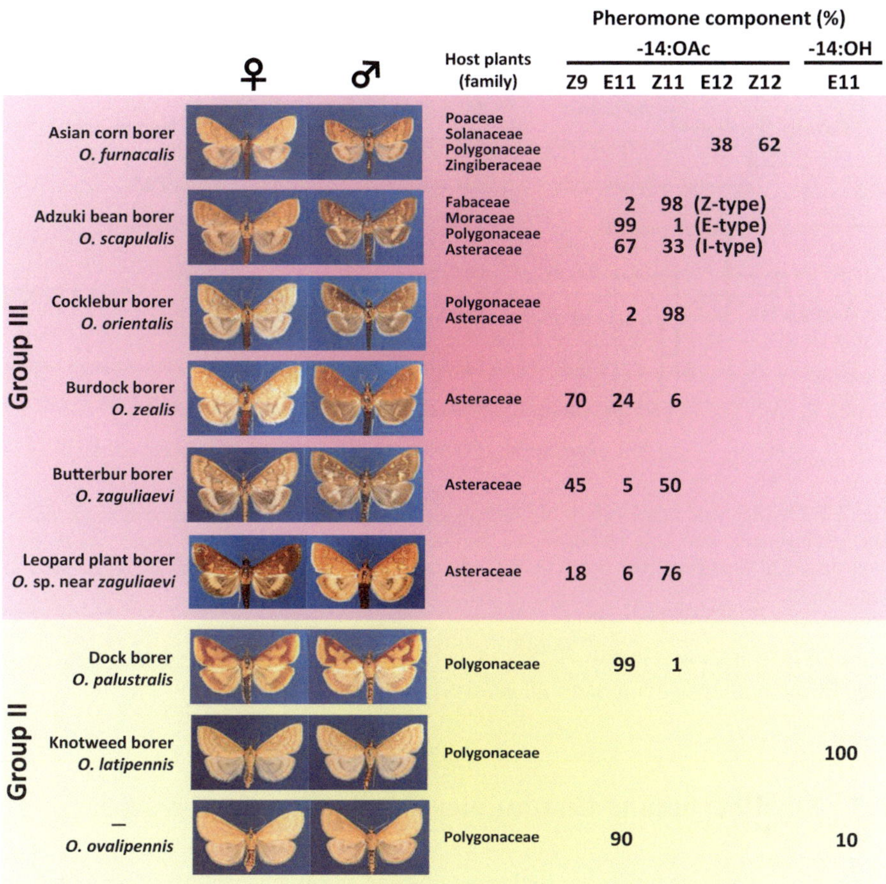

Fig. 7.1 Host plants and sex pheromones of *Ostrinia* species inhabiting Japan (left: female; right: male). Based on the morphology of genitalia, *Ostrinia* species in Japan are classified into Group II and Group III. Only representative host plant families are presented for two polyphagous species, *O. furnacalis* and *O. scapulalis*. The composition of the sex pheromones is expressed by percentages (Huang et al. 1997, 1998a, b, c, 2002; Ishikawa et al. 1999a, b; Takanashi et al. 2000, 2005; Tabata and Ishikawa 2005; Tabata et al. 2003, 2006, 2008; Fu et al. 2004, 2005a, b)

that identification of species in this group based only on the external morphology is difficult. This large classification of *Ostrinia* is supported by the molecular phylogenetic studies using mitochondrial CO II gene sequences (Fig. 7.2; Kim et al. 1999). It should be noted that the two subgroups in Group II, *O. palustralis* and *O. latipennis*/*O. ovalipennis*, are not very closely related. Regarding the relationship between ACB and ECB, the two maize feeders, ACB is distantly related not only to ECB but also to other members of the Group III (Fig. 7.2). Although ECB is geographically isolated from Japanese *Ostrinia*, this species forms a compact clade together with *O. scapulalis* and *O. orientalis*, nonmaize feeders. *Ostrinia* sp. near

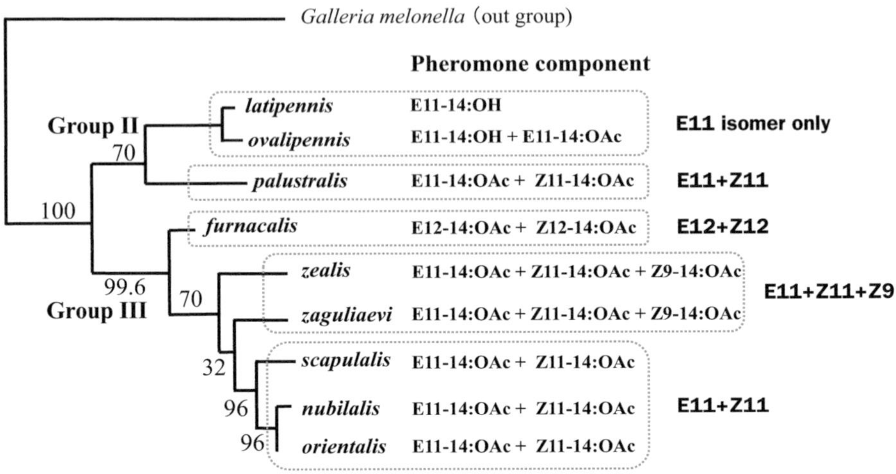

Fig. 7.2 Phylogenetic relationships of *Ostrinia* species along with their pheromone components. The phylogenetic tree was reconstructed by the neighbor-joining method using CO-II gene sequences. The European corn borer, *O. nubilalis*, which is not distributed in Japan, is included for comparison

zaguliaevi (not shown in Fig. 7.2) is closely related to *O. zaguliaevi*, and these two species form a loose clade with *O. zealis* (Fig. 7.2).

7.3 Sex Pheromone Communication Systems in *Ostrinia*

Six compounds have been reported to be utilized as sex pheromone components in *Ostrinia* (Fig. 7.1). They are (*Z*)-9-tetradecenyl acetate (Z9-14:OAc), (*E*)-11-tetradecenyl acetate (E11-14:OAc), (*Z*)-11-tetradecenyl acetate (Z11-14:OAc), (*E*)-12-tetradecenyl acetate (E12-14:OAc), (*Z*)-11-tetradecenyl acetate (Z12-14:OAc), and (*E*)-11-tetradecenol (E11-14:OH). The following points are readily noticeable: (i) all compounds, except for one alcohol component, E11-14:OH, are acetate esters of monoene fatty alcohols with 14-carbon-length chains. (ii) E11-14:OAc and Z11-14:OAc are most commonly utilized as pheromone components among *Ostrinia*. (iii) E11-14:OH is specifically utilized by the *O. latipennis*/*O. ovalipennis* subgroup in Group II. (iv) *O. latipennis*/*O. ovalipennis* use the *E* isomer only, whereas all other species, including *O. palustralis*, which belongs to the same group II, utilize both *E* and *Z* isomers. (v) ACB is exceptional in that only this species utilizes the positional isomers E12-14:OAc and Z12-14:OAc as their pheromone components. (vi) *O. zealis*, *O. zaguliaevi*, and *Ostrinia* sp. near *zaguliaevi* utilize Z9-14:OAc, in addition to E11-14:OAc and Z11-14:OAc. (vii) In *O. scapulalis*, as is known for ECB, three phenotypes are identified in terms of the blend of two pheromone components, E11-14:OAc and Z11-14:OAc, that is, E-type

(Z:E ≈ 1:99), Z-type (Z:E ≈ 97:3), and I-type (Z:E ≈ 36:64) (Fig. 7.1). One possible scenario for the evolution of sex pheromone communication systems in *Ostrinia* is depicted by juxtaposing the CO II phylogenetic tree of this genus with the sex pheromone components of each species (Fig. 7.2).

7.4 Sex Pheromone Biosynthetic Pathways

The biosynthetic pathways for the six pheromone components in *Ostrinia*, depicted based on the knowledge accumulated to date, are shown in Fig. 7.3. All steps of sex pheromone biosynthesis are considered to take place in the PG, which is a modified intersegmental membrane between the 8th and 9th abdominal segments (Fukuzawa et al. 2006; Ma and Roelofs 2002). Palmitoyl-CoA (16:Acyl; CoA is omitted for clarity), which is de novo synthesized from acetyl-CoA via common fatty acid biosynthetic pathways, is the starting material for the pheromone components. Enzymes

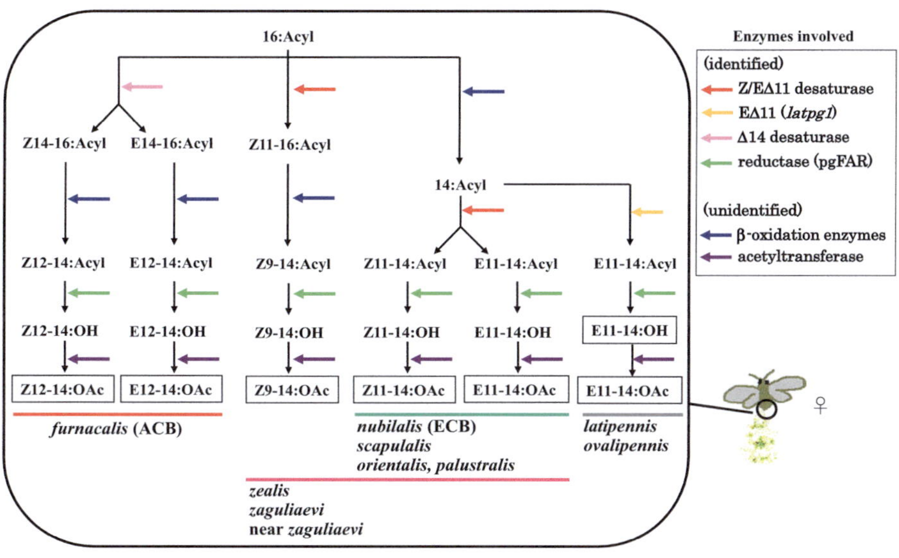

Fig. 7.3 Biosynthetic pathways of six sex pheromone components in *Ostrinia*. Sex pheromone components are biosynthesized in the pheromone gland of a female, which is a modified intersegmental membrane between the 8th and 9th abdominal segments. All components are biosynthesized from palmitoyl-CoA (16:Acyl) by the function of desaturases, reductases, limited β-oxidation enzymes, and acetyltransferase.
The chemical formulae are symbolized as follows. Z and E represent the (*Z*) and (*E*) configurations of a double bond, respectively. The numbers before the hyphen and colon indicate the position of a double bond and the carbon chain length, respectively. OH, OAc, and Acyl stand for primary alcohol, acetate of alcohol, and acyl moiety, respectively. Fatty acids are usually conjugated with Coenzyme A (CoA), but this conjugation is omitted in this figure. (Drawn based on Roelofs et al. (2002), Xue et al. (2007), Sakai et al. (2009), and Fujii et al. (2011))

involved in the biosynthesis are fatty acyl-CoA desaturase (hereafter, "desaturase"), limited β-oxidation enzymes, fatty acyl-CoA reductase ("reductase" or pgFAR), and acetyltransferase. Desaturase, which introduces a double bond into fatty acyl-CoA, and reductase, which converts fatty acyl pheromone precursors to corresponding alcohols, have been identified at the molecular level. Both desaturases and reductases form multiple gene families, members of which exhibit different tissue expression profiles and different substrate and product specificities. Molecular entities of limited β-oxidation enzymes, a group of enzymes that shorten fatty acyl-CoA by 2 carbon atoms for each cycle of function, and acetyltransferase, an enzyme that acetylates fatty alcohol, have not yet been clarified in *Ostrinia* or any other lepidopteran species.

7.4.1 Moth Desaturases

As shown in Fig. 7.3, desaturases function in the early steps of sex pheromone biosynthesis and play an important role in selective production of pheromone components. Molecular studies on desaturases involved in the biosynthesis of sex pheromone components were started in as early as the 1980s, and the knowledge acquired to date is abundant (Wolf and Roelofs 1986; Rodriguez et al. 1992; Knipple et al. 2002). Here, the classification of insect desaturases is first briefly overviewed.

Insect Desaturases Molecular phylogenetic studies of insect desaturases revealed the presence of three clades, which introduce a double bond at positions 9 (Δ9-desturase), 11 (Δ11-desaturase), and 14 (Δ14-desaturase), respectively (Fig. 7.4). Desaturases often exhibit a preference for the carbon chain length of the substrate (Moto et al. 2004; Liu et al. 2004; Fujii et al. 2011). In particular, the Δ9 clade is subdivided into those preferring palmitoyl-CoA to stearoyl-CoA (C16 > C18), those with reversed preference (C18 > C16), and those take fatty acyl-CoA with a carbon chain length between 14 and 26 (C14–C26) as a substrate (Fig. 7.4). Orthologues of Δ11 and Δ14 desaturases have not been found in organisms other than moths, suggesting that desaturases in these groups have evolved specifically to produce sex pheromone components (Roelofs and Rooney 2003). In general, genes encoding desaturases involved in pheromone production are expressed specifically in the PG (Moto et al. 2004; Liu et al. 2004; Fujii et al. 2011). Although the Δ11-desaturase clade also includes Δ3-, Δ6-, and Δ10-desaturases, reports on the functional analyses of these genes are limited (Wang et al. 2010; Foster and Roelofs 1988).

Desaturases Functioning in the PG of ***Ostrinia*** Desaturases functioning in the PG of ECB and ACB exhibit different regiospecificity, catalyzing the introduction of a double bond at positions Δ11 and Δ14, respectively (Fig. 7.3). Desaturase genes functioning in the PG of ECB and ACB, *Z/EΔ11* and *Z/EΔ14*, respectively, were first identified and functionally characterized by Roelofs et al. (2002). In the

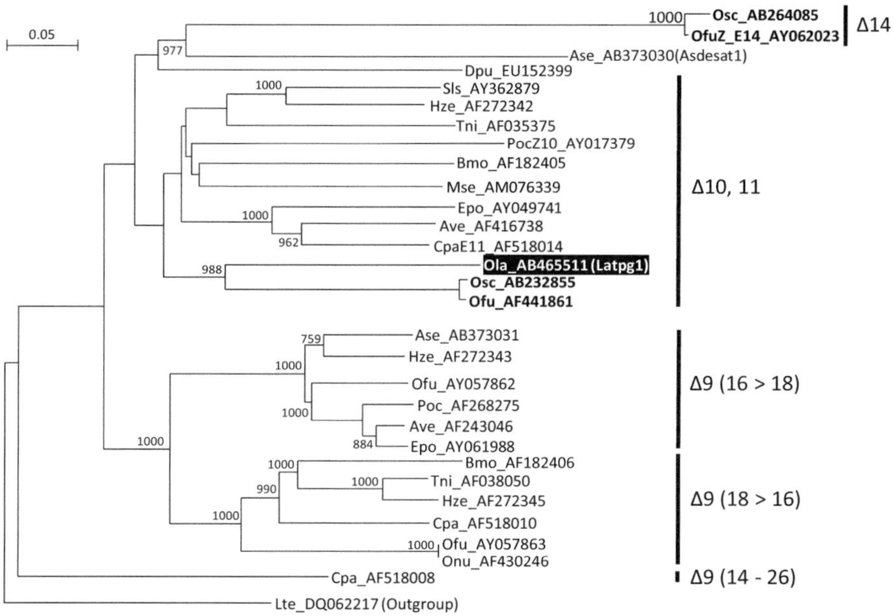

Fig. 7.4 Molecular phylogenetic tree of lepidopteran desaturases. Desaturases expressed in the pheromone gland of moths are presented along with their accession numbers. Desaturases with similar catalytic activity form clades. Functions of some sequences in clade Δ10,11 need to be determined by functional assays of heterologously expressed proteins (Liu et al. 2004)

functional assays using heterologous expression systems, Z/EΔ11 protein produced mainly Z11-16:Acyl, along with a mixture of Z11- and E11-14:Acyl. In contrast, Z/EΔ14 produced a mixture of Z/E14-16:Acyl, with no Z11-16:Acyl. Of note, transcripts almost identical to *Z/EΔ11* were detected in the PG of ACB, and conversely, transcripts almost identical to *Z/EΔ14* were detected in the PG of ECB. A desaturase gene functioning in the PG of *O. latipennis*, the pheromone of which is characterized by the absence of the Z isomer, was subsequently identified and functionally characterized by Fujii et al. (2011). The gene, *latpg1* (*EΔ11*), specifically introduced a double bond of E configuration at carbon position 11 in 14:Acyl. Unlike Z/EΔ11, LATPG1 did not take 16:Acyl as a substrate and, thus, did not produce Z11-16:Acyl. Furthermore, when it took 14:Acyl as the substrate, it produced E11-14:Acyl only; no trace of Z11-14:Acyl was detected. Phylogenetic analysis revealed that LATPG1 is not closely related to Z/ΔE11 (Fig. 7.4). The relationship between *latpg1*, *Z/EΔ11*, and other Δ11-desaturase genes in *Ostrinia* will be discussed later in Sect. 7.6.

7.4.2 Selective Production of Pheromone Components

In this section, molecular bases for the selective production of sex pheromone components in *Ostrinia* are discussed from the standpoint of why pheromone components utilized by other congeners are *not* produced in the species in question. For the convenience of explanation, *Ostrinia* species are grouped into four groups that use the same set of pheromone components: the ACB group (ACB only), ECB group (ECB, *O. scapulalis*, *O. orientalis*, and *O. palustralis*), *O. zealis* group (*O. zealis*, *O. zaguliaevi*, and *Ostrinia* sp. near *zaguliaevi*), and *O. latipennis* group (*O. latipennis* and *O. ovalipennis*). As explained below, selective production of pheromone components depends on the order of enzymes that work on the substrate and the substrate and product specificities of enzymes involved.

In *Ostrinia*, crossing between some combinations of Group-III species is possible and viable offspring are obtained. Postzygotic isolation has not yet been established between these species, probably due to the short time after the split of Group-III species, which is estimated at <0.5 million years ago. Analyses of the sex pheromone components and their precursors in the PG of hybrids between two species (or strains of a species) provided important information about the genetic basis of the variation in sex pheromone components (Sakai et al. 2009; Tabata and Ishikawa 2005, 2011).

Nonproduction of E/Z11-14:OAc in ACB and E/Z12-14:OAc in the ECB Group Information about the genetic basis that controls the nonproduction of E/Z11-14:OAc in ACB and E/Z12-14:OAc in the ECB group was obtained through analyses of pheromone components and their precursors in the hybrid between ACB and *O. scapulalis* (Sakai et al. 2009). The F1 hybrid produced both parents' sex pheromone components (Z/E11-14:OAc and Z/E12-14:OAc). Although the two species have both Δ11-desaturase (*Z/EΔ11*) and Δ14-desaturase (*Z/EΔ14*) genes, transcription of *Z/EΔ14* was only observed in ACB, as was transcription of *Z/EΔ11* in *O. scapulalis* (Fig. 7.5; Fig. 7.6a, b). In F1, both genes were transcribed into mRNA. The production/nonproduction of Z/E11-14:OAc and Z/E12-14:OAc in F1, F2, and backcross progenies were explained by an autosomal transcriptional factor that specifically controls transcription of either *Z/EΔ11* or *Z/EΔ14* (Fig. 7.5; Sakai et al. 2009). Here, a question remains: why Z9-14:OAc is not produced in the ECB group because Δ11-desaturase in *O. nubilalis* is known to catalyze desaturation of 16:Acyl to generate Z11-16:Acyl, a precursor of Z9-14:OA (Fig. 7.6b).

Nonproduction of Z9-14:OAc in Species Other Than *O. zealis* Group Crossing experiments between *O. zealis* and *O. scapulalis* (E type and Z type; see Fig. 7.1) were performed to infer the genetic basis for the selective production of E11-14:OAc (E11), Z11-14:OAc (Z11), and Z9-14:OAc (Z9) in these species (Fig. 7.7; Tabata and Ishikawa 2005). The sex pheromone was analyzed based on the proportions of Z9 and E11 in the total (Z9 + E11 + Z11). E- and Z-type *O. scapulalis* used in this study produced a pheromone with 100% E11 and 0–10% E11, respectively, with no Z9 (Fig. 7.7a). *O. zealis* produced a three-component pheromone: the % Z9 ranged

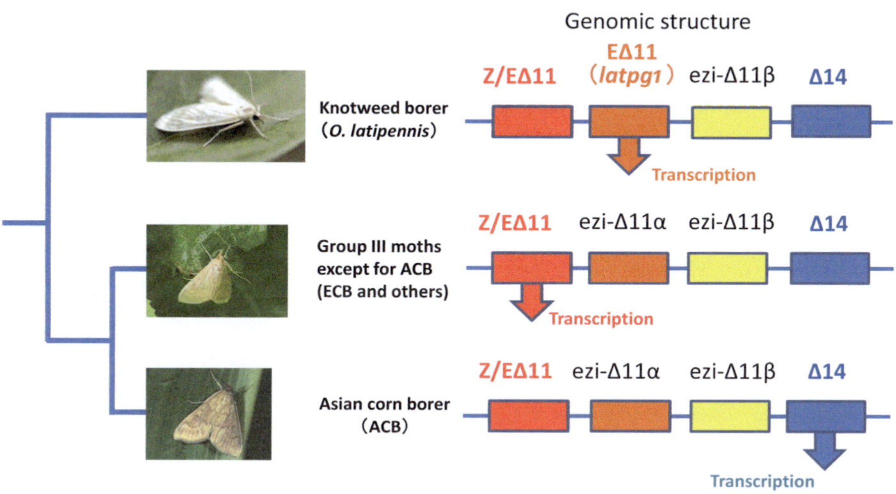

Fig. 7.5 Schematic representation of the selective use of desaturase genes in *Ostrinia* moths. The presence of a mechanism that permits expression of single desaturase gene in a species is assumed. (Redrawn based on Xue et al. (2007) and Fujii et al. (2011, 2015))

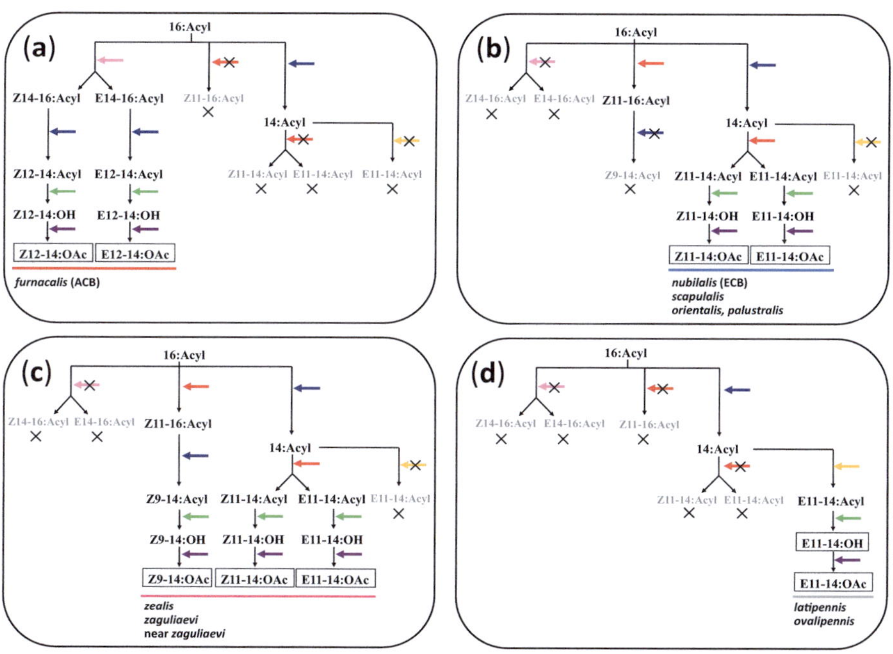

Fig. 7.6 Explanation for the production of a particular set of pheromone components in the (**a**) ACB group (*O. furnacalis* only), (**b**) ECB group (*O. nubilalis*, *O. scapulalis*, *O. orientalis*, and *O. palustralis*), (**c**) *O. zealis* group (*O. zealis*, *O. zaguliaevi*, and *Ostrinia* sp. near *zaguliaevi*), and (**d**) *O. latipennis* group (*O. latipennis* and *O. ovalipennis*)

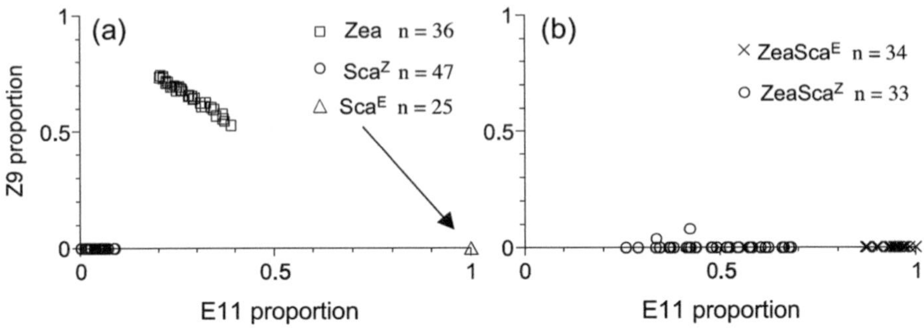

Fig. 7.7 Distribution of sex pheromone blends in the female progeny of *O. zealis*, E-type *O. scapulalis*, and Z-type *O. scapulalis* (**a**), and their F1 hybrids (**b**). Zea: *O. zealis*, Sca^E: E-type *O. scapulalis*, Sca^Z: Z-type *O. scapulalis*. E11: (*E*)-11-tetradecenyl acetate, Z11: (*Z*)-11-tetradecenyl acetate, Z9: (*Z*)-9-tetradecenyl acetate. Abscissa: E11/(Z9 + E11 + Z11), Ordinate: Z9/(Z9 + E11 + Z11)

from 50 to 75 and % E11 from 20 to 40 (Fig. 7.7a). All hybrid F1 females produced sex pheromones lacking Z9 (Fig. 7.7b), suggesting that desaturation of 14:Acyl by Δ9-desaturase is not involved in the biosynthesis of Z9 (Z9 should be present in the pheromone of F1 hybrids if Δ9-desaturase is involved). Here, it should be noted that the proportions of Z9 and E11 in *O. zealis* pheromone were linearly negatively correlated (Fig. 7.7a; r = 0.99). Such a linear relationship is not expected if two desaturase enzymes, one (Δ9-desaturase) for the production of Z9 and the other (Δ11-desaturase) for E11 and Z11, are functioning independently. As Δ11-desaturase of *O. nubilalis* is known to produce Z11-16:Acyl as a "major" by-product of a mixture of Z11- and E11-14:Acyls (Roelofs et al. 2002), we inferred that Z9 in the *O. zealis* group is produced via chain shortening of Z11-16:Acyl, as depicted in Fig. 7.6c. In this scheme, production of Z9 in the *O. zealis* group is assumed to be controlled by an unidentified autosomal recessive gene, the dominant allele of which is involved in the blockage of the chain shortening of Z11-16:Acyl to Z9-14:Acyl in the ECB group (Fig. 7.6b). Although not presented here, the results of the analyses of pheromones of backcross and F2 progenies were consistent with this hypothesis (Tabata and Ishikawa 2005).

Nonproduction of Z Isomer in ***O. latipennis*** **Group** Sex pheromones of *O. latipennis* and *O. ovalipennis* are characterized by the use of only E isomers (Figs. 7.1 and 7.2). Fujii et al. (2011) found that Δ11-desaturase in *O. latipennis*, *latpg1* (*EΔ11*), is specifically expressed in the PG of *O. latipennis*, but not in species of the ECB group or *O. zealis* group (Fig. 7.5). Therefore, nonproduction of the Z isomer in the *O. latipennis* group is fully explainable by the strict product specificity of LATPG1 (EΔ11), which produces only E11-14:Acyl from its specific substrate, 14:Acyl (Fig. 7.6d).

7.5 Production of Species-Specific Blends of Pheromone Components

Thus far, we have described how sex pheromone components used in a certain species are selectively produced. In *Ostrinia* species, however, not only the combination of compounds but also their blend ratio is essential for conferring species specificity to the sex pheromone. For example, although all species in the *O. zealis* group use the same set of compounds, E11-14:OAc, Z11-14:OAc, and Z9-14:OAc, their blend ratios differ species specifically (Fig. 7.1; Ishikawa et al. 1999a; Tabata et al. 2006, 2008). It is necessary to understand how the control of the species-specific blend ratio of pheromone components is achieved. Here, crossing experiments again provided important information on the control of the blend ratio (Tabata and Ishikawa 2011). The differences between *O. zealis* and *O. zaguliaevi* blends, and *O. zealis* and *Ostrinia* sp. near *zaguliaevi* blends, were explainable by a single autosomal locus with three alleles. Furthermore, analyses of the fatty acyl precursors of pheromone components suggested that these genetic factors function in either of the last two steps of pheromone biosynthesis, reduction or acetylation. As acetylation has long been known to exhibit low substrate specificity (Zhao et al. 1995; Zhu et al. 1996), the genetic factor is considered to be involved in reduction.

One of the conspicuous features of the *Ostrinia* sex pheromone is the presence of extreme intraspecific polymorphism in two species, ECB and *O. scapulalis* (Fig. 7.1; Lassance 2016). It has long been known that these variations are controlled by an autosomal locus with two alleles, A^E and A^Z (Takanashi et al. 2005). In the biosynthetic pathways, A^E and A^Z have been suggested to be involved in the reduction step, which follows desaturation, because the E/Z ratio of the desaturase products is approximately 7:3 irrespective of the race (Roelofs et al. 1987; Wolf and Roelofs 1987). However, the molecular entity of reductase remained unclarified for a long time not only in *Ostrinia* but also in all moths. The reductase involved in sex pheromone production was first identified from *B. mori* by Moto et al. (2003). They isolated a cDNA clone encoding a protein homologous to a FAR from the jojoba plant, which was specifically expressed in the PG of *B. mori*. Functional expression of this gene confirmed the FAR activity with a distinct substrate specificity to the pheromone precursor, (*E,Z*)-10,12-hexadecadienoic acid (Moto et al. 2003). Identification of the FAR involved in the biosynthesis of sex pheromones in other moths has been difficult, but by screening of 13 FAR candidates based on the PG-specific expression, Antony et al. (2009) succeeded in identifying FAR in *O. scapulalis*. Subsequently, Lassance et al. (2010) cloned and characterized the orthologues of this gene, *pgFAR-Z* and *pgFAR-E*, from the Z and E strains of *O. nubilalis*. pgFAR-Z and pgFAR-E, which had 7.5% amino-acid divergence (3.8% nucleotide divergence), exhibited strict substrate specificity to Z11-14:Acyl and E11-14:Acyl, respectively (Lassance et al. 2010). Thus, the difference in the pheromone blend between the Z and E races of *O. nubilalis* is clearly explainable by the variation in the substrate specificity of pgFAR. Subsequently, functional analyses of pgFARs isolated from seven additional *Ostrinia* species have demonstrated that the blend ratio of the sex pheromone in these *Ostrinia* species is explainable by the substrate specificity of the pgFAR and the composition of fatty acyl pheromone precursors in the PG (Lassance et al. 2013).

To understand the mechanism that has caused changes in the substrate specificity of pgFAR, it is essential to clarify which mutation in the amino acid sequence of pgFARs is responsible for the variation in the activity of pgFAR. Lassance et al. (2013) constructed several single-codon mutants and examined their substrate specificities. Among these mutants, they found that mutation of amino acid no. 453 of pgFAR affects its substrate specificity to Z11-14:Acyl and Z12-14:Acyl (Fig. 7.8). A single-codon mutation from ttt (phenylalanine) to tgt (cysteine) in the pgFAR of *O. furnacalis* altered its pattern of substrate specificity similar to that of *O. nubilalis*. Conversely, a single-codon change from tgt to ttt in the pgFAR of *O. nubilalis* altered its pattern similar to that of *O. furnacalis*. Here, it is notable that pgFAR of *O. nubilalis* exhibits some activity toward E12-14:Acyl, and conversely, pgFAR of *O. furnacalis* has substantial preference to Z11-14:Acyl. As previously explained, the absence of Z/E11-14:OAc in *O. furnacalis* and Z/E12-14:OAc in *O. nubilalis* is explained by the absence of relevant precursor fatty acyls.

7.6 Suggestions for Future Studies

Lastly, we would like to point out what is remaining to be investigated for a complete understanding of the sex pheromone biosynthetic pathways in *Ostrinia*.

Acetyltransferase Although the involvement of acetyltransferase in the biosynthesis of acetate pheromones has long been recognized, even now its molecular entity remains unknown in all moth species. DNA sequences of the acetyltransfer-

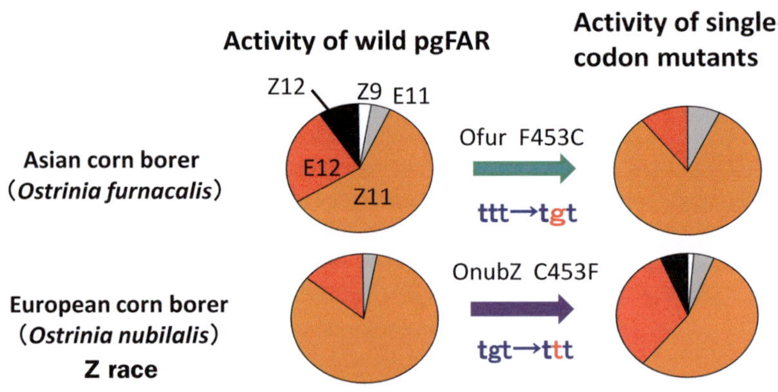

Fig. 7.8 Functional expression assays of parental and single-codon mutant pgFARs. Ratios of fatty alcohol products for wild-type (WT) and mutant pgFARs. Functional impact of the reverse mutation C453F on *O. nubilalis* ZpgFAR activity compared with the *O. furnacalis* WT. The pie charts represent the relative abundances of alcohols (n = 4)

ases of plants and fungi are available in public databases; however, no homologous sequence was found in the transcripts of the PG of *Ostrinia* (Ishikawa, unpublished data). Furthermore, although analysis of the *Agrotis segetum* PG transcriptome was able to nominate as many as 34 putative acetyltransferase sequences, none of them expressed in the heterologous expression system exhibited acetyltransferase activity (Ding and Löfstedt 2015). The low substrate specificity of acetyltransferase was demonstrated by tracing deuterium-labeled fatty alcohols topically applied to the PG of *O. furnacalis* and *O. nubilalis* (Zhao et al. 1995; Zhu et al. 1996); however, information on other biochemical properties of this enzyme is limited. Characterization of the biochemical properties of acetyltransferase protein may help to identify the gene encoding this enzyme. In *Ostrinia*, the control of the expression of acetyltransferase in two closely related species, *O. latipennis* and *O. ovalipennis*, the former of which does not produce acetate pheromone component (Fig. 7.1), is of particular interest.

Limited β-Oxidation Enzymes Limited β-oxidation enzymes play an essential role in adjusting the aliphatic carbon chain length of the pheromone components; however, the molecular entities of enzymes involved have not been clarified in any moth species. Limited β-oxidation is considered to be performed by at least four enzymes, acyl-CoA oxidases, enoyl-CoA hydratase, L-3-hydroxyacyl-CoA dehydrogenase, and thiolase (Ding and Löfstedt 2015). Ding and Löfstedt (2015) were able to find candidate genes for each enzyme, but their functions remain to be verified. As it is impractical to screen candidate genes by simultaneously expressing a set of four enzymes in a heterologous expression system, knockout of genes by gene editing or knockdown by RNAi may be a more practical approach.

Chain-shortening steps in the biosynthetic pathway were demonstrated to be affected in a pheromone mutant strain of *Trichoplusia ni* (Jurenka et al. 1994), and may be important in causing divergence in the pheromone composition. In *Ostrinia*, an autosomal recessive gene of unknown function, not mutations in limited β-oxidation enzymes themselves, is responsible for the control of the chain shortening of Z11-16: Acyl (Fig. 7.6b). Identification of the genes encoding limited β-oxidation enzymes will help to understand the mode of action of the recessive gene in the control of chain shortening.

Evolution of Δ11-Desaturase Genes We demonstrated that production of the E-only pheromone in *O. latipennis* is due to the product specificity of the desaturase (LATPG1 = EΔ11) functioning in this species (Fig. 7.6d; Fujii et al. 2011). Molecular phylogenetic analysis revealed that *latpg1* is an orthologue of the retroposon(ezi)-fused, nonfunctional Δ11-desaturase gene *ezi-Δ11a* in the genomes of *O. nubilalis* and *O. furnacalis*, which was previously reported by Xue et al. (2007). Exploration of the genome of *O. latipennis* revealed two more Δ11-desaturase genes, *latpg2* and *latpg3*, which were orthologous to *ezi-Δ11b* and *Z/EΔ11*, respectively (Fujii et al. 2015; Xue et al. 2007). We found that an *ezi*-like element was integrated in intron 1 of *latpg1*, suggesting that the nonfunctionality of *ezi*-inserted desaturase genes in *O. nubilalis* and *O. furnacalis* is not a direct consequence of the insertion of *ezi* (Fujii et al. 2015).

Another notable finding is the PG-specific expression of a Δ11-desaturase-like gene (*Asdesat1*) in *Ascotis selenaria* (Geometridae), which exclusively utilizes type-II sex pheromones; therefore, no desaturation step is involved in its pheromone biosynthetic pathway (Fujii et al. 2013). The presumed transmembrane domains were found to be degenerated in Asdesat1. Phylogenetic analysis demonstrated that Asdesat1 anciently diverged from the lineage of Δ11-desaturases. These results suggest that an ancestral Δ11-desaturase became dysfunctional in *A. selenaria* after a shift in the use of sex pheromone from type I to type II. Further studies on Δ11-desaturase-like genes in moths that utilize type-II sex pheromones may provide insights into the evolution of desaturases and sex pheromone communication systems in moths.

Alternative Transcription of Desaturase Genes Moth sex-pheromone desaturase genes are considered to be evolving under a birth-and-death process, in which new genes are created through gene duplication (Nei and Rooney 2005; Roelofs and Rooney 2003). Genomic analyses revealed that at least three Δ11-desaturase genes and a Δ14-desaturase gene are commonly present in the genome of *Ostrinia* species (Fig. 7.5; Fujii et al. 2011, 2015; Xue et al. 2007). Through studies on the expression of desaturase genes in the PG of *Ostrinia*, it has become clear that only one desaturase gene is exclusively transcribed in the PG of a species (Fig. 7.5). The presence of a transcription factor is assumed, but the mechanism that enables PG- and species-specific transcription of a specific desaturase gene is unknown at present. Detailed interspecific comparisons of the promoter regions of the Δ11- and Δ14-desaturase genes may provide a clue to better understand the transcriptional control of these genes.

Acknowledgments We are grateful to our numerous collaborators, without whose help it would have been impossible to continue studies on a single group of moths, *Ostrinia*, for more than 30 years.

References

Antony B, Fujii T, Moto K, Matsumoto S, Fukuzawa M, Nakano R, Tatsuki S, Ishikawa Y (2009) Pheromone-gland-specific fatty-acyl reductase in the adzuki bean borer, *Ostrinia scapulalis* (Lepidoptera: Crambidae). Insect Biochem Mol Biol 39:90–95

Ding B, Löfstedt C (2015) Analysis of the *Agrotis segetum* pheromone gland transcriptome in the light of sex pheromone biosynthesis. BMC Genomics 16:711. https://doi.org/10.1186/s12864-015-1909-2

Foster SP, Roelofs WL (1988) Sex pheromone biosynthesis in the leafroller moth *Planotortix excessana* by Δ10 desaturation. Arch Insect Biochem Physiol 8:1–9

Fu X, Tabata J, Huang Y, Takanashi T, Ohno S, Honda H, Tatsuki S, Ishikawa Y (2004) Female sex pheromone of *Ostrinia orientalis* – throwing a light on the relationships between *O. orientalis* and the European corn borer *O. nubilalis*. Chemoecology 14:175–180

Fu X, Fukuzawa M, Tabata J, Tatsuki S, Ishikawa Y (2005a) Sex pheromone biosynthesis in *Ostrinia zaguliaevi*, a congener of the European corn borer moth *O. nubilalis*. Insect Biochem Mol Biol 35:621–626

Fu X, Tatsuki S, Hoshizaki S, Ishikawa Y (2005b) Study of the genetics of female sex pheromone production and male behavioral response in a moth, *Ostrinia orientalis*. Entomol Sci 8:363–369

Fujii T, Ito K, Tatematsu M, Shimada T, Katsuma S, Ishikawa Y (2011) A sex pheromone desaturase functioning in a primitive *Ostrinia* moth is cryptically conserved in congeners' genomes. Proc Natl Acad Sci U S A 108:7102–7106

Fujii T, Suzuki MG, Katsuma S, Ito K, Rong Y, Matsumoto S, Ando T, Ishikawa Y (2013) Discovery of a disused desaturase gene from the pheromone gland of the moth *Ascotis selenaria*, which secretes an epoxyalkenyl sex pheromone. Biochem Biophys Res Comm 441:849–855

Fujii T, Yasukochi Y, Rong Y, Matsuo T, Ishikawa Y (2015) Multiple Δ11-desaturase genes selectively used for sex pheromone biosynthesis are conserved in *Ostrinia* moth genomes. Insect Biochem Mol Biol 61:62–68. https://doi.org/10.1016/j.ibmb.2015.04.007

Fukuzawa M, Fu X, Tatsuki S, Ishikawa Y (2006) cDNA cloning and *in situ* hybridization of Δ11-desaturase, a key enzyme of pheromone biosynthesis in *Ostrinia scapulalis*. J Insect Physiol 52:430–435

Groot AT, Dekker T, Heckel DG (2016) The genetic basis of pheromone evolution in moths. Annu Rev Entomol 61:99–117. https://doi.org/10.1146/annurev-ento-010715-023638

Huang Y, Tatsuki S, Kim C-G, Hoshizaki S, Yoshiyasu Y, Honda H, Ishikawa Y (1997) Identification of sex pheromone of adzuki bean borer, *Ostrinia scapulalis*. J Chem Ecol 23:2791–2802

Huang Y, Tatsuki S, Kim C-G, Hoshizaki S, Ishikawa Y (1998a) Identification of the sex pheromone of *Ostrinia palustralis*. Entomol Expl Appl 86:313–318

Huang Y, Honda H, Yoshiyasu Y, Takanashi T, Hoshizaki S, Tatsuki S, Ishikawa Y (1998b) Geographic variation in sex pheromone of Asian corn borer, *Ostrinia furnacalis*, in Japan. J Chem Ecol 24:2079–2088

Huang Y, Honda H, Yoshiyasu Y, Hoshizaki S, Tatsuki S, Ishikawa Y (1998c) Sex pheromone of the butterbur borer, *Ostrinia zaguliaevi*. Entomol Exp Appl 89:281–287

Huang Y, Takanashi T, Hoshizaki S, Tatsuki S, Ishikawa Y (2002) Female sex pheromone polymorphism in adzuki bean borer, *Ostrinia scapulalis*, is similar to that in European corn borer, *O. nubilalis*. J Chem Ecol 28:533–539

Ishikawa Y, Takanashi T, Huang Y (1999a) Comparative studies on the sex pheromones of *Ostrinia* spp. in Japan: the burdock borer, *Ostrinia zealis*. Chemoecology 9:25–32

Ishikawa Y, Takanashi T, Kim C, Hoshizaki S, Tatsuki S, Huang Y (1999b) *Ostrinia* spp. in Japan: their host plants and sex pheromones. Entomol Exp Appl 91:237–244

Jurenka RA, Haynes KF, Adlof RO, Bengtsson M, Roelofs WL (1994) Sex pheromone component ratio in the cabbage looper moth altered by a mutation affecting the fatty acid chain-shortening reaction in the pheromone biosynthetic pathway. Insect Biochem Mol Biol 24:373–381

Kim C, Hoshizaki S, Huang Y, Tatsuki S, Ishikawa Y (1999) Usefulness of mitochondrial COII gene sequences in examining phylogenetic relationships in the Asian corn borer, *Ostrinia furnacalis*, and allied species (Lepidoptera: Pyralidae). Appl Entomol Zool 34:405–412

Knipple DC, Rosenfield CL, Nielsen R, You KM, Jeong SE (2002) Evolution of the integral membrane desaturase gene family in moths and flies. Genetics 162:1737–1752

Kojima W, Fujii T, Suwa M, Miyazawa M, Ishikawa Y (2010) Physiological adaptation of the Asian corn borer *Ostrinia furnacalis* to chemical defenses of its host plant, maize. J Insect Physiol 56(9):1349–1355

Lassance JM (2010) Journey in the *Ostrinia* world: from pest to model in chemical ecology. J Chem Ecol 36:1155–1169

Lassance JM (2016) The European corn borer *Ostrinia nubilalis*: exotic pest and model system to study pheromone evolution and speciation. In: Pheromone communication in Moths: evolution, behavior, and application. University of California Press, Berkeley, p 233
Lassance JM, Groot AT, Liénard MA, Antony B, Borgwardt C, Andersson F, Hedenström E, Heckel DG, Löfstedt C (2010) Allelic variation in a fatty-acyl reductase gene causes divergence in moth sex pheromones. Nature 466(7305):486
Lassance JM, Liénard MA, Antony B, Qian S, Fujii T, Tabata J, Ishikawa Y, Löfstedt C (2013) Functional consequences of sequence variation in the pheromone biosynthetic gene *pgFAR* for *Ostrinia* moths. Proc Natl Acad Sci U S A 110:3967–3972
Liu W, Rooney AP, Xue B, Roelofs WL (2004) Desaturases from the spotted fireworm moth (*Choristoneura parallela*) shed light on the evolutionary origins of novel moth sex pheromone desaturases. Gene 342:303–311
Ma PWK, Roelofs WL (2002) Sex pheromone gland of the female European corn borer moth, *Ostrinia nubilalis* (Lepidoptera, Pyralidae): ultrastructural and biochemical evidences. Zool Sci 19:501–511
Moto K, Yoshiga T, Yamamoto M, Takahashi S, Okano K, Ando T, Nakata T, Matsumoto S (2003) Pheromone gland-specific fatty-acyl reductase of the silkmoth, *Bombyx mori*. Proc Natl Acad Sci U S A 100:9156–9161
Moto K, Suzuki GM, Hull JJ, Kurata R, Takahashi S, Yamamoto M, Okano K, Imai K, Ando T, Matsumoto S (2004) Involvement of a bifunctional fatty-acyl desaturase in the biosynthesis of the silkmoth, *Bombyx mori*, sex pheromone. Proc Natl Acad Sci U S A 101:8631–8636
Mutuura A, Munroe E (1970) Taxonomy and distribution of the European corn borer and allied species: genus *Ostrinia* (Lepidoptera: Pyralidae). Mem Entomol Soc Canada 102(71):1–112
Nei M, Rooney AP (2005) Concerted and birth-and death evolution of multigene families. Annu Rev Genet 39:121–152
Ohno S, Hoshizaki S, Tatsuki S, Ishikawa Y (2003) New records of *Ostrinia ovalipennis* (Lepidoptera: Crambidae) from Hokkaido, and morphometric analyses for species identification and geographic variation. Appl Entomol Zool 38:529–536
Ohno S, Ishikawa Y, Tatsuki S, Hoshizaki S (2006) Variation in mitochondrial COII gene sequences among two species of Japanese knotweed-boring moths, *Ostrinia latipennis* and *O. ovalipennis* (Lepidoptera: Crambidae). Bull Entomol Res 96:243–249
Phuong TTT, Yamamoto M, Fujii T, Kojima W, Matsuo T, Ishikawa Y (2016) Comparison of the ability to catabolize DIMBOA, a maize antibiotic, between *Ostrinia furnacalis* and *Ostrinia scapulalis* (Lepidoptera: Crambidae), with reference to their hybrids. Appl Entomol Zool 51:143–149
Phuong TTT, Yamamoto M, Matsuo T, Fujii T, Ishikawa Y (2018) *In vitro* analysis of DIMBOA catabolism in the Asian corn borer *Ostrinia furnacalis* (Lepidoptera: Crambidae). Appl Entomol Zool 53:223–227
Rodriguez F, Hallahan DL, Pickett JA, Camps F (1992) Characterization of the Δ11-palmitoyl-CoA-desaturase from *Spodoptera littoralis* (lepidoptera: noctuidae). Insect Biochem Mol Biol 22:143–148
Roelofs WL, Rooney AP (2003) Molecular genetics and evolution of pheromone biosynthesis in Lepidoptera. Proc Natl Acad Sci U S A 100:9179–9184
Roelofs W, Glover T, Tang XH, Sreng I, Robbins P, Eckenrode C, Löfstedt C, Hansson BS, Bengtsson BO (1987) Sex pheromone production and perception in European corn borer moths is determined by both autosomal and sex-linked genes. Proc Natl Acad Sci U S A 84:7585–7589
Roelofs WL, Liu W, Hao G, Jiao H, Rooney AP, Linn-Jr CE (2002) Evolution of moth sex pheromones via ancestral genes. Proc Natl Acad Sci U S A 99:13621–13626
Sakai R, Fukuzawa M, Nakano R, Tatsuki S, Ishikawa Y (2009) Alternative suppression of transcription from two desaturase genes is the key for species-specific sex pheromone biosynthesis in two *Ostrinia* moths. Insect Biochem Mol Biol 39:62–67
Tabata J, Ishikawa Y (2005) Genetic basis to divergence of sex pheromones in two closely related moths, *Ostrinia scapulalis* and *O. zealis*. J Chem Ecol 31:1111–1124

Tabata J, Ishikawa Y (2011) Genetic basis regulating the sex pheromone blend in *Ostrinia zealis* (Lepidoptera: Crambidae) and its allies inferred from crossing experiments. Annal Entomol Soc Amer 104:326–336

Tabata J, Ishikawa Y (2016) Divergence of the sex pheromone systems in "Oriental" *Ostrinia* species. In: Allison JD, Carde RT (eds) Pheromone communication in Moths: evolution, behavior, and application. University of California Press, Berkeley, p 245

Tabata J, Takanashi T, Ishikawa Y (2003) Pheromone analysis of wild female moths with a PBAN C-terminal peptide injection for an estimation of assortative mating in adzuki bean borer, *Ostrinia scapulalis*. J Chem Ecol 28:2749–2759

Tabata J, Hoshizaki S, Tatsuki S, Ishikawa Y (2006) Heritable pheromone blend variation in a local population of the butterbur borer moth *Ostrinia zaguliaevi*. Chemoecology 16:123–128

Tabata J, Huang Y, Ohno S, Yoshiyasu Y, Sugie H, Tatsuki S, Ishikawa Y (2008) Sex pheromone of *Ostrinia* sp. newly found on the leopard plant *Farfugium japonicum*. J Appl Entomol 132:566–574

Takanashi T, Ohno S, Huang Y, Tatsuki S, Honda H, Ishikawa Y (2000) A sex pheromone component novel to *Ostrinia* identified from *Ostrinia latipennis* (Lepidoptera: Crambidae). Chemoecology 10:143–147

Takanashi T, Huang Y, Takahasi R, Hoshizaki S, Tatsuki S, Ishikawa Y (2005) Genetic analysis and population survey of sex pheromone variation in adzuki bean borer moth, *Ostrinia scapulalis*. Bio J Linn Soc 84:143–160

Wang HL, Liénard MA, Zhao CH, Wang CZ, Löfstedt C (2010) Neofunctionalization in an ancestral insect desaturase lineage led to rare Δ6 pheromone signals in the Chinese tussah silkworm. Insect Biochem Mol Biol 40:742–751

Wolf WA, Roelofs WL (1986) Properties of the Δ11-desaturase enzyme used in cabbage looper moth sex pheromone biosynthesis. Arch Insect Biochem Physiol 3:45–52

Wolf WA, Roelofs WL (1987) Reinvestigation confirms action of Δ11-desaturases in spruce budworm moth sex pheromone biosynthesis. J Chem Ecol 13:1019–1027

Xue B, Rooney AP, Kajikawa M, Okada N, Roelofs WL (2007) Novel sex pheromone desaturases in the genomes of corn borers generated through gene duplication and retroposon fusion. Proc Natl Acad Sci U S A 104:4467–4472

Zhao CH, Lu F, Bengtsson M, Löfstedt C (1995) Substrate specificity of acetyltransferase and reductase enzyme systems used in pheromone biosynthesis by Asian corn borer *Ostrinia furnacalis*. J Chem Ecol 21:1495–1510

Zhu JW, Zhao CH, Bengtsson M, Löfstedt C (1996) Reductase specificity and the ratio regulation of E/Z isomers in the pheromone biosynthesis of the European corn borer, *Ostrinia nubilalis* (Lepidoptera: Pyralidae). Insect Biochem Mol Biol 26:171–176

Chapter 8
Epoxidases Involved in the Biosynthesis of Type II Sex Pheromones

Takeshi Fujii, Yu Rong, and Yukio Ishikawa

Abstract In moth species that utilize alkenyl (type II) sex pheromones, selective epoxidation of double bonds in the alkene pheromone components confers further diversity on their chemical structures. Two arctiids, the fall webworm *Hyphantria cunea* and the mulberry tiger moth *Lemyra imparilis*, use the same epoxyalkene, Z3,Z6,epo9-21:H, as the main pheromone component. In these species, we recently identified cytochrome P450s (CYPs) belonging to the CYP341 family as enzymes involved in the specific epoxidation of a Z9 double bond of the pheromone precursor Z3,Z6,Z9-21:H. Furthermore, a cytochrome P450 belonging to a different family, CYP340, was identified as an enzyme responsible for the specific epoxidation of a Z3 double bond of the pheromone precursor Z3,Z6,Z9-19:H in the Japanese giant looper *Ascotis selenaria*, which uses epo3,Z6,Z9-19:H as the main pheromone component. These findings suggest that epoxidases (CYPs) with different regio-specificities evolved independently.

Keywords Epoxidases · Cytochrome P450 · Type II sex pheromones

8.1 Introduction

An epoxide is a three-membered cyclic ether consisting of two carbon atoms and one oxygen atom. This ring assumes an equilateral triangle, which makes it strained, and hence highly reactive (Parker and Isaacs 1959). This high reactivity makes epoxides versatile intermediates in the production of a variety of industrially useful materials. Insects are notable in that they have successfully exploited epoxidation for detoxification of plant allelochemicals (Niu et al. 2011; Schuler 2011; Schuler and Berenbaum 2013), the biosynthesis of biologically active compounds, including hormones and pheromones (Ando et al. 2004; Daimon et al. 2012; Després et al.

T. Fujii (✉) · Y. Rong · Y. Ishikawa
Laboratory of Applied Entomology, Graduate School of Agricultural and Life Sciences, The University of Tokyo, Tokyo, Japan

Y. Ishikawa (ed.), *Insect Sex Pheromone Research and Beyond*, Entomology Monographs, https://doi.org/10.1007/978-981-15-3082-1_8

2007; Helvig et al. 2004; Niwa et al. 2004), and acquirement of resistance to insecticides (Daborn et al. 2007; Feyereisen 2012).

In applied chemistry, epoxides are usually generated by treating olefin with peroxyacids, which are potentially explosive and require careful handling. It is remarkable that organisms have evolved epoxidases that enable mild and regio-/stereo-specific epoxidation of a variety of molecules in life. We recently identified epoxidases involved in the biosynthesis of type II sex pheromones in moths as cytochrome P450s (CYPs) belonging to the CYP340 and CYP341 families, the function of which has long been unknown. In this chapter, we first overview general knowledge of the biosynthetic pathways of type II pheromones, and then describe our recent identification of pheromone gland (PG)-specific CYPs involved in the production of epoxyalkene pheromone components. The evolution of PG-specific CYPs in moths utilizing type II pheromones will be discussed in the context of the evolution of moths.

8.2 Biosynthetic Pathways of Type II Sex Pheromones

Moth sex pheromones have been classified into type I, type II, and miscellaneous based on their chemical structures (Ando et al. 2004). Type I pheromones consist of unsaturated aliphatic compounds with a C_{10}–C_{18} straight chain and terminal functional group such as hydroxyl, acetoxyl, or formyl group. On the other hand, type II pheromones are composed of C_{17}–C_{23} hydrocarbons with two or three *cis* double bonds at the 3, 6, or 9 positions, and their corresponding epoxy derivatives (Ando et al. 2004; Millar 2000, 2010). Although an increasing number of enzymes involved in the biosynthesis of type I pheromones have been clarified at the molecular level (Jurenka 2004; Matsumoto et al. 2007; Chap. 7 of this book), little is known about the molecular entities of enzymes involved in the biosynthesis of type II pheromones.

Biosynthetic pathways of type II sex pheromones have been extensively investigated using the Japanese giant looper *Ascotis selenaria* (Geometridae), which secretes *cis*-3,4-epoxy-*(Z,Z)*-6,9-nonadecadiene (epo3,Z6,Z9-19:H) as the main sex pheromone component (Ando et al. 1997; Fujii et al. 2007, 2013; Matsuoka et al. 2006; Miyamoto et al. 1999; Wei et al. 2003, 2004). The two marked differences between the biosynthetic pathways of type I and type II sex pheromones are (i) the starting materials and (ii) presence of the transport/incorporation step. Type I sex pheromone components are de novo biosynthesized in the PG from acetyl-coenzyme A (CoA), whereas type II pheromones are derived from dietary essential fatty acids (EFAs) such as linoleic acid and α-linolenic acid (Fig. 8.1). As shown in Fig. 8.1, the chemical structure characteristic to type II sex pheromones, that is, 2–3 *cis* double bonds, originates from EFAs. Previous studies demonstrated that EFAs, ingested from the diet during the larval stage, typically undergo chain elongation, desaturation (optional), and decarboxylation into alkenes in specialized cells called oenocytes, which are associated with the epidermis of the abdominal integument of adult female moths. These EFA-derived alkenes are then loaded onto

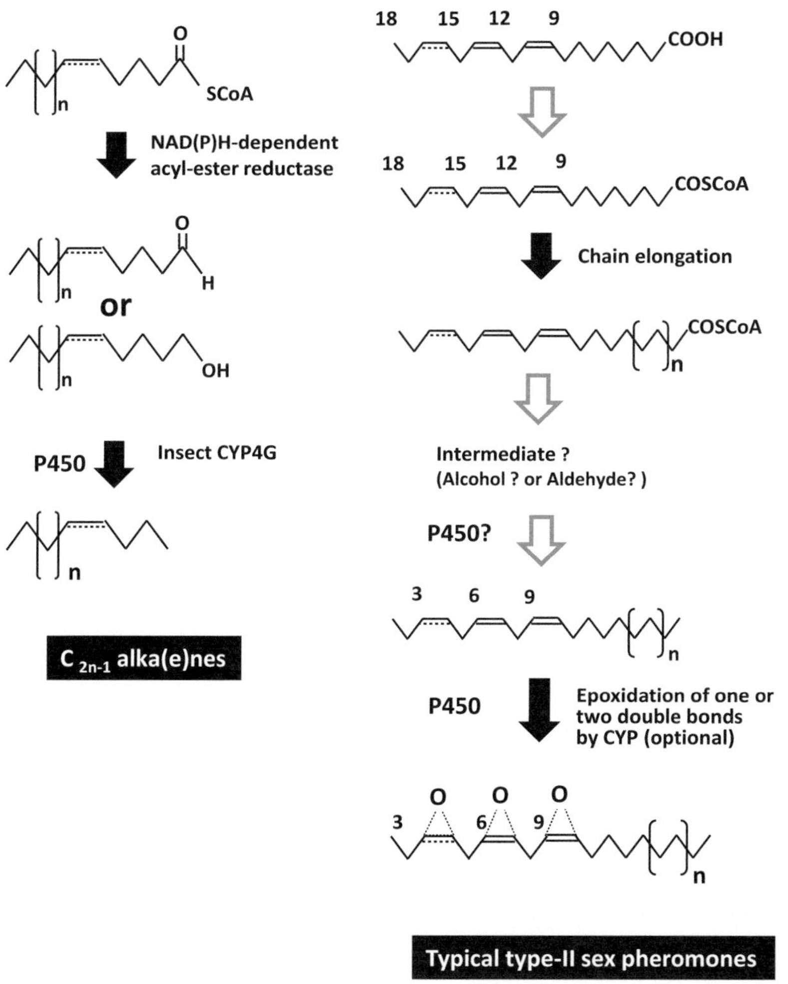

Fig. 8.1 Comparison of the biosynthetic pathways of cuticular hydrocarbons in *Drosophila* (left) and type II sex pheromones with the putative pathway indicated by the open arrow (right). Two series of CYPs are presumed to be involved in the biosynthesis of type II sex pheromones: one is CYP4G functioning in oenocytes and the other is CYP340/341 functioning in the pheromone gland

hemolymph lipophorin, transported to the PG located at the abdominal tip, and secreted without further modification or after epoxidation (Chino 1985; Matsuoka et al. 2006; Schal and Sevala 1998; Wei et al. 2003). Accordingly, EFA-derived alkenes can be experimentally detected in the hemolymph extract of the female moths by gas chromatography–mass spectrometry (GC-MS) analysis (Blomquist 2010; Fujii et al. 2015; Jurenka and Subchev 2000; Matsuoka et al. 2006; Schal and Sevala 1998; Wei et al. 2003).

In the biosynthesis of cuticular hydrocarbons in insects, decarboxylation of precursor fatty acyl-CoA is considered to occur in two steps, that is, the formation of fatty aldehyde/alcohol by reductase and subsequent conversion into hydrocarbon by the cytochrome P450 belonging to CYP4G (Fig. 8.1; MacLean et al. 2018; Qiu et al. 2012). Although not yet experimentally confirmed, similar enzymes are considered to function in the production of EFA-derived hydrocarbons in oenocytes.

Regarding the incorporation of EFA-derived alkenes into the PG, analysis of the hemolymph of the cabbage moth *Mamestra brassicae*, a moth utilizing type I pheromone, revealed the presence of EFA-derived alkenes; however, no trace of them was detected in the PG (Fujii et al. 2015). This suggests the presence of a mechanism that controls selective uptake of EFA-derived alkenes into the PG in type I pheromone users.

Many species in Erebidae and Geometridae secrete alkenes with one or two epoxy ring(s) at positions 3, 6, or 9 as sex pheromone components. Regarding the position of epoxy rings in mono-epoxy pheromone components, position 9 is most common (45%), followed by position 6 (32%) and position 3 (23%). Survey of the database of lepidopteran sex pheromones (Ando and Yamamoto 2018) revealed the presence of phylogenetic bias in the utilization of epoxy positional isomers (Table 8.1). Namely, the position of the epoxy ring in Arctiinae is limited to position 9, with the exception of only one species. The single exceptional species within Arctiinae utilizes atypical type II pheromone components that are not derived from EFAs. The exclusive use of 9epo compounds in Arctiinae suggests the presence of genetic constraint. In contrast, the position of epoxy rings in Geometridae is approximately evenly distributed among positions 3, 6, and 9.

8.3 Sequence Information of Epoxidases in Erebidae and Geometridae

As described above, epoxyalkenes are often used as pheromone components in moth species in Erebidae and Geometridae. We selected four species that use mono-epoxy alkenes with different epoxy positions as the main pheromone component to identify epoxidases responsible for the epoxidation of a double bond at a specific position: the fall webworm *Hyphantria cunea*, the mulberry tiger moth *Lemyra imparilis*, the Japanese giant looper *Ascotis selenaria*, and the giant geometer moth *Biston robustum* (Fig. 8.2).

Rong et al. (2014) succeeded in identifying the enzyme involved in the epoxidation of pheromone precursors in *H. cunea* by focusing on cytochrome P450s, which are known to catalyze the oxidation of diverse compounds. A P450-like sequence (*Hc_epo1*), which was specifically expressed in the PG (Fig. 8.3), was identified in the complementary DNA (cDNA) library prepared from the PG of *H. cunea*. Phylogenetic analysis of the sequence demonstrated that Hc_epo1 belongs to the CYP341B clade in the CYP341 family and was given the name CYP341B14

Table 8.1 The positions of epoxy rings in type II sex pheromones[a]

Superfamily	Family	Subfamily	Position of the epoxy rings							Non-epo
			Mono-epo				Di-epo			
			3	6	9	Others[b]	3,6	6,9	Unknown[c]	
Geometroidea	Geometridae		30	29	21	1			2	48
Noctuoidea[d]	Noctuidae[d]				1					
	Erebidae[d]	Erebinae		1	13					13
		Calpinae	1	5	3		1			5
		Hypeninae			4					
		Herminiinae		8		1			1	
		Lymantriinae[e]		1	2	9	1	1		6
		Arctiinae[e]			17	1				21
Pyraloidea	Crambidae					1				
Tischerioidea	Tischeriidae									1
Yponomeutoidea	Lyonetiidae									4
Total			31	44	61	13	2	1	3	98

[a]As of September 2018 (excerpted from https://lepipheromone.sakura.ne.jp/lepi_phero_lis)
[b]These pheromone components (4epo, 7epo, and 11epo) are not derived from essential fatty acids
[c]The epoxy ring position in the pheromone component is unidentified
[d]Some subfamilies formerly placed in Noctuidae have been transferred to the family Erebidae (Regier et al. 2017; Zahiri et al. 2012)
[e]These two subfamilies were formerly treated as two independent families (Lymantridae and Arctiidae) in the superfamily Noctuoidea

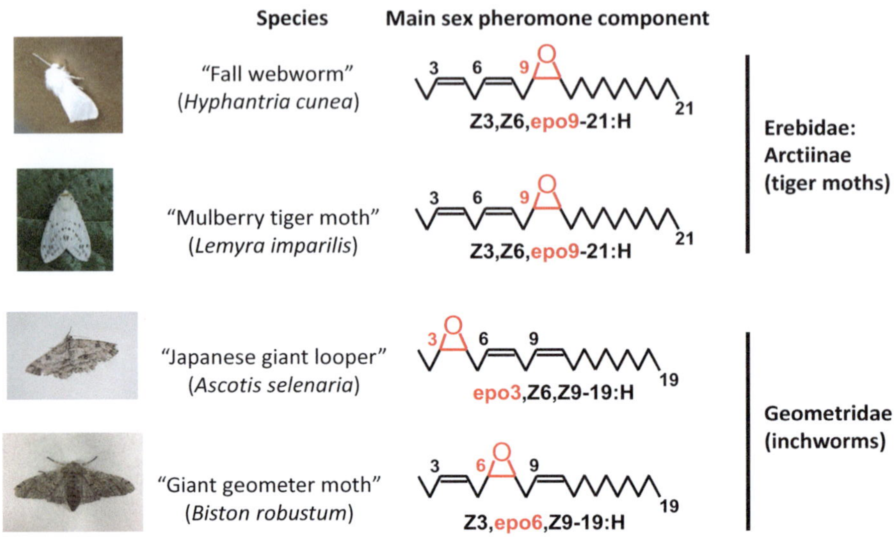

Fig. 8.2 Moth species of interest for the study of epoxidase involved in regio-selective epoxidation of the precursors of type II pheromones

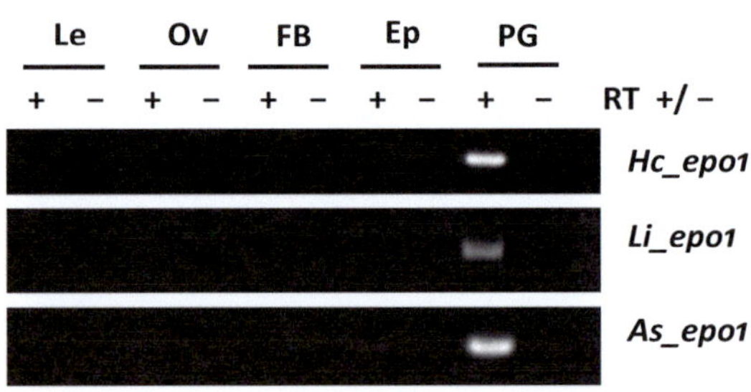

Fig. 8.3 Tissue distribution analysis of transcripts encoding epoxidase functioning in the epoxidation of sex pheromone precursors in the pheromone gland. RT-PCR analysis demonstrated transcripts of the epoxidase localized in the pheromone gland. The pattern of tissue distribution is identical to that of transcripts of key enzymes in type I sex pheromone generation such as desaturase and reductase. *Le* legs, *Ov* ovary, *FB* fat body, *Ep* epidermis with cuticle, *PG* pheromone gland

(Fig. 8.4). A subsequent functional assay using Sf-9 cells transiently expressing Hc_epo1 demonstrated that this P450 protein can specifically epoxidize a (*Z*)-double bond at the 9th position in the pheromone precursor (3Z,6Z,9Z)-3,6,9-henicosatriene (Z3,Z6,Z9-21:H) (Rong et al. 2014).

Rong et al. (2019a, b) also succeeded in identifying two more genes encoding epoxidases, *Li_epo1* and *As_epo1*, from *L. imparilis* (Erebidae) and *A. selenaria* (Geometridae), respectively. The former gene (*Li_epo1*) was cloned from the PG of *L. imparilis* via amplification of the fragment of *Hc_epo1* homolog by reverse transcription polymerase chain reaction (RT-PCR) with degenerate primers (Rong et al.

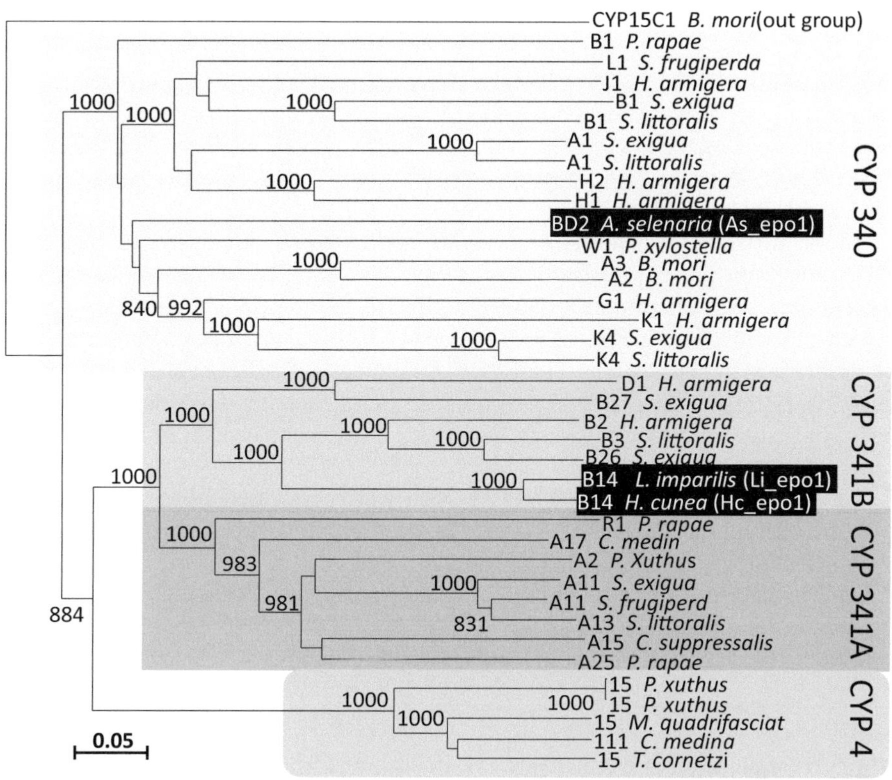

Fig. 8.4 Phylogenetic relationships of Hc_epo1 (Accession No.: AB795798), Li_epo1 (Accession No.: LC326250), As_epo1 (Accession No.: LC424537), and sequences related to these three CYPs. The phylogenetic tree was reconstructed using the neighbor-joining method. CYP15C1 of *Bombyx mori* (Accession No.: AB124839) was used as an out-group. Bootstrap values after 1000 replications are shown near the branches. The branch length is drawn to the genetic distance. *CYP* cytochrome P450

2019a), whereas the latter gene (*As_epo1*) was identified via screening of transcriptomes obtained by RNA sequencing (RNA-seq) of the PG of *A. selenaria* (Rong et al. 2019b). The deduced amino acid sequence of Li_epo1 shared a high identity (88.5%) with that of Hc_epo1, and phylogenetic analysis revealed that Li_epo1 is an ortholog of Hc_epo1 (Fig. 8.4). This is consistent with (1) the close relationship between *L. imparilis* and *H. cunea*, both of which belong to the same subfamily Arctiinae, and (2) the use of the same compound, Z3,Z6,epo9-21:H, as the main component of their sex pheromones. In contrast, the deduced amino acid sequence of As_epo1 had low identity (9.5%) with Hc_epo1, and phylogenetic analysis demonstrated that As_epo1 belongs to CYP340, a CYP family different from CYP341 to which Hc_epo1 and Li_epo1 belong (Fig. 8.4). To identify the PG-specific epoxidase in *A. selenaria*, we initially investigated whether Z3-specific epoxidase in *A. selenaria* also belongs to CYP341B. However, no band of the expected size was obtained by RT-PCR using the degenerate primers designed based on the amino acid sequences of the CYP341B subfamily in the public database. The failure of RT-PCR with degenerate primers is consistent with the low identity between As_epo1 and Hc_epo1 sequences.

Tissue distribution analysis of *Li_epo1* and *As_epo1* demonstrated that both genes are transcribed specifically in the PG (Fig. 8.3). In the female moths that secrete type I sex pheromones, the transcripts of desaturase and reductase genes, key enzymes involved in the sex pheromone production, have been demonstrated to be localized in the PG (Antony et al. 2009; Liénard et al. 2010; Liu et al. 2004; Moto et al. 2003, 2004). The expression of a set of genes encoding enzymes involved in the sex pheromone production may therefore be under the control of a single transcription factor.

Functional assays of *Li_epo1* and *As_epo1* were performed using the Sf9 insect cell line-baculovirus expression system. Li_epo1 exhibited epoxidase activity with strict selectivity to the double bond at position 9 of two trienes, Z3,Z6,Z9-21:H and Z3,Z6,Z9-23:H, precursors of epoxy diene sex pheromone components in *L. imparilis*. On the other hand, functional assay of As_epo1 demonstrated that it specifically epoxidizes the Z3 double bond in the pheromone precursor Z3,Z6,Z9-19:H.

8.4 Phylogeny of Epoxidases Involved in Sex Pheromone Production

Insect P450 families are classified into four clans: clans 2, 3, 4, and mitochondrial (Fig. 8.5; Feyereisen 2012; Yu et al. 2015). Both CYP341 and CYP340, to which epoxidases involved in type II pheromone biosynthesis belong, are members of clan 4. CYP families with experimentally confirmed epoxidase activity, which are indicated by bold letters in Fig. 8.5, were newly found in clan 4. With the addition of CYP341 and CYP340, CYP families with epoxidase activity are now found in all four clans. In other words, the epoxidase activity is not limited to CYPs in particular clans. CYPs with epoxidase activity in different clans may have different substrate preferences and, as a whole, may play a role in expanding the repertory of substrates.

When we reconstructed the phylogenetic tree shown in Fig. 8.4, we collected sequences related to Hc_epo1, Li_epo1, and As_epo1 using their amino acid sequences as queries in the Basic Local Alignment Search Tool: Protein BLAST (BLASTP) search of the National Center for Biotechnology Information (NCBI) nonredundant protein sequence database. We quickly noted that the sequences with significant homology to the queries were only those of lepidopterans. As no insect or animal sequences other than those of lepidopterans exhibited significant homology to the queries, we currently consider CYP341 and CYP340 to be Lepidoptera-

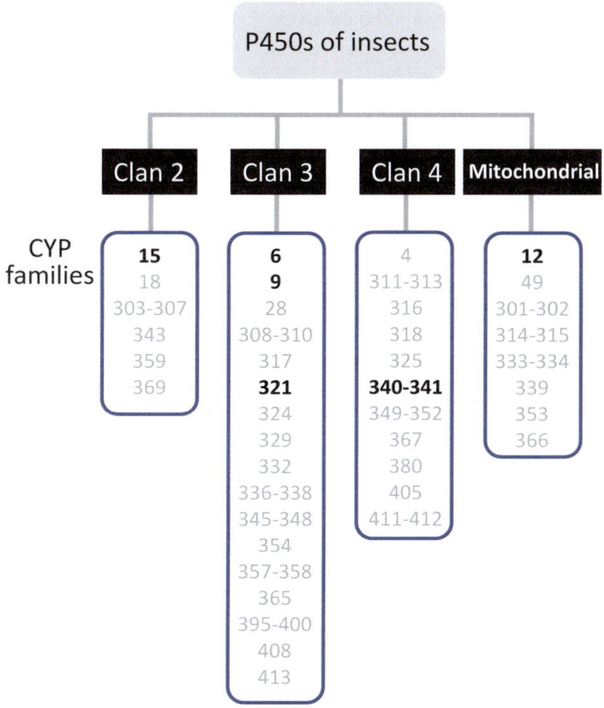

Fig. 8.5 Classification of insect P450 families. Insect P450 families are classified into four clans (Feyereisen 2012; Yu et al. 2015). Numbers indicate the name of a family, for example, "4" designates the CYP4 family. Bold characters indicate families, members of which were experimentally demonstrated to exhibit epoxidase activity

specific (Fig. 8.4). Furthermore, many CYP341 and CYP340 sequences have been reported from type I moths, such as *Helicoverpa armigera* and *Spodoptera littoralis*, which do not utilize epoxyalkene sex pheromones, and epoxidation is therefore not included in their pheromone biosynthetic pathways. Of note, CYP341 and CYP340 sequences were also reported from butterflies (*Pieris rapae* and *Papilio xuthus*). Here, it should be noted that as most of these CYP sequences were deduced from genome sequences, neither the functions nor tissue expression patterns of these genes are known. The sequences and functions of CYPs in insects are markedly diversified by repeated duplication events and the subsequent acquirement of new functions by these duplicated genes (Calla et al. 2017; Feyereisen 2012). The presence of CYP340/341 genes not only in type II moths, but also in type I moths and butterflies, may reflect these processes in lepidoptera. As moth species belonging to relatively recently diverged groups (Geometridae and Erebidae) utilize type II pheromones, the production of type II pheromones and thus the utilization of epoxidase in pheromone production are considered to be derived traits. Information on the roles of CYP340/341 genes in type I moths and butterflies may provide insight into how these CYP genes became involved in the biosynthesis of type II pheromones.

8.5 Future Direction of the Study

Through our studies, CYPs involved in Z3- and Z9-specific epoxidation of EFA-derived hydrocarbons have been identified. The Z3 double bond-specific epoxidase, As_epo1, belongs to the CYP340 family, whereas the Z9 double bond-specific epoxidases, Hc_epo1 and Li_epo1, belong to the CYP341 family (Fig. 8.4). The two CYP families were estimated to have diverged long before the divergence of moth families. To elucidate the evolution of the utilization of CYPs in the biosynthesis of moth sex pheromones, the identification of many more epoxidases in type II moths, especially those specific for Z6 double bonds, is needed. We aimed to clarify the CYP involved in Z6-specific epoxidation using the giant geometer moth *B. robustum*, which utilizes Z3,epo6,Z9-19:H as the major pheromone component (Fig. 8.2). As even a single amino acid substitution may affect the substrate specificity or catalytic activity of CYP (Ortiz de Montellano 2015), CYPs in the same subfamily may have different regio-specificities in terms of epoxidation. In addition to moths that utilize mono-epoxy alkenes, some moths in Lymantriinae were reported to secrete alkenes with two epoxy rings at positions 3 + 6 or 6 + 9 (Table 8.1). It is of interest to know whether a single CYP introduces two epoxy rings or whether two independent CYPs function in the formation of two epoxy rings.

Among the enzymes involved in the biosynthesis of type II pheromones, only epoxidases have been identified at the molecular level. Fatty acyl elongase, fatty acyl-CoA reductase (FAR), and CYP4G functioning in the oenocytes of moths remain to be identified (Fig. 8.1). As the sequences of elongase and CYP4G involved in the formation of cuticular hydrocarbons have been reported for *D. melanogaster* and some other species (MacLean et al. 2018; Qiu et al. 2012; Yu et al. 2016), homologs of these enzymes in moth species may be relatively easily found. Regarding fatty acyl-CoA reductases (FARs), those in oenocytes have been assumed to produce aldehyde that is decarbonylated by CYP4G (Fig. 8.1; Qiu et al. 2012); however, FARs that produce aldehyde have not yet been identified in any insect, including *D. melanogaster* (Finet et al. 2019). Recently, MacLean et al. (2018) found that CYP4G in the mountain pine beetle *Dendroctonus ponderosae* takes both aldehyde and alcohol as substrates, with a greater preference for alcohol, and converts them to hydrocarbon. When CYP4G takes alcohol as substrate, an aldehyde is presumed to be transiently formed as an intermediate and immediately converted into hydrocarbon (MacLean et al. 2018). Thus, identification and functional assay of FAR specifically expressed in the oenocytes are awaited to clarify the products of these enzymes.

Acknowledgment We thank Prof. Tetsu Ando, Dr. Masanobu Yamamoto, and Dr. Masataka G. Suzuki for continuous support of this study.

References

Ando T, Yamamoto M (2018) Internet Database. https://lepipheromone.sakura.ne.jp/lepi_phero_list

Ando T, Ohtani K, Yamamoto M, Miyamoto T, Qi XR, Witjaksono (1997) Sex Pheromone of Japanese giant looper, *Ascotis selenaria cretacea*: identification and field tests. J Chem Ecol 23:2413–2423

Ando T, Inomata S, Yamamoto M (2004) Lepidopteran sex pheromones. Top Curr Chem 239:51–96

Antony B, Fujii T, Moto K, Matsumoto S, Fukuzawa M, Nakano R, Tatsuki S, Ishikawa Y (2009) Pheromone-gland-specific fatty-acyl reductase in the adzuki bean borer, *Ostrinia scapulalis* (Lepidoptera: Crambidae). Insect Biochem Mol Biol 39:90–95

Blomquist GJ (2010) Biosynthesis of cuticular hydrocarbons. In: Blomquist GJ, Bagnères AG (eds) Insect hydrocarbons: biology, biochemistry, and chemical ecology. Cambridge University Press, New York, pp 35–52

Calla B, Noble K, Johnson RM, Walden KK, Schuler MA, Robertson HM, Berenbaum MR (2017) Cytochrome P450 diversification and hostplant utilization patterns in specialist and generalist moths: birth, death and adaptation. Mol Ecol 26:6021–6035

Chino H (1985) Lipid transport: biochemistry of hemolymph lipophorin. In: Kerkut GA, Gilbert LI (eds) Comprehensive insect physiology, biochemistry and pharmacology, vol. 10. Pergamon Press, New York, pp 115–135

Daborn PJ, Lumb C, Boey A, Wong W, Batterham P (2007) Evaluating the insecticide resistance potential of eight *Drosophila melanogaster* cytochrome P450 genes by transgenic over-expression. Insect Biochem Mol Biol 37:512–519

Daimon T, Kozaki T, Niwa R, Kobayashi I, Furuta K (2012) Precocious metamorphosis in the juvenile hormone-deficient mutant of the silkworm, *Bombyx mori*. PLoS Genet 8:e1002486

Després L, David JP, Gallet C (2007) The evolutionary ecology of insect resistance to plant chemicals. Trends Ecol Evol 22:298–307

Feyereisen R (2012) Insect CYP genes and P450 enzymes. In: Gilbert LI (ed) Insect molecular biology and biochemistry. Academic, New York, pp 236–316

Finet C, Slavik K, Pu J, Carroll SB, Chung H (2019) Birth-and-death evolution of the fatty acyl-CoA reductase (FAR) gene family and diversification of cuticular hydrocarbon synthesis in *Drosophila*. Genome Biol Evol 11:1541–1551. https://doi.org/10.1101/509588

Fujii T, Suzuki MG, Kawai T, Tsuneizumi K, Ohnishi A, Kurihara M, Matsumoto S, Ando T (2007) Determination of the pheromone-producing region that has epoxidation activity in the abdominal tip of the Japanese giant looper, *Ascotis selenaria cretacea* (Lepidoptera: Geometridae). J Insect Physiol 53:312–318

Fujii T, Suzuki MG, Katsuma S, Ito K, Rong Y, Matsumoto S, Ando T, Ishikawa Y (2013) Discovery of a disused desaturase gene from the pheromone gland of the moth *Ascotis selenaria*, which secretes an epoxyalkenyl sex pheromone. Biochem Biophys Res Comm 441:849–855

Fujii T, Yamamoto M, Nakano R, Nirazawa T, Rong Y, Dong SL, Ishikawa Y (2015) Alkenyl sex pheromone analogs in the hemolymph of an arctiid *Eilema japonica* and several non-arctiid moths. J Insect Physiol 82:109–113

Helvig C, Koener JF, Unnithan GC, Feyereisen R (2004) CYP15A1, the cytochrome P450 that catalyzes epoxidation of methyl farnesoate to juvenile hormone III in cockroach corpora allata. Proc Natl Acad Sci U S A 101:4024–4029

Jurenka R (2004) Insect pheromone biosynthesis. In: The chemistry of pheromones and other semiochemicals I. Springer, Berlin/Heidelberg, pp 97–132

Jurenka R, Subchev M (2000) Identification of cuticular hydrocarbons and the alkene precursor to the pheromone in hemolymph of the female gypsy moth, *Lymantria dispar*. Arch Insect Biochem Physiol 43:108–115

Liénard MA, Hagström ÅK, Lassance JM, Löfstedt C (2010) Evolution of multicomponent pheromone signals in small ermine moths involves a single fatty-acyl reductase gene. Proc Natl Acad Sci U S A 107:10955–10960

Liu W, Rooney AP, Xue B, Roelofs WL (2004) Desaturases from the spotted fireworm moth (*Choristoneura parallela*) shed light on the evolutionary origins of novel moth sex pheromone desaturases. Gene 342:303–311

MacLean M, Nadeau J, Gurnea T, Tittiger C, Blomquist GJ (2018) Mountain pine beetle (*Dendroctonus ponderosae*) CYP4Gs convert long and short chain alcohols and aldehydes to hydrocarbons. Insect Biochem Mol Biol 201:11–20

Matsumoto S, Hull JJ, Ohnishi A, Moto KI, Fónagy A (2007) Molecular mechanisms underlying sex pheromone production in the silkmoth, *Bombyx mori*: characterization of the molecular components involved in bombykol biosynthesis. J Insect Physiol 53:752–759

Matsuoka K, Tabunoki H, Kawai T, Ishikawa S, Yamamoto M, Sato R, Ando T (2006) Transport of a hydrophobic biosynthetic precursor by lipophorin in the hemolymph of geometrid female moth which secretes an epoxyalkenyl sex pheromone. Insect Biochem Mol Biol 36:576–583

Millar JG (2000) Polyene hydrocarbons and epoxides: a second major class of lepidopteran sex attractant pheromones. Annu Rev Entomol 45:575–604

Millar JG (2010) Polyene hydrocarbons, epoxides, and related compounds as components of lepidopteran pheromone blends. In: Blomquist GJ, Bagnères AG (eds) Insect Hydrocarbons. Cambridge University Press, New York, pp 390–447

Miyamoto T, Yamamoto M, Ono A, Ohtani K, Ando T (1999) Substrate specificity of the epoxidation reaction in sex pheromone biosynthesis of the Japanese giant looper (Lepidoptera: Geometridae). Insect Biochem Mol Biol 29:63–69

Moto K, Yoshiga T, Yamamoto M, Takahashi S, Okano K, Ando T, Nakata T, Matsumoto S (2003) Pheromone gland-specific fatty-acyl reductase of the silkmoth, *Bombyx mori*. Proc Natl Acad Sci U S A 100:9156–9161

Moto K, Suzuki GM, Hull JJ, Kurata R, Takahashi S, Yamamoto M, Okano K, Imai K, Ando T, Matsumoto S (2004) Involvement of a bifunctional fatty-acyl desaturase in the biosynthesis of the silkmoth, *Bombyx mori*, sex pheromone. Proc Natl Acad Sci U S A 101:8631–8636

Niu G, Rupasinghe SG, Zangerl AR, Siegel JP, Schuler MA, Berenbaum MR (2011) A substrate-specific cytochrome P450 monooxygenase, CYP6AB11, from the polyphagous navel orangeworm (*Amyelois transitella*). Insect Biochem Mol Biol 41:244–253. https://doi.org/10.1016/j.ibmb.2010.12.009

Niwa R, Matsuda T, Yoshiyama T, Namiki T, Mita K, Fujimoto Y, Kataoka H (2004) CYP306A1, a cytochrome P450 enzyme, is essential for ecdysteroid biosynthesis in the prothoracic glands of *Bombyx* and *Drosophila*. J Biol Chem 279:35942–35949

Ortiz de Montellano PR (2015) Substrate oxidation by cytochrome P450 enzymes. In: Ortiz de Montellano PR (ed) Cytochrome P450. Springer, Boston, pp 111–176

Parker RE, Isaacs NS (1959) Mechanisms of epoxide reactions. Chem Rev 59:737–799

Qiu Y, Tittiger C, Wicker–Thomas C, Le Goff G, Young S, Wajnberg E, Fricauxc T, Taquetc N, Blomquist GJ, Feyereisen R (2012) An insect–specific P450 oxidative decarbonylase for cuticular hydrocarbon biosynthesis. Proc Natl Acad Sci USA 109:14858–14863

Regier JC, Mitter C, Mitter K, Cummings MP, Bazinet AL, Hallwachs W, Janzen DH, Zwick A (2017) Further progress on the phylogeny of Noctuoidea (Insecta: Lepidoptera) using an expanded gene sample. Syst Entomol 42(1):82–93

Rong Y, Fujii T, Katsuma S, Yamamoto M, Ando T, Ishikawa Y (2014) CYP341B14: a cytochrome P450 involved in the specific epoxidation of pheromone precursors in the fall webworm *Hyphantria cunea*. Insect Biochem Mol Biol 54:122–128

Rong Y, Fujii T, Naka H, Yamamoto M, Ishikawa Y (2019a) Functional characterization of the epoxidase gene, Li_epo1 (CYP341B14), involved in generation of epoxyalkene pheromones in the mulberry tiger moth *Lemyra imparilis*. Insect Biochem Mol Biol 107:46–52

Rong Y, Fujii T, Ishikawa Y (2019b) CYPs in different families are involved in the divergent regio-specific epoxidation of alkenyl sex pheromone precursors in moths. Insect Biochem Mol Biol 108:9–15

Schal C, Sevala V (1998) Novel and highly specific transport of a volatile sex pheromone by hemolymph lipophorin in moths. Naturwissenschaften 85:339–342

Schuler MA (2011) P450s in plant-insect interactions. Biochim Biophys Acta 1814:36–45. https://doi.org/10.1016/j.bbapap.2010.09.012

Schuler MA, Berenbaum MR (2013) Structure and function of cytochrome P450s in insect adaptation to natural and synthetic toxins: insights gained from molecular modeling. J Chem Ecol 39:1232–1245. https://doi.org/10.1007/s10886-013-0335-7

Wei W, Miyamoto T, Endo M, Murakawa T, Pu G-Q, Ando T (2003) Polyunsaturated hydrocarbons in the hemolymph: biosynthetic precursors of epoxy pheromones of geometrid and Arctiid moths. Insect Biochem Mol Biol 33:397–405

Wei W, Yamamoto M, Asato T, Fujii T, Pu GQ, Ando T (2004) Selectivity and neuroendocrine regulation of the precursor uptake by pheromone glands from hemolymph in geometrid female moths, which secrete epoxyalkenyl sex pheromones. Insect Biochem Mol Biol 34:1215–1224

Yu L, Tang W, He W, Ma X, Vasseur L, Baxter SW, Yang G, Huang S, Song F, You M (2015) Characterization and expression of the cytochrome P450 gene family in diamondback moth, *Plutella xylostella* (L.). Sci Rep 5:8952

Yu Z, Zhang X, Wang Y, Moussian B, Zhu KY, Li S, Ma E, Zhang J (2016) LmCYP4G102: an oenocyte-specific cytochrome P450 gene required for cuticular waterproofing in the migratory locust, *Locusta migratoria*. Sci Rep 6:29980

Zahiri R, Holloway JD, Kitching IJ, Donald Lafontaine J, Mutanen M, Wahlberg N (2012) Molecular phylogenetics of Erebidae (Lepidoptera, Noctuoidea). Syst Entomol 37(1):102–124

Part III
Reception of Sex Pheromones

Chapter 9
Molecular Mechanisms of Sex Pheromone Reception in Moths

Yusuke Shiota and Takeshi Sakurai

Abstract Male moths locate their mates using species-specific sex pheromones emitted by conspecific females. One striking feature of sex pheromone detection in male moths is the high degree of specificity and sensitivity at all levels from sensory to behavior. In recent years, significant progress has been made elucidating the molecular mechanisms underlying the reception of sex pheromones, which involve several molecular components, such as pheromone-binding proteins, olfactory receptor co-receptor proteins, sex pheromone receptor proteins, and sensory neuron membrane proteins. In this chapter, we focus on these latest advances and discuss what they unraveled about underlying mechanisms of specific and sensitive detection of sex pheromones in moths.

Keywords Sex pheromone reception · Pheromone-binding proteins · Pheromone receptors · Sensory neuron membrane proteins

9.1 Introduction

Male moths utilize sex pheromones emitted by conspecific females to identify and locate their mating partner. For successful mating, male moths have evolved with a highly sensitive olfactory system that can detect minute amounts of sex pheromones dispersed in the air (Leal 2013; Sakurai et al. 2014; Fleischer and Krieger 2018). The sex pheromone detection system in male moths is also highly species specific,

Y. Shiota
Reseach Center for Advanced Science and Technology, The University of Tokyo, Tokyo, Japan

T. Sakurai (✉)
Department of Agricultural Innovation for Sustainable Society, Faculty of Agriculture, Tokyo University of Agriculture, Atsugi, Japan
e-mail: ts206448@nodai.ac.jp

Y. Ishikawa (ed.), *Insect Sex Pheromone Research and Beyond*, Entomology Monographs, https://doi.org/10.1007/978-981-15-3082-1_9

such that hybridization between different species is prevented (Baker 2008). Mechanisms underlying such a sensitive and specific pheromone detection system have attracted the interest of many researchers and have been a major topic in the field of insect olfaction.

Analysis of the physiological properties of the male olfactory system started when Schneider (1957) developed electroantennogram (EAG) recording, which enabled the detection of electrophysiological responses of antennae to odor stimulation. Subsequent development of single sensillum recording (SSR) has enabled the analysis of electrophysiological responses of single olfactory receptor neurons (ORNs), which constitute the physiological basis of odor and pheromone discrimination at the peripheral level (Schneider and Boeckh 1962). These studies have revealed the marked ability of sex pheromone detection systems in male moths. For example, pheromone-sensitive ORNs of male silkmoth antennae can theoretically detect a single bombykol molecule (Kaissling 1987).

In addition to high sensitivity and specificity, mechanisms that control the temporal response properties of pheromone-sensitive ORNs are an important subject. In the natural environment, pheromone plumes consist of discontinuous intermittent odor filaments (Murlis and Jones 1981; Murlis et al. 1992, 2000). The pheromone source orientation behavior of male moths must be adapted to track these discontinuously distributed pheromones, making detection of intermittent odor filaments with high temporal resolution essential for pheromone source localization (Baker and Kuenen 1982; Baker et al. 1985; Kanzaki et al. 1992, Kramer 1992; Vickers and Baker 1994, 1996; Mafra-Neto and Cardé 1995).

In the last decade, identification and functional analyses of molecular components in pheromone reception, such as pheromone-binding protein (PBP), sex pheromone receptor (PR), sensory neuron membrane protein (SNMP), and pheromone-degrading enzyme (PDE), have shed light on how sex pheromones are detected and discriminated at a molecular level. In this chapter, we first summarize the basic knowledge of pheromone detection by male antennae and then review recent advances in studies on the molecular mechanisms of sex pheromone detection in male moths.

9.2 Structure of Antennae

Male moths can locate their mating partners by sensing and tracking sex pheromones emitted by conspecific females from even 4 km away (Collins and Potts 1932). To orient toward the pheromone source from such a long distance, male moths need to detect sex pheromones with high sensitivity and selectivity by their antennae. Two antennal structures, bipectinate (comb-like) and simple filamentous, are typically observed in moths. Bipectinate antennae have many pairs of side branches that stem from the antennal stalk (Fig. 9.1a). These structures and their arrangement serve as an efficient odorant "sieve" that facilitates capturing odorants in the air (Kohel 2006). Although filamentous antennae have a simple structure, a

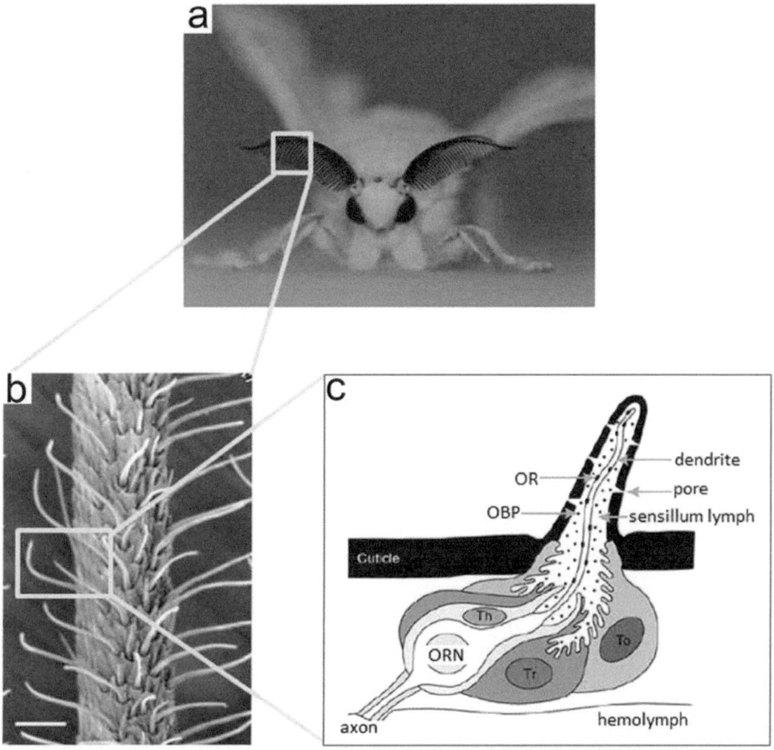

Fig. 9.1 Sensory structure for odorant detection in the silkmoth *Bombyx mori*
(**a**) A male silkmoth with a pair of antennae for odorant detection. (**b**) Scanning electron micrograph of an antenna side branch displaying the external morphology of trichodea sensilla. Scale bar: 25 μm. (**c**) Schematic diagram of an olfactory sensillum showing the detailed configuration of an ORN and accessory cells, and components in the sensillum lymph space. (**a**) and (**b**) modified from Sakurai et al. 2014. (c) modified from Jacquin-Joly and Merlin 2004

recent computational fluid dynamics study using the antennal structure of Heliozelidae moths suggested that arrangements of antennal scales affect fluid dynamics around the antennae, which enhances the efficacy of pheromone detection (Wang et al. 2018). These studies revealed that the structures of male moth antennae are sophisticatedly designed for efficient pheromone detection.

On the surface of an antenna are more than thousands of small cuticular specializations termed olfactory sensilla. These sensilla have numerous minute pores (ten to several tens of nanometers in diameter) through which odorant molecules enter inside sensilla (Fig. 9.1b; Ernst 1969; Slifer 1970; Steinbrecht 1970, 1973, Keil 1982). Olfactory sensilla typically house one to four ORNs, which are surrounded by accessory cells. Dendrites of ORNs innervate into the sensillum lymph space, whereas axons of ORNs project into the brain (Fig. 9.1c). Olfactory sensilla are categorized into several types based on morphology such as trichodea, basiconic, and coeloconic sensilla. In general, trichodea sensilla on male moth antennae house pheromone-sensitive ORNs and exclusively detect pheromones. In the silkmoth *Bombyx mori*, trichodea sensilla are subdivided into long and medium types. Long trichodea sensilla are specifically tuned to pheromone components (Kaissling et al. 1978). Basiconic sensilla detect general odorants such as plant volatiles (Anderson et al. 1995; Pophof 1997). Coeloconic sensilla were reported to detect plant odorants (Pophof 1997), and are speculated to house a sensory neuron for detection of temperature and humidity (Shields and Hildebrand 2001).

9.3 Overview of Pheromone Reception

Sex pheromone molecules in the air are first adsorbed on the cuticular surface of olfactory sensilla, diffuse in the cuticular wax layer, and then enter inside olfactory sensilla through olfactory pores (Kaissling 1987). Sex pheromones are normally highly hydrophobic and difficult to dissolve into sensillum lymph. Here, pheromone-binding proteins (PBPs) in sensillum lymph assist in the solubilization and movement of pheromone molecules in the lymph. When transported to the vicinity of the dendritic membrane of ORNs, pheromones are released from PBPs and detected by PR proteins on the dendritic membrane. Insect odorant receptors (ORs), including PRs, form heteromeric complexes with olfactory receptor co-receptor (Orco). These complexes function as ligand-gated non-selective cation channels, and cause cation influx when a sex pheromone molecule binds to a specific PR. This influx depolarizes the membrane potential that leads to the generation of action potentials in ORNs. After activating PRs, pheromones are rapidly degraded by pheromone-degrading enzymes (PDEs) and/or inactivated via unknown mechanisms (Kaissling 2009). In the following sections, we will review the functions of the above-mentioned molecular components in the context of specificity, sensitivity, and temporal properties of pheromone responses.

9.4 Pheromone-Binding Proteins

PBPs, which belong to the insect OBP (odorant-binding protein) family, are thought to play a role in the efficient solubilization of pheromone molecules into aqueous sensillum lymph, and enhance the selectivity and sensitivity of responses to sex pheromones. The first PBP was identified in 1981 from the giant silkmoth *Antheraea polyphemus* (Vogt and Riddiford 1981). To date, PBPs have been identified from more than 16 lepidopteran species (Fig. 9.2a; Gong et al. 2009; Gu et al. 2013; Krieger et al. 1996; Sun et al. 2013; Liu et al. 2013; Vogt et al. 2015). PBPs are small (15–20 kDa) soluble proteins secreted from trichogen and tormogen cells, and are highly enriched in the sensillum lymph space of trichoid sensilla (Klein 1987; Pelosi et al. 2006; Forstner et al. 2006).

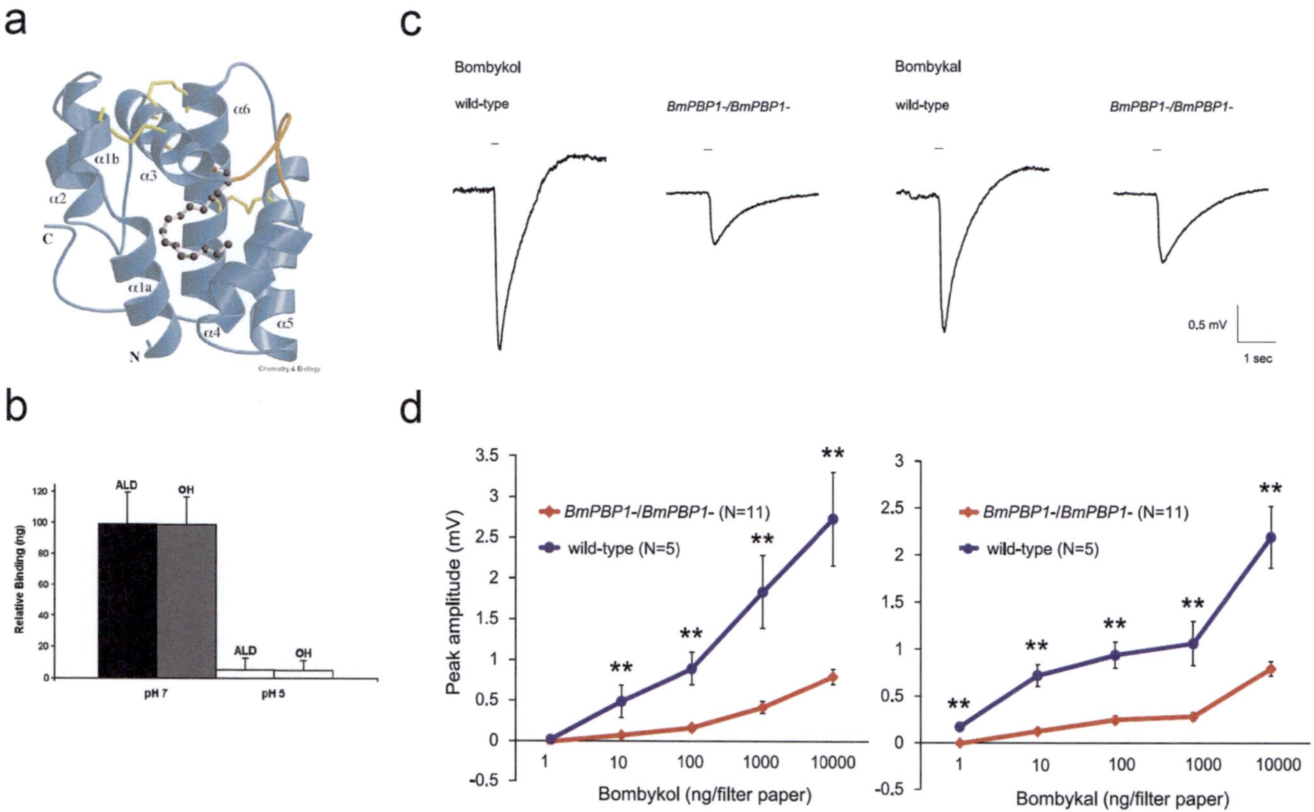

Fig. 9.2 Structure, binding assay, and functional characterization of BmPBP1
(**a**) Overall structure of BmPBP1. Three disulfide bridges are shown in yellow and the bombykol binding pocket is shown in orange (modified from Sandler et al. 2000). (**b**) Binding affinity of bombykol and bombykal to BmPBP1 (modified from Gräter et al. 2006). (**c**) Representative EAG responses of antennae from *BmPBP1-/BmPBP1-* and wild-type male moths to 1000 ng of bombykol. (**d**) Dose-dependent EAG responses of *BmPBP1-/BmPBP1-* and wild-type male moth antennae to bombykol (left) and bombykal (right). (**c**) and (**d**) modified from Shiota et al. 2018

Previous studies suggested that PBPs (1) transport and/or solubilize hydrophobic pheromone molecules and act as their carriers in sensillum lymph (Vogt and Riddiford 1981), (2) protect pheromone molecules from enzymatic degradation during transportation (Vogt and Riddiford 1981; Vogt et al. 1985; Ishida and Leal 2005), and (3) are involved in the deactivation of pheromone molecules after activation of ORNs (Ziegelberger 1995; Kaissling 2001, 2009, 2013).

As the binding selectivity and affinity of PBPs are very important to understand whether PBPs affect the specificity and sensitivity of pheromone reception, in vitro binding analyses have been carried out using PBPs of several moth species. In the case of *B. mori*, female moths emit (*E,Z*)-10,12-hexadecadien-1-ol (bombykol) as a sex pheromone along with a small amount of its analog (*E,Z*)-10,12-hexadecadienal (bombykal), which negatively modulates the behavioral activity of bombykol (Butenandt et al. 1959; Kaissling et al. 1978).

Although one PBP gene (BmPBP1) and two PBP-like genes (BmPBP2, BmPBP3) are reported in *B. mori*, only BmPBP1 is expressed in sensillum lymph of trichodea sensilla (Forstner et al. 2006). Gräter et al. (2006) found that BmPBP1 binds both bombykol and bombykal with nearly the same affinity, whereas Hooper et al. (2009) reported that BmPBP1 binds bombykol with a higher affinity than bombykal (Fig. 9.2b). This inconsistency in binding selectivity between the two studies may be resolved by in vivo functional analysis using recently developed genome editing techniques (TALENs, CRISPR). *B. mori* is one of the model insects amenable to genome editing (Takasu et al. 2013; Daimon et al. 2014). Using transcription activator-like nuclease (TALEN)-mediated gene targeting, *BmPBP1*-knockout male moths were generated to examine its functional role by both behavioral and physiological experiments (Fig. 9.2c; Shiota et al. 2018). The peak EAG responses of *BmPBP1*-knockout male antennae to both bombykol and bombykal were significantly lower than those of wild-type male antennae, indicating that BmPBP1 functions in the sensitive detection of pheromones, but not in the discrimination of the two compounds. Likewise, in the cotton bollworm *Helicoverpa armigera*, EAG responses of the antennae of *PBP1*-knockout males to the three sex pheromone components were reduced by the same degree (Ye et al. 2017). We will further discuss the molecular mechanisms of highly specific pheromone recognition in moths in the *Specificity of pheromone receptor* section below.

9.5 Sex Pheromone Receptors

After delivery by PBPs, pheromones are detected by PRs expressed on the dendritic membrane of ORNs. PRs belong to the insect olfactory receptor family (OR). Insect OR genes were first identified from the fruit fly *Drosophila melanogaster* using a large-scale subtractive cloning strategy and mining the draft version of *Drosophila* genome database in 1999 (Clyne et al. 1999; Vosshall et al. 1999; Gao and Chess 1999). Although insect ORs have 7 transmembrane domains like those in vertebrates and the nematode *Caenorhabditis elegans*, which belong to the G protein-coupled receptor (GPCR) superfamily (reviewed by Touhara and Vosshall 2009), insect ORs form a unique gene family with no notable similarity to other ORs. Indeed, insect ORs have a reversed membrane topology compared with GPCRs with their amino terminal inside and carboxy terminal outside of the cell (Benton et al. 2006; Lundin et al. 2007; Jordan et al. 2009). In *B. mori*, a pair of ORNs housed in a trichodea sensillum respond to bombykol and bombykal, respectively (Kaissling et al. 1978). Moth PRs were first identified from *B. mori* by screening male antennae-specific ORs (Fig. 9.3a, b; Sakurai et al. 2004). Two PRs, BmOR1 and BmOR3, specifically responded to bombykol and bombykal, respectively, when coexpressed with BmOrco (formerly called BmOR2) in a *Xenopus* oocyte heterologous expression system (Fig. 9.3c, d; Nakagawa et al. 2005). These two PRs are mutually exclusively expressed in a pair of ORNs in pheromone-sensitive sensilla, suggesting that specific responses of pheromone-sensitive ORNs are conferred by strict ligand selectivity of PRs (Nakagawa et al. 2005). Some other studies, however, reported that interaction between PRs and PBPs is important for specific responses to sex pheromone components (Große-Wilde et al. 2006; see the *Specificity determinant of pheromone receptors* section for details). In addition to BmOR1 and BmOR3, three male-specific or -predominant ORs with amino acid sequence similarity to BmOR1 and BmOR3 were found in the genome of *B. mori*. These ORs, however, did not respond to bombykol and bombykal in heterologous expression systems (Nakagawa et al. 2005; data not shown), and ligands for these ORs remain unknown.

To date, many candidate PR genes have been identified from several moth species. Molecular phylogenetic analyses revealed that moth PRs form an isolated

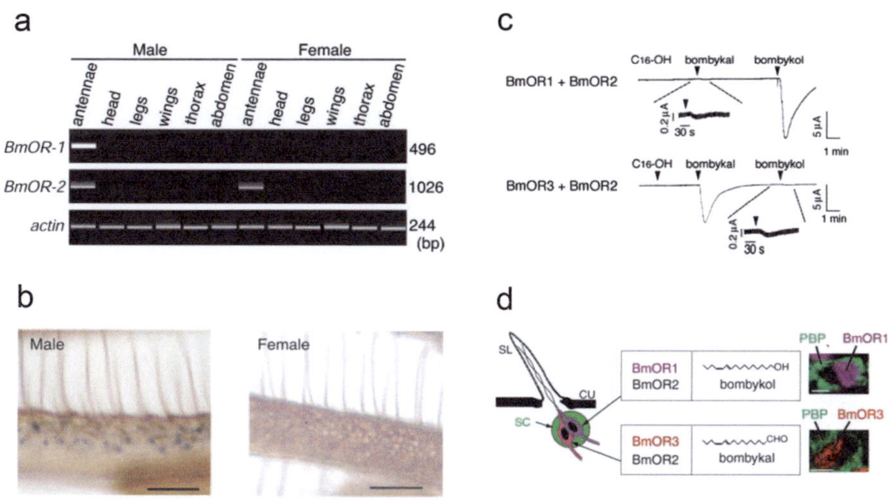

Fig. 9.3 Identification and functional characterization of sex pheromone receptors
(**a**) Tissue- and sex-specific expression of BmOR1 and BmOR2, which is currently known as BmOrco. (**b**) Male-specific expression pattern of BmOR1 by whole-mount in situ hybridization. (**c**) Ligand specificity of BmOR1 and BmOR3 to bombykol and bombykal. (**d**) Schematic drawing of a pheromone-sensitive sensillum (SL) housing BmOR1 and BmOR3, and its ligand. (**a**) and (**b**) modified from Sakurai et al. 2004. (**c**) and (**d**) modified from Nakagawa et al. 2005

clade within the insect OR family, suggesting that PRs have evolved from a single ancestral gene (Mitsuno et al. 2008). Some candidate PRs have been functionally characterized as sex pheromone receptors (Engsontia et al. 2014). In principle, a PR in a moth species is tuned to a specific sex pheromone component of the species, although some PRs respond to multiple pheromone components. It should be noted that not all ORs in the PR clade are tuned to sex pheromones. For example, one OR in the PR clade in the light brown apple moth *Epiphyas postvittana*, EpOR1, responds to plant-derived odorants but not to sex pheromone components of this species (Jordan et al. 2009).

9.6 Sensory Neuron Membrane Protein

Sensory neuron membrane proteins (SNMPs), which belong to the CD36 scavenger family, are thought to function in pheromone detection in moths (Vogt et al. 2009). SNMP was first identified from *A. polyphemus* (Rogers et al. 1997) and then from *M. sexta* (Robertson et al. 1999). In moths, SNMP consists of two subfamilies, SNMP1 and SNMP2. In *Heliothis virescens*, HvSNMP1 is expressed in an ORN in pheromone-sensitive sensilla along with HvOR13, and HvSNMP2 is expressed in supporting cells that surround the ORN (Forstner et al. 2008). This suggests that SNMP1 is related to pheromone detection, but the function of SNMP2 is unclear. Similar expression profiles were reported for *A. Polyphemus*, suggesting that the functions of SNMP1 and SNMP2 are conserved among moth species.

Functions of SNMP family genes were first reported for *D. melanogaster*. In an in vivo functional study, an SNMP mutant fly did not respond to *cis*-vaccenyl acetate (cVA), one of the sex pheromones for this fly (Benton et al. 2007; Jin et al. 2008), demonstrating that SNMP is essential for pheromone detection (Fig. 9.4a). SNMP is needed for functional reconstruction of a *H. virescens* PR (HvOR13) in cVA-sensitive ORNs of *D. melanogaster* (Benton et al. 2007), suggesting the necessity of SNMP for sex pheromone reception in moths (Figs. 9.4b–e). On the other hand, expression of SNMP was not necessary for the functional reconstruction of moth sex pheromone receptors in a heterologous expression system using *Xenopus* oocytes or Sf21 cells; coexpression of PR and Orco conferred cell responsiveness to sex pheromones in the absence of SNMPs (Nakagawa et al. 2005; Mitsuno et al. 2015).

SNMP1 was reported to affect the response kinetics of *Xenopus* oocytes expressing BmOR1 and BmOrco; coexpression of BmSNMP1 with BmOrco and BmOR1 resulted in faster activation and termination of electrical responses to bombykol than in cells not expressing BmSNMP1 (Fig. 9.4f, g; Li et al. 2014). This suggests that SNMP1 modulates the temporal resolution of responses to sex pheromones. Further studies on SNMPs using different methods, such as in vivo functional characterization, may clarify the function of SNMPs.

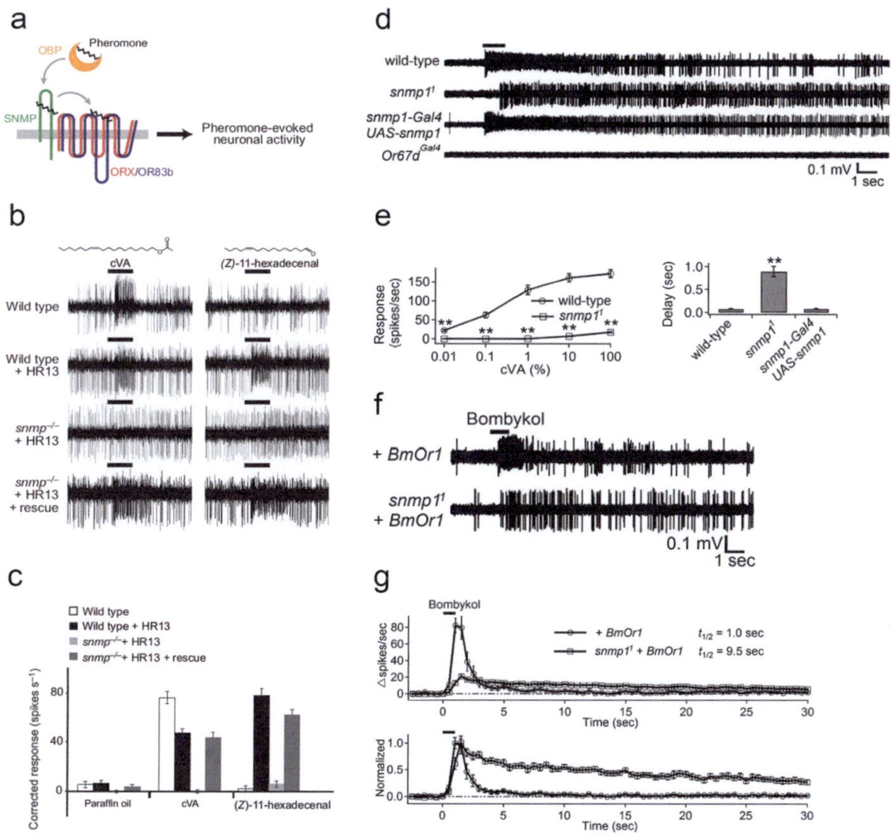

Fig. 9.4 Functional characterization of SNMP in insects
(**a**) Proposed schematic diagram of SNMP. (**b**) Representative responses of OR67d neurons in wild-type, HR13-expressing OR67d neurons, HR13-expressing OR67d neurons in SNMP-/SNMP-, and HR13-expressing OR67d neurons in SNMP rescue animals. (**c**) Responses to cVA and (*Z*)-11-hexadecenal. (**a**)–(**c**) modified from Benton et al. (2007). (**d**) Representative responses of OR67d neurons in wild-type, OR67d SNMP mutant, SNMP rescue animals, and OR67d mutant. (**e**) Quantification of peak firing rate (left) and onset delay (right) of responses. (**f**) Representative responses of OR67d neurons expressing BmOR1 and of OR67d neurons expressing BmOR1 in SNMP mutant to bombykol. (**g**) Response dynamics of OR67d neurons expressing BmOR1 and of OR67d neurons expressing BmOR1 in SNMP mutant to bombykol (upper graph). Lower graph shows the normalized response dynamics by respective peak spiking rates. (**d**)–(**g**) modified from Li et al. (2014)

9.7 Pheromone-Degrading Enzymes

It is essential for odor source localization in moths to detect odor signals with high temporal resolution because odor plumes in nature consist of discontinuous intermittent odor filaments. Moths need to instantly behave in response to intermittent odor filaments during odor source localization. For this, quick inactivation of pheromone molecules that once activated pheromone receptors is highly important. Indeed, male moth antennae can rapidly degrade sex pheromone molecules. Seventeen percent of pheromone molecules that enter inside olfactory sensilla are degraded as rapidly as within 3 ms; the half-life of pheromone molecules is 4–5 min (Kasang 1971, 1973; Kasang and Kaissling 1972; Kasang et al. 1988, 1989a, b).

Pheromone-degrading enzymes (PDEs) are thought to play a central role in this rapid degradation of sex pheromone molecules. A PDE was originally reported for *A. polyphemus* in 1981 along with the discovery of PBP (Vogt and Riddiford 1981), and later, this enzyme was isolated from the antennae (Vogt et al. 1985). Biochemical analysis of *A. polyphemus* PDE revealed that it plays a role in the rapid degradation of the pheromone (*E,Z*)-6,11-hexadecadienyl acetate, with a half-life of 15 ms (Vogt et al. 1985). The PDE gene was cloned by Ishida and Leal (2005). Recombinant PDE exhibited enzymatic activity similar to that by the enzyme expressed in the antennae. Subsequently, two carboxylesterase (CXE) genes, SexiCEX13 and SlituCXE13, were identified from the antennae of *Spodoptera exigua* and *Spodoptera litura*, respectively (He et al. 2014). Both CEXs degraded sex pheromone components of their own species (Z9E12-14:Ac and Z9-14:Ac for *S. exigua*; Z9E12-14:Ac, Z9-14:Ac, and Z9E11-14:Ac for *S. litura*) as well as plant volatiles such as pentyl acetate, hexyl acetate, (*Z*)-3-hexenyl acetate, and (*E*)-2-hexenyl acetate.

In *B. mori*, alcohol oxidase (AOX) plays a key role in sex pheromone degradation: AOX in the antennae of male silkmoths oxidize bombykol into bombykal, and then bombykal into an inactive compound (Rybczynski et al. 1990). Two candidate AOX genes (AOX1 and AOX2) were cloned (Pelletier et al. 2007). Of the two, AOX2 may function in pheromone degradation because only AOX2 is expressed in the pheromone-sensitive long trichodea sensilla of male moths (Pelletier et al. 2007).

Although candidate PDEs have been isolated from several moth species, functional characterization using heterologous expression systems or in vivo analysis of the function of these genes is limited. To clarify how PDEs regulate the temporal resolution of pheromone responses, both identification and in vivo characterization of PDEs from many moth species are indispensable.

9.8 Chemoelectrical Signal Transduction of Odorants and Sex Pheromones

The function of ORs is to transduce chemical signals (odorants) into electrical signals. In contrast to vertebrate ORs, which belong to the GPCR superfamily and transduce signals via the G-protein-coupled second messenger system, insect ORs coupled with Orco function as an odorant-gated non-selective cation channel. Orco was originally isolated as a member of the OR family in *D. melanogaster*, and formerly named Or83b (Vosshall et al. 2000). However, Orco has two characteristics that distinguish it from other conventional ORs: (1) Orco is expressed in most ORNs, whereas conventional ORs are expressed in a small subset of ORNs and (2)

Orco is well conserved among insect species, whereas conventional ORs are highly divergent between and within species. These characteristics suggest that Orco does not encode a receptor for a specific odorant but plays essential roles in odorant detection. Indeed, knockout analysis using *D. melanogaster* revealed that Orco has chaperon-like functions, namely, to locate conventional ORs to the dendritic membrane of ORNs (Larsson et al. 2004; Benton et al. 2006). Analyses of antennal and behavioral responses of Orco-knockout *Ostrinia furnacalis*, *B. mori*, and *Spodoptera littoralis* confirmed that Orco is necessary for the detection of sex pheromones as well as plant-derived odorants by moths (Koutroumpa et al. 2016; Yang et al. 2016; Liu et al. 2017).

The detailed functions of Orco in odorant/sex pheromone detection were investigated by analysis of cultured cells expressing Orco and conventional ORs (Sato et al. 2008; Wicher et al. 2008). Analyses of electrophysiological and pharmacological properties of cells coexpressing BmOR and BmOrco revealed that these two form a heteromeric complex that functions as odorant-/pheromone-activated non-selective cation channel (Fig. 9.5a; Sato et al. 2008). Additionally, they found no evidence of elevation of 2nd messengers after odorant stimulation of these cells, suggesting that this channel complex plays a role in the primary pheromone/odorant transduction pathway in insects. Wicher et al. (2008), on the other hand, found that HEK293 cells expressing *Drosophila* Orco and OR exhibited two currents with different timescales: rapid transient and slow prolonged responses. The authors proposed a transduction model where fast transient responses result from direct activation of Orco by activated OR, whereas slow prolonged responses were mediated by a metabotropic pathway involving G-protein and adenylyl cyclase (Fig. 9.5b). Observations by both Sato et al. (2008) and Wicher et al. (2008) are consistent with the hypothesis that odorant signals are mediated by the odorant-induced channel activity of ORs/Orco complexes or Orco. However, further studies are necessary to build a consensus of the roles of G-protein-mediated 2nd messenger cascades in the reception of odorants and pheromones.

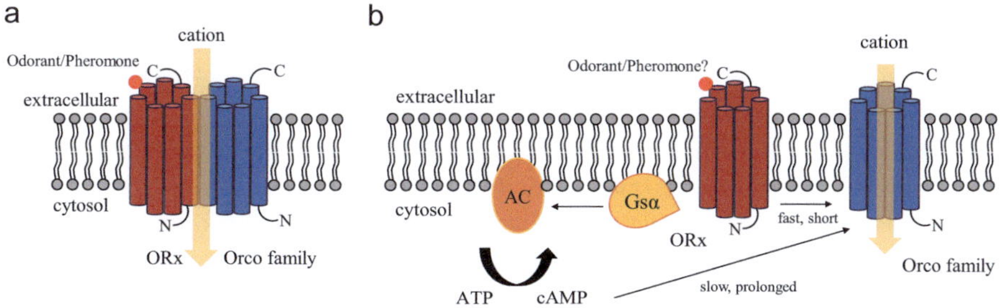

Fig. 9.5 Proposed signal transduction mechanisms in insects
(**a**) Model of the heteromeric pheromone/odorant-gated non-selective cation channel formed by the odorant receptor with an odorant receptor co-receptor family protein. (**b**) Model of two signal transduction pathways. The ionotropic pathway via direct activation of Orco followed by ligand binding to an OR, inducing a fast transient cation influx. The metabotropic pathway is coupled to the G protein, inducing slow prolonged cation currents

9.9 Specificity of Pheromone Receptors

The mechanism that confers specificity to sex pheromone recognition systems in moths has been one of the major subjects in insect olfaction, and several different molecular mechanisms have been proposed. Nakagawa et al. (2005) reported that *Xenopus* oocytes expressing BmOR1/BmOrco respond to bombykol but not to bombykal, and those expressing BmOR3/BmOrco respond to bombykal but not to bombykol. Regarding the ligand specificity of BmOR1 and BmOR3, the same results were obtained in a study using the Sf21 cell line expression system (Mitsuno et al. 2015). These results suggested that the PR determines the specificity of responses. In contrast, Xu et al. (2012) demonstrated that *Xenopus* oocytes expressing BmOR1/BmOrco respond to both bombykol and bombykal, suggesting that selective responses to bombykol are not determined only by BmOR1. In addition, Große-Wilde et al. (2006) reported that HEK293T cells expressing BmOR1 responded to both bombykol and bombykal when these compounds were dissolved in the culture medium using dimethyl sulfoxide (DMSO), whereas these cells selectively responded to bombykol when these compounds were solubilized using BmPBP1. BmPBP1 was considered to function in the specificity of the pheromone receptor by selectively binding bombykol and presenting it to PR. However, we argue against this notion because the response of *BmPBP1*-knockout male antennae to both bombykol and bombykal was reduced to a similar extent, indicating that BmPBP1 enhances the detection of both bombykol and bombykal in *B. mori* (Shiota et al. 2018). Our results strongly support that BmOR1 is the determinant of specificity of the pheromone reception system. Examination of the responses of bombykol- and bombykal-sensitive ORNs in *BmPBP1*-KO antennae by single sensillum recordings may fortify our conclusion.

9.10 Projection Patterns of Pheromone-Sensitive ORNs

Odorants are detected by ORNs and transduced to electrical signals (action potentials). These signals are conveyed to the antennal lobe (AL), the primary olfactory center in the insect brain (Fig. 9.6a; Hildebrand and Shepherd 1997). The AL of the insect brain is a sphere-shaped region in the deutocerebrum and consists of spheroidal structures of neuropils termed glomeruli. The AL includes three types of neurons: ORNs, projection neurons (PNs), and local interneurons (LNs). In male moths, relatively enlarged glomeruli, referred to as the macroglomerular complex (MGC), exclusively receive axonal projections of pheromone-sensitive ORNs (Hildebrand and Shepherd 1997). In general, one ORN expresses one OR (PR) gene, and ORNs expressing the same OR (PR) convergently project to a single glomerulus (Gao and Chess 1999; Vosshall et al. 2000; Sakurai et al. 2011). Single-cell staining revealed that ORNs tuned to a particular pheromone component project to the same subdivision of the MGC (Fig. 9.6b; Hansson et al. 1992; Anton and Hansson 1995). Genetic

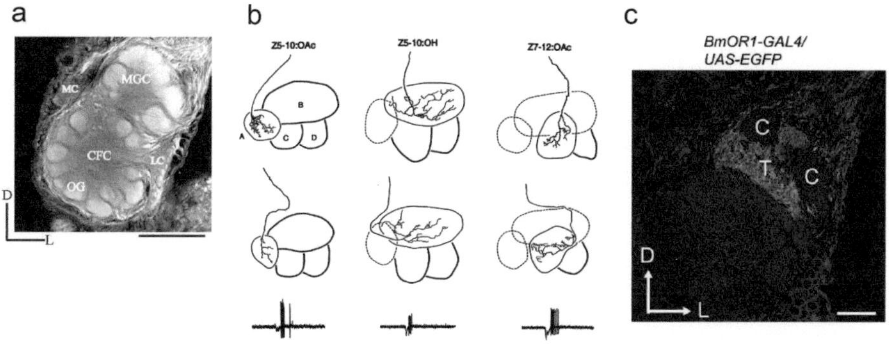

Fig. 9.6 Antennal lobe structure
(**a**) Cell clusters and neural structure of the AL. The AL consists of the macroglomerular complex (MGC), ordinary glomeruli (OG), medial cell cluster (MC), and lateral cell cluster (LC). The central fiber core (CFC) is a region densely innervated by neuropils from AL cells (modified from Iwano and Kanzaki 2005). (**b**) ORNs that respond to different pheromone components innervate different regions in the MGC (modified from Hansson et al. 1992). (**c**) A single population of ORNs expressing one type of OR innervates a specific region of the MGC (modified from Sakurai et al. 2011)

labeling of ORNs expressing a particular PR in *B. mori* also revealed that ORNs expressing the same PR convergently project into the same subdivision of the MGC (Fig. 9.6c; Sakurai et al. 2011). These studies helped to form a topographic map of pheromone component information in the MGC.

9.11 Hard Wiring Between ORNs and Behavioral Responses

Specific responses of ORNs to sex pheromones and resultant induction of sexual behavior have long suggested that sex pheromone information is processed via a specific neural pathway in the brain and coded by the so-called labeled line. Recent studies using silkmoths provided experimental evidence for this long-held assumption (Fig. 9.7a–c; Sakurai et al. 2011, 2015). As silkmoth males exhibit pheromone source orientation behavior only to bombykol, we hypothesized that activation of a single population of ORNs that express BmOR1 is necessary and sufficient to drive pheromone source orientation behavior. To confirm this hypothesis, we generated transgenic silkmoths that express exogenous PR in BmOR1-expressing ORNs. When PxOR1, a PR of the diamondback moth *Plutella xylostella,* tuned to its major pheromone component Z11-hexadecenal (Z11-16:Ald; Mitsuno et al. 2008) was expressed in these ORNs, Z11-16:Ald evoked responses from these ORNs and triggered pheromone source searching behavior in male moths (Fig. 9.7d). Furthermore, optogenetic activation of BmOR1-expressing ORNs mediated by ectopic expression of a blue-light gated ion channel, channel rhodopsin 2 (ChR2), triggered typical pheromone source searching behavior in response to blue light stimulation (Tabuchi et al. 2013). We also generated BmOR1 gene knockout moths using the

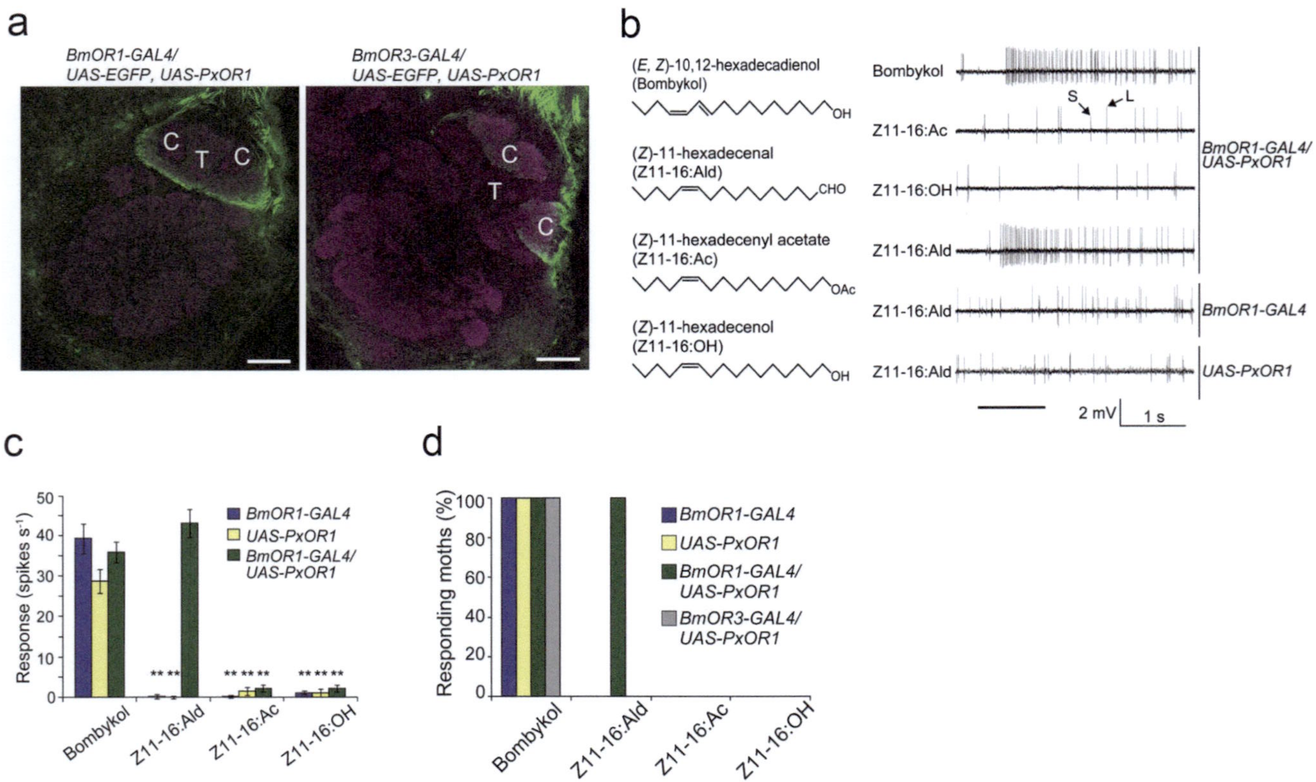

Fig. 9.7 A single subset of ORNs determines response specificity
(**a**) Axon terminations of bombykol or bombykal receptor neurons with ectopic expression of PxOR1. (**b**) Representative physiological profiles of PxOR1-expressing bombykol receptor neurons. (**c**) The responses of PxOR1-expressing bombykol receptor neurons to bombykol and pheromone components of *P. xylostella*. (**d**) Behavioral responses of PxOR1-expressing male moths (modified from Sakurai et al. 2011)

genome-editing tool TALEN. BmOR1-knockout male moths completely lost both antennal and behavioral responses to bombykol (Sakurai et al. 2015). These studies provided proof for labeled line coding of sex pheromone information in the silkmoth.

Given that ligand specificity of PRs not only determines ORN selectivity but also pheromone preference at the behavioral level, understanding of the effects of mutations in PR on ligand specificity will help understand the evolution of pheromone preferences in male moths. In *Ostrinia* species, mutation of a single amino acid residue of a PR protein markedly altered its ligand specificity, suggesting that change of only a single amino acid residue can affect sex pheromone preference (Leary et al. 2012). PRs from two related species *Helicoverpa assulta* (HassOR14b) and *H. armigera* (HarmOr14b) respond to Z9-16:Ald and Z9-14:Ald, respectively (Yang et al. 2017). Replacement of only two amino acids in these PRs accounted for the difference in ligand specificity. These studies suggest that ligand selectivity of PRs can be changed by mutations of a few amino acid residues, which can lead to alteration of the pheromone preference in male moths.

9.12 Effects of Nonpheromonal Odorants

In nature, male moths often encounter pheromones and plant odors simultaneously. Plant odors influence and modulate the sensitivity of mating behavior and odor quality recognition in moths. Synergism between host volatiles and pheromones is reported in many moths (Reddy and Guerrero 2010): linalool and (*Z*)-3-hexenol (plant volatiles) and (*Z*)-11-hexadecenal (pheromone) in *Helicoverpa zea* (Ochieng et al. 2002), linalool and (*Z,E*)-9,11-tetradecadienyl acetate in *S. littoralis* (Party et al. 2009), and a mulberry leaf volatile *cis*-3-hexen-1-ol and bombykol in *B. mori* (Namiki et al. 2008). In the field trap tests, a greater number of *H. zea* and *Cydia pomonella* were captured by pheromone traps when green leaf volatiles, such as (*Z*)-3-hexenyl-acetate, were added to the pheromones (Light et al. 1993).

From a physiological standpoint, in *H. zea*, stimulation with a mixture of (*Z*)-11-hexadecenal and linalool significantly increased the firing rate of pheromone-sensitive neurons compared with stimulation by the major pheromone component alone, whereas the plant volatile linalool alone did not elicit responses from pheromone-sensitive neurons (Ochieng et al. 2002). Similar synergistic effects of plant volatiles on pheromones were reported for *B. mori* (Namiki et al. 2008). When male *B. mori* moths were exposed to a mixture of bombykol and *cis*-3-hexen-1-ol, the spike responses of PNs were significantly increased compared with those to bombykol alone. Furthermore, the behavior sensitivity was higher when male moths were exposed to a mixture of bombykol and *cis*-3-hexen-1-ol, indicating that moths become more sensitive at both the behavioral and primary brain center levels due to synergistic interaction between host plant odors and pheromones. Of note, these synergistic effects were observed only when bombykol and *cis*-3-hexen-1-ol were applied to the same region of the antennae. This suggests that interaction of the two odorants occurred at the antennal level.

At the level of sex pheromone receptors, attenuating effects of plant odorants occur in *H. virescens*. A mixture of main sex pheromone component, (Z)-11-hexadecenal, and a plant volatile, such as linalool, inhibited responses of the cumulus in the MGC (Pregitzer et al. 2012). Similar inhibition by plant volatiles was observed for responses of HEK293 cells expressing HR13 to (Z)-11-hexadecenal, demonstrating that plant odorants interfere with the response to sex pheromones at the pheromone receptor level. These results suggest that moths can interpret both plant odors and pheromone information at the same time.

9.13 Closing Remarks

Since the first identification of sex pheromone chemicals from the silkmoth by Butenandt et al. in 1959, mechanisms of sex pheromone reception have been one of the main topics in the field of insect olfactory research. Studies in recent years have greatly advanced our understanding of the molecular mechanisms underlying the sensitive and specific detection of sex pheromones by male moth antennae. However, despite such advancements, important unanswered questions remain. For example, the molecular determinants of pheromone specificity and the signal transduction mechanisms of pheromone detection are still controversial. In addition, the molecular mechanisms that control the temporal properties of pheromone-sensitive ORNs that are required for efficient pheromone source localization are not fully understood. For a full understanding of the molecular process of pheromone reception, a systematic approach that can examine how each molecular component involved in pheromone reception functions in the context of an in vivo molecular network will be important.

References

Anderson P, Hansson BS, Löfqvist J (1995) Plant-odour-specific receptor neurones on the antennae of female and male *Spodoptera littoralis*. Physiol Entomol 20(3):189–198

Anton S, Hansson BS (1995) Sex pheromone and plant-associated odor processing in antennal lobe interneurons of male *Spodoptera littoralis* (Lepidoptera: Noctuidae). J Comp Physiol A 176:773–789

Baker TC (2008) Balanced olfactory antagonism as a concept for understanding evolutionary shifts in moth sex pheromone blends. J Chem Ecol 34(7):971–981

Baker TC, Kuenen LP (1982) Pheromone source location by flying moths: a supplementary non-anemotactic mechanism. Science 216(4544):424–427

Baker TC, Willis MA, Haynes KF, Phelan PL (1985) A pulsed cloud of sex pheromone elicits upwind flight in male moths. Physiol Entomol 10(3):257–265

Benton R, Sachse S, Michnick SW, Vosshall LB (2006) Atypical membrane topology and heteromeric function of *Drosophila* odorant receptors *in vivo*. PLoS Biol 4:e20

Benton R, Vannice KS, Vosshall LB (2007) An essential role for a CD36-related receptor in pheromone detection in *Drosophila*. Nature 450:289

Butenandt A, Beckmann R, Stamm D, Hecker E (1959) Über den Sexuallockstoff des Seidenspinners *Bombyx mori*. Reindarstellung und Konstitution. Z Naturforsch 14b:283–284
Clyne PJ, Warr CG, Freeman MC, Lessing D, Kim J, Carlson JR (1999) A novel family of divergent seven-transmembrane proteins: candidate odorant receptors in *Drosophila*. Neuron 22:327–338
Collins CW, Potts SF (1932) Attractants for the flying gypsy moths as an aid in locating new infestations. Techn Bull 336(336):1–44
Daimon T, Kiuchi T, Takasu Y (2014) Recent progress in genome engineering techniques in the silkworm. Bombyx mori Dev Growth Differ 6(1):14–25
Engsontia P, Sangket U, Chotigeat W, Satasook C (2014) Molecular evolution of the odorant and gustatory receptor genes in lepidopteran insects: implications for their adaptation and speciation. J Mol Evol 79:21–39
Ernst KD (1969) Die Feinstruktur von Riechsensillen auf der Antenne des Aaskiifers *Necrophorus* (Coleoptera). Structure 94:72–102
Fleischer J, Krieger J (2018) Insect pheromone receptors – key elements in sensing intraspecific chemical signals. Front Cell Neurosci 12:425
Forstner M, Gohl T, Breer H, Krieger J (2006) Candidate pheromone binding proteins of the silkmoth *Bombyx mori*. Invertebr Neurosci 6(4):177–187
Forstner M, Gohl T, Gondesen I, Raming K, Breer H, Krieger J (2008) Differential expression of SNMP-1 and SNMP-2 proteins in pheromone-sensitive hairs of moths. Chem Senses 33(3):291–299
Gao Q, Chess A (1999) Identification of candidate *Drosophila* olfactory receptors from genomic DNA sequence. Genomics 60(1):31–39
Gong DP, Zhang HJ, Zhao P, Xia QY, Xiang ZH (2009) The odorant binding protein gene family from the genome of silkworm, *Bombyx mori*. BMC Genomics 10:332
Gräter F, Xu W, Leal W, Grubmüller H (2006) Pheromone discrimination by the pheromone-binding protein of *Bombyx mori*. Structure 14(10):1577–1586
Große-Wilde E, Svatos A, Krieger J (2006) A pheromone-binding protein mediates the bombykol-induced activation of a pheromone receptor *in vitro*. Chem Senses 31:547–555
Gu SH, Zhou JJ, Wang GR, Zhang YJ, Guo YY (2013) Sex pheromone recognition and immunolocalization of three pheromone binding proteins in the black cutworm moth *Agrotis ipsilon*. Insect Biochem Mol Biol 43(3):237–251
Hansson BS, Ljungberg H, Hallberg E, Lofstedt C (1992) Functional specialization of olfactory glomeruli in a moth. Science 256(5061):1313–1315
He P, Li ZQ, Liu CC, Liu SJ, Dong SL (2014) Two esterases from the genus *Spodoptera* degrade sex pheromones and plant volatiles. Genome 57(4):201–208
Hildebrand JG, Shepherd GM (1997) Mechanisms of olfactory discrimination: converging evidence for common principles across phyla. Annu Rev Neurosci 20:595–631
Hooper AM, Dufour S, He X, Muck A, Zhou J, Almeida R, Field LM, Svatos A, Pickett JA (2009) High-throughput ESI-MS analysis of binding between the *Bombyx mori* pheromone-binding protein BmorPBP1, its pheromone components and some analogues. Chem Commun 38:5725–5727
Ishida Y, Leal WS (2005) Rapid inactivation of a moth pheromone. Proc Natl Acad Sci 102:14075–14079
Iwano M, & Kanzaki R. (2005). Immunocytochemical identification of neuroactive substances in the antennal lobe of the male silkworm moth Bombyx mori. Zoological Science 22(2):199–211
Jacquin-Joly E, Merlin C (2004) Insect olfactory receptors: contributions of molecular biology to chemical ecology. J Chem Ecol 30:2359–2397
Jin X, Ha TS, Smith DP (2008) SNMP is a signaling component required for pheromone sensitivity in *Drosophila*. Proc Natl Acad Sci 105(31):10996–11001
Jordan MD, Anderson A, Begum D, Carraher C, Authier A, Marshall SD, Kiely A, Gatehouse LN, Greenwood DR, Christie DL, Kralicek AV, Trowell SC, Newcomb RD (2009) Odorant

receptors from the light brown apple moth (*Epiphyas postvittana*) recognize important volatile compounds produced by plants. Chem Senses 34:383–394
Kaissling KE (1987) R. H. Wright lectures on insect olfaction. Simon Fraser Univ, Burnaby
Kaissling KE (2009) Olfactory perireceptor and receptor events in moths: a kinetic model revised. J Comp Physiol A Neuroethol Sens Neural Behav Physiol 195(10):895–922
Kaissling KE (2013) Kinetics of olfactory responses might largely depend on the odorant-receptor interaction and the odorant deactivation postulated for flux detectors. J Comp Physiol A Neuroethol Sens Neural Behav Physiol 199(11):879–896
Kaissling KE (2001) Olfactory perireceptor and receptor events in moths: a kinetic model. Chem Senses 26(2):125–150
Kaissling KE, Kasang G, Bestmann HJ, Stransky W, Vostrowsky O (1978) A new pheromone of the silkworm moth *Bombyx mori* – sensory pathway and behavioral effect. Naturwissenschaften 65(7):382–384
Kanzaki R, Sugi N, Shibuya T (1992) Self-generated zigzag turning of *Bombyx mori* males during pheromone-mediated upwind walking. Zool Sci 9(3):515–527
Kasang G (1971) Bombykol reception and metabolism on the antennae of the silkmoth *Bombyx mori*. In: Ohloff G, Thomas AF (eds) Gustation and olfaction. Academic, London, pp 245–250
Kasang G (1973) Physikochemische vorgange beim riechen des seidenspinners. Naturwissenschaften 60:95–101
Kasang G, Kaissling KE (1972) Specificity of primary and secondary olfactory processes in *Bombyx* antennae. In: Schneider D (ed) International symposium Olfaction and Taste IV. Wissensch Verlagsgesellsch, Stuttgart, pp 200–206
Kasang G, Von Proff L, Nicholls M (1988) Enzymatic conversion and degradation of sex pheromones in antennae of the male silkworm moth *Antheraea polyphemus*. Zeitschrift Fur Naturforschung Sect C J Biosci 43(3–4):275–284
Kasang G, Nicholls M, Keil T, Kanaujia S (1989a) Enzymatic conversion of sex pheromones in olfactory hairs of the male silkworm moth *Antheraea polyphemus*. Zeitschrift Fur Naturforschung – Sect C J Biosci 44(11–12):920–926
Kasang G, Nicholls M, von Proff L (1989b) Sex pheromone conversion and degradation in antennae of the male silkworm moth *Bombyx mori*. L. Experientia 45:81–87
Keil TA (1982) Contacts of pore tubules and sensory dendrites in antennal chemosensilla of a silkmoth: demonstration of a possible pathway for olfactory molecules. Tissue Cell 14(3):451–462
Klein U (1987) Sensillum-lymph proteins from antennal olfactory hairs of the moth *Antheraea polyphemus* (Saturniidae). Insect Biochem 17(8):1193–1204
Kohel MAR (2006) The fluid mechanics of arthropod sniffing in turbulent odor plumes. Chem Senses 31(2):93–105
Koutroumpa FA, Monsempes C, François MC, de Cian A, Royer C, Concordet JP, Jacquin-Joly E (2016) Heritable genome editing with CRISPR/Cas9 induces anosmia in a crop pest moth. Sci Rep 6:29620
Kramer E (1992) Attractivity of pheromone surpassed by time-patterned application of two nonpheromone compounds. J Insect Behav 5(1):83–97
Krieger J, Nickisch-rosenegk E, Mameli M, Pelosi P, Breer H (1996) Binding proteins from the antennae of *Bombyx mori*. Insect Biochem Mol Biol 26(3):297–307
Larsson MC, Domingos AI, Jones WD, Chiappe ME, Amrein H, Vosshall LB (2004) Or83b encodes a broadly expressed odorant receptor essential for *Drosophila* olfaction. Neuron 43:703–714
Leal WS (2013) Odorant reception in insects: roles of receptors, binding proteins, and degrading enzymes. Annu Rev Entomol 58:373–391
Leary GP, Allen JE, Bunger PL, Luginbill JB, Linn CE, Macallister IE, Kavanaugh MP, Wanner KW (2012) Single mutation to a sex pheromone receptor provides adaptive specificity between closely related moth species. Proc Natl Acad Sci 109(35):14081–14086
Li Z, Ni JD, Huang J, Montell C (2014) Requirement for *Drosophila* SNMP1 for rapid activation and termination of pheromone-induced activity. PLoS Genet 10(9):e1004600

Light DM, Flath RA, Buttery RG, Zalom FG, Rice RE, Dickens JC, Jang EB (1993) Host-plant green-leaf volatiles synergize the synthetic sex pheromones of the corn earworm and codling moth (Lepidoptera). Chemoecology 4(3–4):145–152

Liu NY, Liu CC, Dong SL (2013) Functional differentiation of pheromone-binding proteins in the common cutworm *Spodoptera litura*. Comp Biochem Physiol A Mol Integr Physiol 165(2):254–262

Liu Q, Liu W, Zeng B, Wang G, Hao D, Huang Y (2017) Deletion of the *Bombyx mori* odorant receptor co-receptor (BmOrco) impairs olfactory sensitivity in silkworms. Insect Biochem Mol Biol 86:58–67

Lundin C, Käll L, Kreher SA, Kapp K, Sonnhammer EL, Carlson JR, Von Heijne G, Nilsson I (2007) Membrane topology of the *Drosophila* OR83b odorant receptor. FEBS Lett 581:5601–5604

Mafra-Neto A, Cardé RT (1995) Influence of plume structure and pheromone concentration on upwind flight of *Cadra cautella* males. Physiol Entomol 20(2):117–133

Mitsuno H, Sakurai T, Murai M, Yasuda T, Kugimiya S, Ozawa R, Toyohara H, Takabayashi J, Miyoshi H, Nishioka T (2008) Identification of receptors of main sex-pheromone components of three Lepidopteran species. Eur J Neurosci 28:893–902

Mitsuno H, Sakurai T, Namiki S, Mitsuhashi H, Kanzaki R (2015) Novel cell-based odorant sensor elements based on insect odorant receptors. Biosens Bioelectron 65:287–294

Murlis J, Jones C (1981) Fine-scale structure of odour plumes in relation to insect orientation to distant pheromone and other attractant sources. Physiol Entomol 6:71–86

Murlis J, Elkinton JS, Carde RT (1992) Odor plumes and how insects use them. Annu Rev Entomol 37:505–532

Murlis J, Willis MA, Cardé RT (2000) Spatial and temporal structures of pheromone plumes in fields and forests. Physiol Entomol 25:211–222

Nakagawa T, Sakurai T, Nishioka T, Touhara K (2005) Insect sex-pheromone signals mediated by specific combinations of olfactory receptors. Science 307(5715):1638–1642

Namiki S, Iwabuchi S, Kanzaki R (2008) Representation of a mixture of pheromone and host plant odor by antennal lobe projection neurons of the silkmoth *Bombyx mori*. J Comp Physiol A Neuroethol Sens Neural Behav Physiol 194(5):501–515

Ochieng SA, Park KC, Baker TC (2002) Host plant volatiles synergize responses of sex pheromone-specific olfactory receptor neurons in male *Helicoverpa zea*. J Comp Physiol A Neuroethol Sens Neural Behav Physiol 188(4):325–333

Party V, Hanot C, Sain I, Rochat D, Michel R (2009) Plant terpenes affect intensity and temporal parameters of pheromone detection in a moth. Chem Senses 34:763–774

Pelletier J, Bozzolan F, Solvar M, François MC, Jacquin-Joly E, Maïbèche-Coisne M (2007) Identification of candidate aldehyde oxidases from the silkworm *Bombyx mori* potentially involved in antennal pheromone degradation. Gene 404(1–2):31–40

Pelosi P, Zhou JJ, Ban LP, Calvello M (2006) Soluble proteins in insect chemical communication. Cell Mol Life Sci 63(14):1658–1676

Pophof B (1997) Olfactory responses recorded from sensilla coeloconica of the silkmoth *Bombyx mori*. Physiol Entomol 22(3):239–248

Pregitzer P, Schubert M, Breer H, Hansson BS, Sachse S, Krieger J (2012) Plant odorants interfere with detection of sex pheromone signals by male *Heliothis virescens*. Front Cell Neurosci 6:42

Reddy GVP, Guerrero A (2010) New pheromones and insect control strategies. Vitam Horm 83(C):493–519

Robertson HM, Martos R, Sears CR, Todres EZ, Walden KKO, Nardi JB (1999) Diversity of odourant binding proteins revealed by an expressed sequence tag project on male *Manduca sexta* moth antennae. Insect Mol Biol 8(4):501–518

Rogers ME, Sun M, Lerner MR, Vogt RG (1997) Snmp-1, a novel membrane protein of olfactory neurons of the silk moth *Antheraea polyphemus* with homology to the CD36 family of membrane proteins. J Biol Chem 272(23):14792–14799

Rybczynski R, Vogt RG, Lerner MR (1990) Antennal-specific pheromone-degrading aldehyde oxidases from the moths *Antheraea polyphemus* and *Bombyx mori*. J Biol Chem 265:19712–19715

Sakurai T, Nakagawa T, Mitsuno H, Mori H, Endo Y, Tanoue S, Yasukochi Y, Touhara K, Nishioka T (2004) Identification and functional characterization of a sex pheromone receptor in the silkmoth *Bombyx mori*. Proc Natl Acad Sci 101:16653–16658

Sakurai T, Mitsuno H, Haupt SS, Uchino K, Yokohari F, Nishioka T, Kanzaki R (2011) A single sex pheromone receptor determines chemical response specificity of sexual behavior in the silkmoth *Bombyx mori*. PLoS Genet 7(6):e1002115

Sakurai T, Namiki S, Kanzaki R (2014) Molecular and neural mechanisms of sex pheromone reception and processing in the silkmoth *Bombyx mori*. Front Physiol 5:125

Sakurai T, Mitsuno H, Mikami A, Uchino K, Tabuchi M, Zhang F, Kanzaki R (2015) Targeted disruption of a single sex pheromone receptor gene completely abolishes *in vivo* pheromone response in the silkmoth. Sci Rep 5:11001

Sandler BH, Nikonova L, Leal WS, Clardy J (2000) Sexual attraction in the silkworm moth: structure of the pheromone- binding-protein-bombykol complex. Chem Biol 7(2):143–151

Sato K, Pellegrino M, Nakagawa T, Nakagawa T, Vosshall LB, Touhara K (2008) Insect olfactory receptors are heteromeric ligand-gated ion channels. Nature 452:1002

Schneider D (1957) Electrophysiological investigation on the antennal receptors of the silk moth during chemical and mechanical stimulation. Experientia 13(2):89–91

Schneider D, Boeckh J (1962) Rezeptorpotential und nervenimpulse einzel-ner olfaktorischer sensillen der insektenan-tenne. Z Vergl Physiol 45:405–412

Shields VDC, Hildebrand JG (2001) Recent advances in insect olfaction, specifically regarding the morphology and sensory physiology of antennal sensilla of the female sphinx moth *Manduca sexta*. Microsc Res Tech 55(5):307–329

Shiota Y, Sakurai T, Daimon T, Mitsuno H, Fujii T, Matsuyama S, Sezutsu H, Ishikawa Y, Kanzaki R (2018) *In vivo* functional characterisation of pheromone binding protein-1 in the silkmoth, *Bombyx mori*. Sci Rep 8(1):1–8

Slifer EH (1970) The structure of arthropod chemoreceptors. Annu Rev Entomol 15(1):121–142

Steinbrecht RA (1970) Stimulus transfering tubules in insect olfactory receptors. In: Proceeding of 7th International Congress on Electron Microscopy. Grenoble

Steinbrecht RA (1973) Der Feinbau olfaktorischer Sensillen des Seidenspinners (Insecta, Lepidoptera). Rezeptorforts„tze und reizleitender Apparat. Z Zellforsch 139:533–565

Sun M, Liu Y, Wang G (2013) Expression patterns and binding properties of three pheromone binding proteins in the diamondback moth, *Plutella xylostella*. J Insect Physiol 59(1):46–55

Tabuchi M, Sakurai T, Mitsuno H, Namiki S, Minegishi R, Shiotsuki T, Uchino K, Sezutsu H, Tamura T, Haupt SS, Nakatani K, Kanzaki R (2013) Pheromone responsiveness threshold depends on temporal integration by antennal lobe projection neurons. Proc Natl Acad Sci U S A 110(38):15455–15460

Takasu Y, Sajwan S, Daimon T, Osanai-Futahashi M, Uchino K, Sezutsu H, Tamura T, Zurovec M (2013) Efficient TALEN construction for *Bombyx mori* gene targeting. PLoS One 8:e73458

Touhara K, Vosshall LB (2009) Sensing odorants and pheromones with chemosensory receptors. Annu Rev Physiol 71:307–332

Vickers NJ, Baker TC (1994) Reiterative responses to single strands of odor promote sustained upwind flight and odor source location by moths. Proc Natl Acad Sci U S A 91(13):5756–5760

Vickers NJ, Baker TC (1996) Latencies of behavioral response to interception of filaments of sex pheromone and clean air influence flight track shape in *Heliothis virescens* (F.) males. J Comp Physiol A 178:831–847

Vogt RG, Riddiford LM (1981) Pheromone binding and inactivation by moth antennae. Nature 293:161

Vogt RG, Riddiford LM, Prestwich GD (1985) Kinetic properties of a sex pheromone-degrading enzyme: the sensillar esterase of *Antheraea polyphemus*. Proc Natl Acad Sci 82(24):8827–8831

Vogt RG, Miller NE, Litvack R, Fandino RA, Sparks J, Staples J, Friedman R, Dickens JC (2009) The insect SNMP gene family. Insect Biochem Mol Biol 39(7):448–456

Vogt RG, Große-Wilde E, Zhou JJ (2015) The Lepidoptera odorant binding protein gene family: gene gain and loss within the GOBP/PBP complex of moths and butterflies. Insect Biochem Mol Biol 62:142–153

Vosshall LB, Amrein H, Morozov PS, Rzhetsky A, Axel R (1999) A spatial map of olfactory receptor expression in the *Drosophila* antenna. Cell 96(5):725–736

Vosshall LB, Wong AM, Axel R (2000) An olfactory sensory map in the fly brain. Cell 102:147–159

Wang Q, Shang Y, Hilton DS, Inthavong K, Zhang D, Elgar MA (2018) Antennal scales improve signal detection efficiency in moths. Proc R Soc B Biol Sci 285(1874):20172832

Wicher D, Schäfer R, Bauernfeind R, Stensmyr MC, Heller R, Heinemann SH, Hansson BS (2008) *Drosophila* odorant receptors are both ligand-gated and cyclic-nucleotide-activated cation channels. Nature 452:1007–1011

Xu P, Hooper AM, Pickett JA, Leal WS (2012) Specificity determinants of the silkworm moth sex pheromone. PLoS One 7(9):2–11

Yang B, Fujii T, Ishikawa Y, Matsuo T (2016) Targeted mutagenesis of an odorant receptor co-receptor using TALEN in *Ostrinia furnacalis*. Insect Biochem Mol Biol 70:53–59

Yang K, Huang LQ, Ning C, Wang CZ (2017) Two single-point mutations shift the ligand selectivity of a pheromone receptor between two closely related moth species. elife 6:e29100

Ye ZF, Liu XL, Han Q, Liao H, Dong XT, Zhu GH, Dong SL (2017) Functional characterization of PBP1 gene in *Helicoverpa armigera* (Lepidoptera: Noctuidae) by using the CRISPR/Cas9 system. Sci Rep 7(1):8470

Ziegelberger G (1995) Redox-shift of the pheromone-binding protein in the silkmoth *Antheraea polyphemus*. Eur J Biochem 232(3):706–711

Chapter 10
Evolutionary History of Lepidopteran Genes Associated with Sex Pheromone Recognition

Yuji Yasukochi

Abstract Insect genes encoding odorant receptors (ORs) and odorant-binding proteins (OBPs) comprise Lepidoptera (moths and butterflies)-specific subfamilies named pheromone receptors (PRs) and pheromone-binding proteins (PBPs), respectively. Both subfamilies are monophyletic, and some of the genes in these groups are responsible for sex pheromone recognition. Recent progress in sequencing by next-generation sequencers has facilitated the detection of gains, losses, and translocations of these genes, which have occurred in a wide variety of lepidopteran lineages. The evolutionary history of these genes provides valuable insights into how the sex pheromone recognition system has evolved in Lepidoptera.

Keywords Evolutionary history · Pheromone receptors · Pheromone-binding proteins

10.1 Insect Chemosensory-Related Genes

Sex pheromones in moths are odorants emitted by females, and responses of conspecific males to sex pheromones are released when pheromone molecules are detected by the male's olfactory system (for review, see Leal 2013). The genes encoding chemosensory receptors in insects include three families, that is, odorant receptors (ORs), gustatory receptors (GRs), and ionotropic receptors (IRs).

ORs and GRs share a seven-transmembrane domain structure similar to vertebrate G protein-coupled receptors; however, there are several clear differences such as opposite membrane topologies (for review, Nei et al. 2008). GR-like genes are

Y. Yasukochi (✉)
Institute of Agrobiological Sciences, National Agriculture and Food Research Organization, Tsukuba, Ibaraki, Japan
e-mail: yaskoch@affrc.go.jp

Y. Ishikawa (ed.), *Insect Sex Pheromone Research and Beyond*, Entomology Monographs, https://doi.org/10.1007/978-981-15-3082-1_10

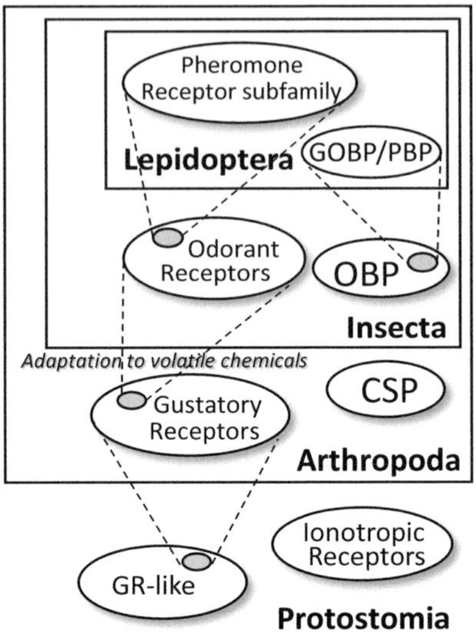

Fig. 10.1 Schematic representation of evolutionary relationships between invertebrate chemosensory proteins. *CSP* chemosensory protein, *GR* gustatory receptor, *OBP* odorant-binding protein, *GOBP* general odorant-binding protein, *PBP* pheromone-binding protein

widely present in animal lineages (Saina et al. 2015; Robertson 2015). OR genes are thought to have evolved from a single lineage of GR genes as an adaptation to detection of volatile chemicals in the terrestrial conditions (Robertson et al. 2003) (Fig. 10.1); however, OR genes are absent in noninsect terrestrial arthropods such as Chelicerata and Myriapoda (Gulia-Nuss et al. 2016; Chipman et al. 2014). Most GRs expanded within each terrestrial arthropod subphylum independently, consistent with their distant evolutionary relationships (Eyun et al. 2017).

ORs function in the form of heterodimers that consist of a ligand-specific subunit and single co-receptor, Orco (Vosshall and Hansson 2011). A recent survey on genomes of noninsect hexapods and basal wingless insects revealed that OR genes are absent in noninsect hexapods and present in basal wingless insects, indicating that ORs are insect-specific (Brand et al. 2018). The Orco gene is not found in the draft genome assembly of a jumping bristletail, *Machilis hrabei*, suggesting that the OR heterodimer system has not yet arisen in the common ancestor of insects, but has evolved after the divergence of jumping bristletails from the other insect lineages (Brand et al. 2018).

Benton et al. (2009) found a large family of divergent ionotropic glutamate receptor-like genes in *Drosophila* and defined them as a novel chemoreceptor family, ionotropic receptors (IRs; for review, see Rytz et al. 2013). IRs have been identified from olfactory and taste sensory organs across protostomes (Croset et al. 2010) (Fig. 10.1), and may play an important role in the olfactory system of noninsect terrestrial arthropods without OR genes (Missbach et al. 2014).

In addition to chemosensory receptor genes, two chemosensory-related gene families, odorant-binding proteins (OBPs) and chemosensory proteins (CSPs), are involved in chemoreception in insects. Both OBPs and CSPs encode small water-soluble proteins abundant in sensillum lymph, which bind and transport odorants to the receptors. OBPs and CSPS are characterized by conserved cysteine residues that form disulfide bridges within the protein (for review, Sánchez-Gracia et al. 2009). CSP is commonly found across Arthropoda, whereas OBP is specific to insects (Vieira and Rozas 2011; Eyun et al. 2017), similar to the relationship between GR and OR (Fig. 10.1). Although the evolutionary relationship between OBP and CSP is not clear, unlike that between GR and OR, they have significantly similar protein structures (Vieira and Rozas 2011).

10.2 Sex Pheromone Receptors in Lepidoptera

10.2.1 Lepidopteran Odorant Receptor Genes and the Sex Pheromone Receptor Subfamily

It is difficult to identify all OR genes by random Sanger sequencing of cDNA clones because antennae are very small organs and the quantity of mRNA prepared from them is too small to achieve sufficient sequence depth. Therefore, determination of the OR gene repertoire in Lepidoptera was first attempted for the silkworm based on the genome sequence (Tanaka et al. 2009).

Identification of odorant receptor genes was accelerated by the advent of next-generation sequencers (NGS). To date, antennal RNA sequencing (RNA-seq) studies have been conducted for a wide variety of lepidopteran lineages including Bombycoidea (Grosse-Wilde et al. 2011), Noctuoidea (Liu et al. 2012), Tortricoidea (Bengtsson et al. 2012), Lasiocampoidea (Zhang et al. 2014), Pyraloidea (Yang et al. 2015), Yponomeutoidea (Yang et al. 2017), and Carposinoidea (Tian et al. 2018). These studies consistently demonstrated that all known moth pheromone receptor genes are included in a single Lepidoptera-specific clade, named the sex pheromone receptor (PR) subfamily, suggesting that the pheromone recognition system in these lineages did not evolve independently multiple times. However, all lepidopteran superfamilies studied to date belong to Ditrysia, which consists of most extant lepidopteran species. Thus, the PR subfamily may have emerged between the divergence of Lepidoptera from Diptera and the radiation of Ditrysia.

Recently, a series of antennal transcriptome analyses provided a clue to the origin of the pheromone recognition system utilized in Ditrysia (Yuvaraj et al. 2017, 2018a, 2018b). Trichoptera (caddisflies) is the closest insect order to Lepidoptera, and no OR genes belonging to the PR subfamily were found in a caddisfly, *Rhyacophila nubile*. A nonditrysian Lepidoptera, *Eriocrania semipurpurella,* which uses short-chain secondary alcohols or ketones as sex pheromone components (Type-0 pheromone), also does not harbor PR subfamily OR genes (Yuvaraj et al. 2017, 2018b). Moreover, *E. semipurpurella* OR genes, which encode ORs responding to Type-0 pheromone, are clearly unrelated to the PR subfamily. In contrast, another nonditrysian Lepidoptera, *Lampronia capitella*, which uses sex pheromone components similar to Type-I pheromone (C10–C18 acetates, alcohols, and aldehydes) utilized in Ditrysia, harbors PR subfamily ORs (Yuvaraj et al. 2018a).

Therefore, the PR subfamily should have been established just prior to the radiation of Ditrysia.

10.2.2 Chromosomal Distribution of Lepidopteran OR Genes

Although transcriptome analysis by NGS is a very high throughput technique, there are several limitations to elucidating the evolutionary history of significantly duplicated chemosensory-related genes. First, bulk RNA derived from the antennae of multiple individuals is used for analysis. Hence, detected sequence variations are very complicated, which makes it difficult to estimate copy-number variations of recently duplicated genes. We previously succeeded in revealing an unknown duplication of OR genes by sequencing introns and intergenic regions using bacteria artificial chromosomes (BAC) (Yasukochi et al. 2011). Second, incomplete coding sequences (CDSs) are often obtained for weakly expressed genes, which may lead to misinterpretation of portions of a single transcript to be independent genes. Lastly, it is impossible to detect nontranscribed CDS, such as pseudogenes, which may be informative for understanding loss-of-function mutations. The order of individual genes in clusters also provides important information regarding gene duplication processes.

Genome annotation is a good method to complement these limitations of transcriptome analyses. However, the length of OR CDS usually reaches nearly 1300 nucleotides, and it is not rare for the total genomic size to exceed 10 kb. Thus, the presence of well-assembled genome sequences is essential for precise gene prediction. Recently, novel sequencing technologies like pac-bio have markedly improved the quality and average length of scaffold sequences. Furthermore, an integrated genome database for Lepidoptera, Lepbase, which facilitates simultaneous similarity searches against nearly 50 species, was recently released (Challis et al. 2016).

As interchromosomal rearrangements rarely occur in Lepidoptera (e.g., Yasukochi et al. 2016), orthologous correspondence of OR genes evolving at a higher rate can be confirmed by their locations relative to neighboring single-copy conserved genes (Yasukochi et al. 2018). Several preceding studies reported the chromosomal distribution of OR genes in the silkworm *Bombyx mori* (Engsontia et al. 2014) and the postman butterfly *Heliconius melpomene* (Briscoe et al. 2013). The comparison of chromosomal OR loci between *B. mori* and *H. melpomene* based on updated genome assemblies of the two species is shown in Fig. 10.2 (SilkBase v2.1 and Lepbase:Hmel 2.5). Locations of OR genes are generally conserved between species, regardless of their distant relationships. One-to-one correspondences of OR genes are even found between Lepidoptera and Trichoptera, suggesting the existence of highly conserved olfactory functions (Yuvaraj et al. 2018b). Lineage-specific losses and expansion of genes are also observed. For example, significant OR gene duplications occurred in *H. melpomene* chromosomes 5 and 6 (Fig. 10.2).

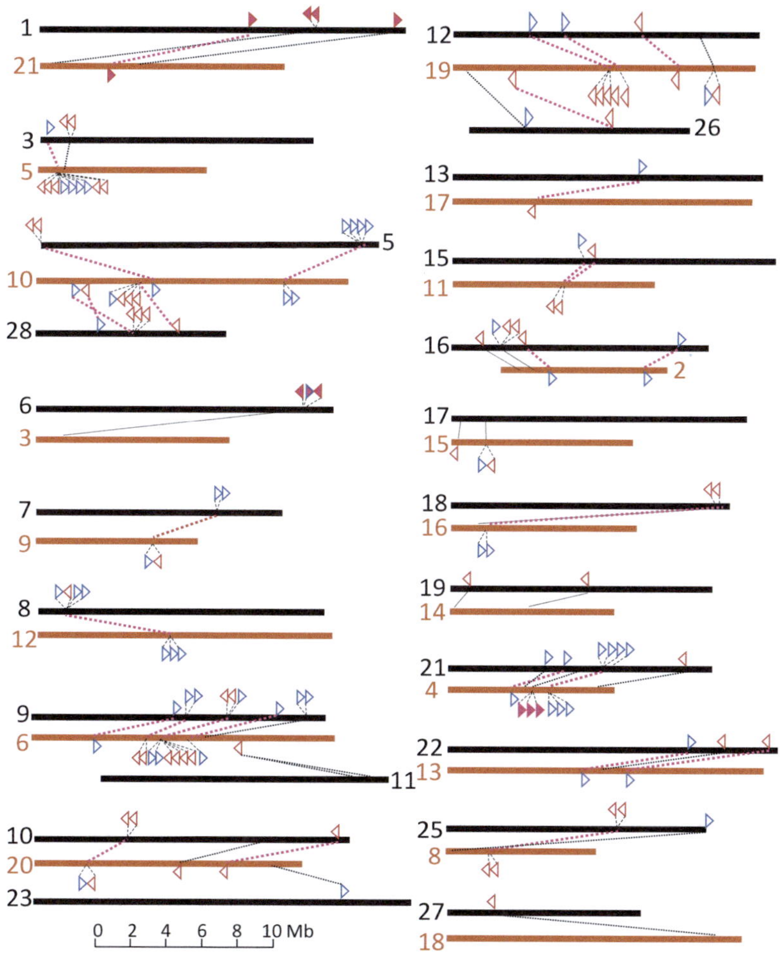

Fig. 10.2 Chromosomal distribution of OR genes in *B. mori* (black) and *H. melpomene* (brown). Chromosome numbers are indicated beside the start point of the chromosomal position. Arrowheads represent the position and transcriptional orientation of OR genes. Magenta arrowheads indicate PR subfamily OR genes. Magenta dotted lines represent the correspondence between *B. mori* and *H. melpomene* OR genes. Black dotted lines represent gene losses in either species

10.2.3 Evolutionary History of PR Subfamily Genes

PR subfamily genes are markedly concentrated on the Z chromosome (Fig. 10.3), suggesting that the common ancestor of PR subfamily genes was localized on the Z chromosome. There are three loci where PR subfamily genes are located in distantly related lineages, presumably because duplication and intrachromosomal translocation had occurred earlier in Ditrysia (Fig. 10.3). Of these, the locus containing *B. mori* OR1 and *H. melpomene* OR3 genes is likely to be that of the ancestral gene

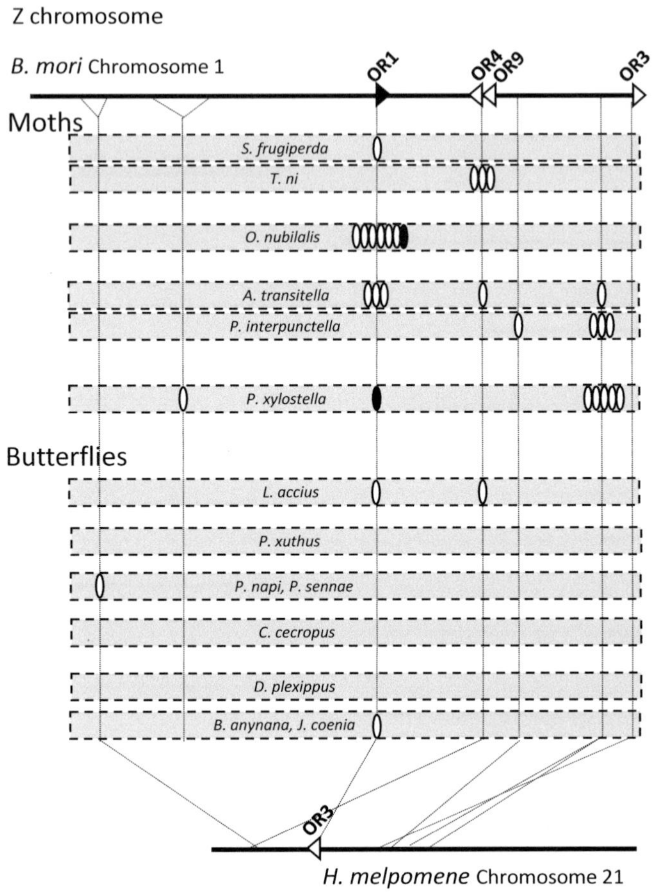

Fig. 10.3 The presence of PR subfamily OR genes in the Z chromosome. Horizontal bars represent *B. mori* (top, Bombycoidea; Bombycidae) and *H. melpomene* (bottom, Papilionoidea; Nymphalidae). Ovals represent PR subfamily OR genes of moths, *S. frugiperda* (Noctuoidea; Noctuidae), *T. ni* (Noctuoidea; Noctuidae), *O. nubilalis* (Pyraloidea; Crambidae), *A. transitella* (Pyraloidea; Pyralidae), *Plodia interpunctella* (Pyraloidea; Pyralidae), *P. xylostella* (Yponomeutoidea; Plutellidae), and butterflies, *Lerema accius* (Hesperiidae), *Papilio xuthus* (Papilionidae), *Pieris napi* (Pieridae), *Phoebis sennae* (Pieridae), *Calycopis cecrops* (Lycaenidae), *Danaus plexippus* (Nymphalidae), *Bicyclus anynana* (Nymphalidae), and *Junonia coenia* (Nymphalidae). Closed ovals represent ORs that respond to sex pheromone components

because OR1 orthologs were reported to respond to respective major sex pheromone components in *B. mori* (Sakurai et al. 2004), the European corn borer *Ostrinia nubilalis* (Wanner et al. 2010), and the diamondback moth *Plutella xylostella* (Sun et al. 2013).

Of note, several butterflies retain PR subfamily genes in this locus (Fig. 10.3), although butterflies do not use female-produced sex pheromones for mate recognition. Sex pheromone biosynthetic pathways are also conserved between moths and butterflies (Liénard et al. 2014), and the male sex pheromone of a butterfly *Bicyclus anynana* increases mating success (Nieberding et al. 2008). Thus, orthologs of moth PR genes may play a role in the recognition of such chemical signals.

PR subfamily genes are also located on autosomes (Fig. 10.4). For example, orthologs of *B. mori* OR6, which is located on *B. mori* chromosome 12, are distributed in several lineages (Fig. 10.4). Thus, these genes are likely to have been translocated to an autosome and gained conserved functions immediately after the split from the other PR subfamily genes. Other putative translocation events to autosomes were likely lineage-specific (Fig. 10.4). Marked gene expansion was previously reported for *P. xylostella* (Engsontia et al. 2014), as this gene cluster was inserted into introns of a cadherin gene in the opposite transcriptional orientation and copy-number variation (10 and 15 copies) due to further gene duplication involving three exons of the cadherin gene was observed between independent genome assemblies (Fig. 10.5).

In some species, PR genes experimentally confirmed to respond to sex pheromones are localized on autosomes. *AtraOR1* of the navel orangeworm *Amyelois transitella* (Xu et al. 2012) is located on a region orthologous to *B. mori* chromosome 4 (Fig. 10.4). Similarly, the trait of male response to sex pheromones in two congeneric Noctuid moths, *Heliothis subflexa* and *Heliothis virescens*, was mapped to chromosome 27, which corresponds to *B. mori* chromosome 14 (Gould et al. 2010). Therefore, *H. virescens* OR6 and OR14–OR16 genes located on chromosome 27 are candidates responsible for the trait, although OR13 located on the Z chromosome has been reported to respond to the most abundant sex pheromone component in *H. virescens* (Gohl and Krieger 2006). Genes orthologous to these genes have been found in other Noctuid moths, *Spodoptera frugiperda* and *Trichoplusia ni* (Fig. 10.4).

Ostrinia OR1 and OR3 genes, located on chromosome 23 (Fig. 10.4) (Yasukochi et al. 2011), exhibit male-biased expression and responses to several sex pheromone components (Miura et al. 2009; Wanner et al. 2010); however, the male response trait was found to be Z-linked (Dopman et al. 2004). Thus, the exact roles of many pheromone receptors remain unknown.

10.3 Pheromone-Binding Proteins in Lepidoptera

10.3.1 Lepidopteran Odorant-Binding Protein Genes and the GOBP/PBP Subfamily

OBP genes have been classified into three subfamilies, classical, Plus-C, and Minus-C, based on the number and position of cysteines in the molecule. According to the transcriptome analyses referred to in this chapter, only Plus-C is monophyletic across Insecta and Minus-C is monophyletic within Lepidoptera. In addition, the GOBP/PBP subfamily, which consists of the GOBP (general odorant-binding

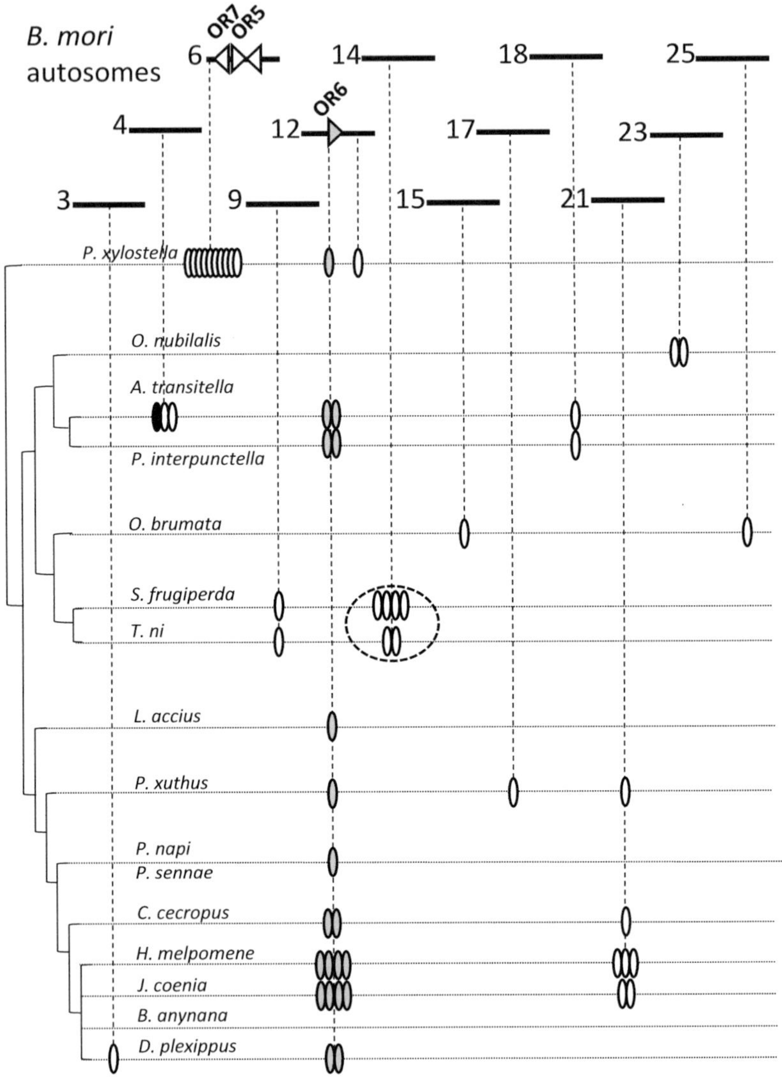

Fig. 10.4 Putative translocation of PR subfamily OR genes to autosomes. Horizontal bars represent *B. mori* autosomes. Numbers indicate chromosome numbers. Ovals represent PR subfamily OR genes, and closed ovals represent ORs that respond to sex pheromone components. Gray ovals represent OR genes belonging to the *B. mori* OR6 clade. Ovals enclosed by a dotted circle indicate orthologs of *H. virescens* OR6, 14–16 genes. *O. brumata*: *Operophtera brumata* (Geometroidea; Geometridae). See the legend to Fig. 10.3 for other species names. Phylogeny of the species is based on Regier et al. (2013)

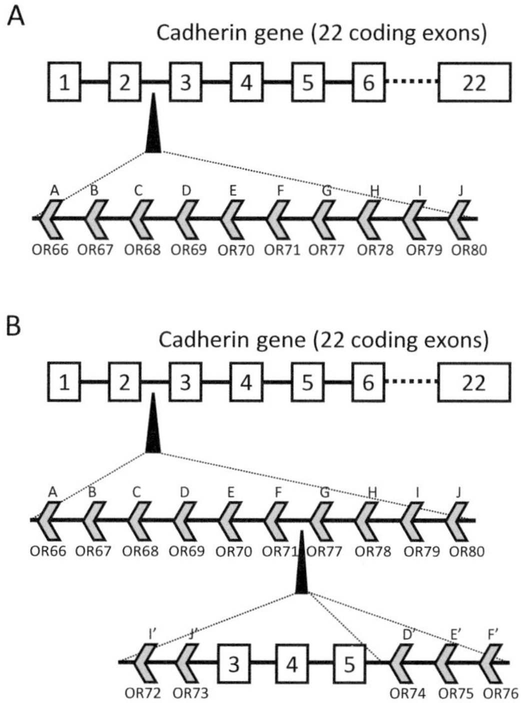

Fig. 10.5 Insertion of duplicated OR genes into a cadherin gene (GenBank Gene ID 105388727) in scaffold sequences (panel A: unitig_15513|quiver; panel B: NW_011952229.1) of independent *P. xylostella* genome assemblies (panel A: pacbiov1; panel B: DBM FJ v 1.1). A–J show types of nearly identical genes. Gene names are according to Engsontia et al. (2014)

protein)-clade and PBP (pheromone-binding protein)-clade, exists in Lepidoptera (Gong et al. 2009). These clades are further classified into six subgroups, GOBP1, GOBP2, PBP-A, PBP-B, PBP-C, and PBP-D, arrayed in the conserved gene order (Vogt et al. 2015). OBPs belonging to the GOBP/PBP subfamily are not found in the caddisfly *R. nubile* (Yuvaraj et al. 2018b). Only GOBP2 genes are found in *E. semipurpurella*, which uses Type-0 pheromones, whereas GOBP1, GOBP2, and PBP genes exist in *L. capitella*, consistent with its usage of Type-1 pheromones (Yuvaraj et al. 2018b). Although transcriptome analysis does not provide solid evidence of the absence of a gene, the emergence of the PBP-clade is likely closely linked with that of Type-1 pheromones.

10.3.2 Chromosomal Distribution of Lepidopteran OBP Genes

The OBP loci in *B. mori* and *H. melpomene* are presented in Fig. 10.6. Compared with OR genes, gene losses and translocations have not occurred so frequently, and not only the position but also the transcriptional orientation is conserved (Fig. 10.6). Plus-C and GOBP/PBP subfamilies are exclusively located on a single chromosome (Fig. 10.6), suggesting that these subfamilies were generated by gene duplications within the chromosomes. Although one lineage-specific gene expansion was found

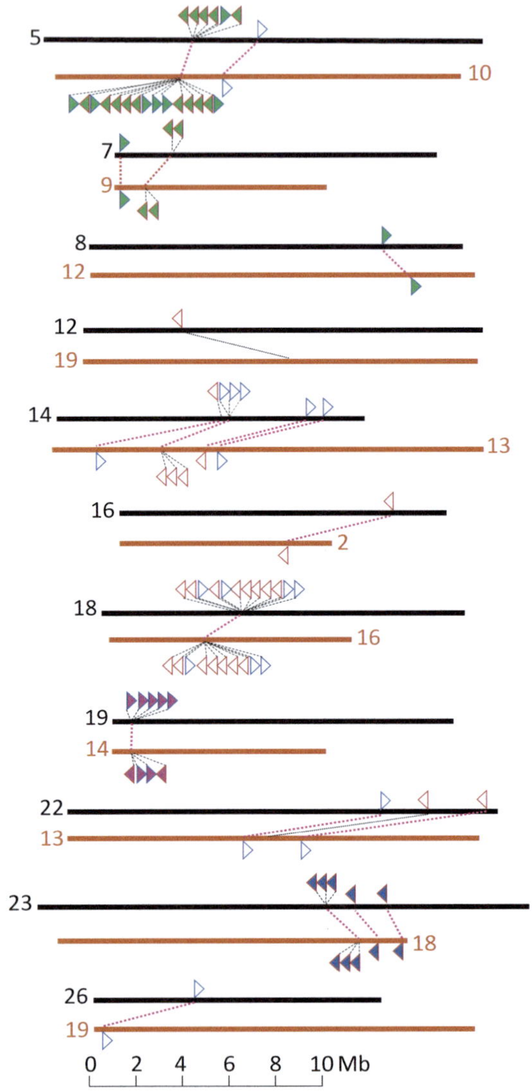

Fig. 10.6 Chromosomal distribution of OBP genes in *B. mori* (black) and *H. melpomene* (brown). Chromosome numbers are indicated beside the start point of the chromosomal position. Arrowheads represent the position and transcriptional orientation of OR genes. Green, magenta, and blue arrowheads represent Minus-C, GOBP/PBP, and Plus-C clades, respectively. Magenta dotted lines represent the correspondence between *B. mori* and *H. melpomene* OBP genes. Black dotted lines represent gene losses in either species

in a Minus-C gene cluster on *H. melpomene* chromosome 5 (Fig. 10.6), other clusters may be an assembly of individual conserved genes in the same order.

The detailed comparison of the GOBP/PBP complex in different lepidopteran lineages is shown in Fig. 10.7. Although overall gene orders are well conserved, lineage-specific gains, losses, inversions, and translocations are frequent (Yasukochi et al. 2018). Of these, PBP-A genes have been specifically lost in many butterfly lineages (Vogt et al. 2015; Yasukochi et al. 2018), suggesting that PBP-A genes are specialized in sex pheromone recognition. As described above, it is likely that basal moths using Type-0 pheromones possess several GOBP2 genes (Yuvaraj et al.

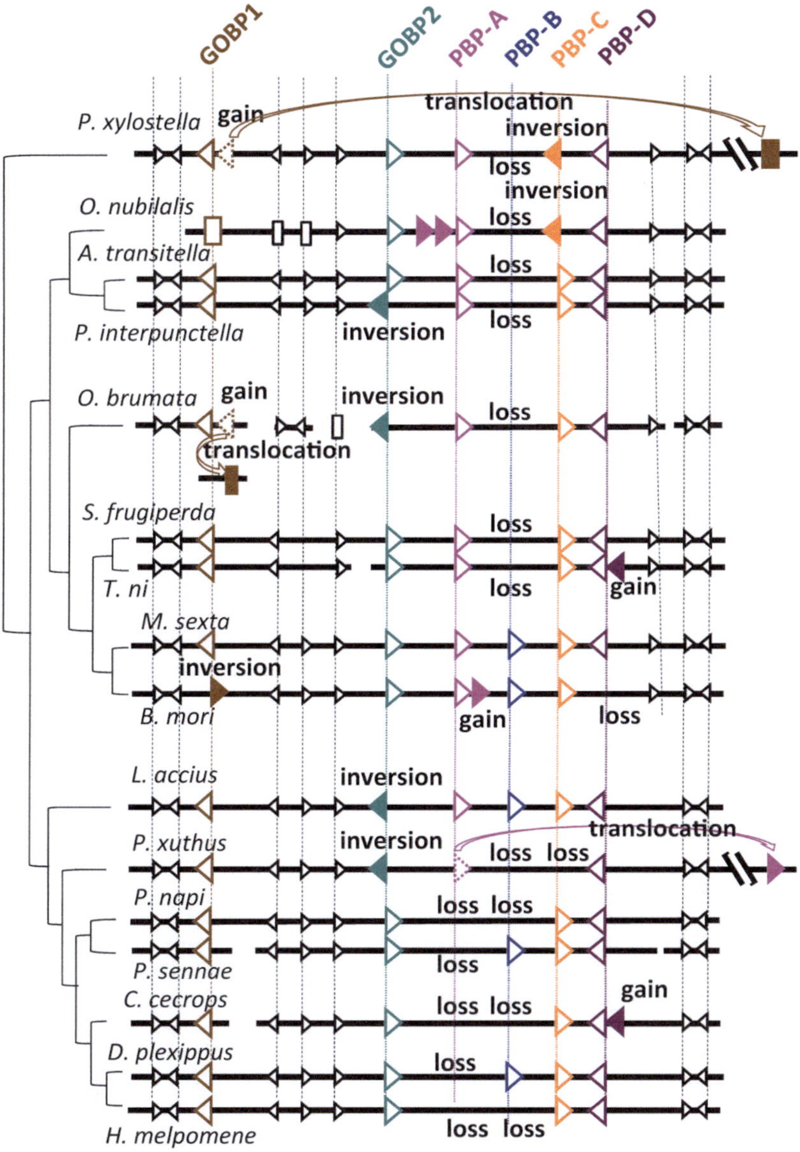

Fig. 10.7 Schematic representation of the order and transcriptional orientation of the GOBP/PBP complex of species listed above and *Manduca sexta* (Bombycoidea; Sphingidae). Arrowheads represent CDSs for which the transcriptional orientation is identified. Squares represent CDSs for which the transcriptional orientation is not identified. Closed arrowheads and squares represent lineage-specific gains and inversions, respectively. Brown, GOBP1; green, GOBP2; magenta, PBP-A; blue, PBP-B; orange, PBP-C; purple, PBP-D; black, neighboring non-OBP genes

2018b). Thus, it is plausible that the conversion of one of the GOBP2 genes to a PBP-A gene was the first step for adaptation to Type-I pheromones, and other PBP genes were generated by further gene duplication and gain of novel functions. Recent duplications of PBP-A genes were observed in the Crambidae lineage (Fig. 10.7). Of note, the first copy of the GOBP2 gene exhibits typical male-biased expression, whereas the third copy of the PBP-A gene does not (Yasukochi et al. 2018).

10.4 Conclusion

The PR subfamily of OR genes and the GOBP/PBP subfamily of OBP genes emerged and immediately evolved after the split of Lepidoptera and Trichoptera until the radiation of Ditrysia. This evolutionary process may be closely associated with the shift from Type-0 to Type-1 pheromones. However, both subfamilies include genes not associated with sex pheromone recognition that gained novel functions.

Acknowledgments I greatly appreciate Yukio Ishikawa for providing me the opportunity to publish this chapter and important contributions to our articles referred to in this chapter. I also thank Ken Sahara, Nami Miura, Bin Yang, Ryo Nakano, and Takashi Matsuo for their helpful assistance.

References

Bengtsson JM, Trona F, Montagne N et al (2012) Putative chemosensory receptors of the codling moth, *Cydia pomonella*, identified by antennal transcriptome analysis. PLoS One 7:e31620

Benton R, Vannice KS, Gomez-Diaz C et al (2009) Variant ionotropic glutamate receptors as chemosensory receptors in *Drosophila*. Cell 136:149–162

Brand P, Robertson HM, Lin W et al (2018) The origin of the odorant receptor gene family in insects. elife 7:e38340

Briscoe AD, Macias-Munoz A, Kozak KM et al (2013) Female behaviour drives expression and evolution of gustatory receptors in butterflies. PLoS Genet 9:E1003620

Challis RJ, Kumar S, Dasmahapatra KK et al (2016) Lepbase: the Lepidopteran genome database. bioRxiv preprint. https://doi.org/10.1101/056994

Chipman AD, Ferrier DEK, Brena C et al (2014) The first myriapod genome sequence reveals conservative arthropod gene content and genome organisation in the centipede *Strigamia maritima*. PLoS Biol 12:e1002005

Croset V, Rytz R, Cummins SF et al (2010) Ancient protostome origin of chemosensory ionotropic glutamate receptors and the evolution of insect taste and olfaction. PLoS Genet 6:e1001064

Dopman EB, Bogdanowicz SM, Harrison RG (2004) Genetic mapping of sexual isolation between E and Z pheromone strains of the European corn borer (*Ostrinia nubilalis*). Genetics 167:301–309

Engsontia P, Sangket U, Chotigeat W (2014) Molecular evolution of the odorant and gustatory receptor genes in lepidopteran insects: implications for their adaptation and speciation. J Mol Evol 79:21–39

Eyun SI, Soh HY, Posavi M et al (2017) Evolutionary history of chemosensory-related gene families across the arthropoda. Mol Biol Evol 34:1838–1862

Gohl T, Krieger J (2006) Immunolocalization of a candidate pheromone receptor in the antenna of the male moth, *Heliothis virescens*. Invertebr Neurosci 6:13–21

Gong D, Zhang H, Zhao P et al (2009) The odorant binding protein gene family from the genome of silkworm, *Bombyx mori*. BMC Genomics 10:332

Gould F, Estocka M, Hillierb NK et al (2010) Sexual isolation of male moths explained by a single pheromone response QTL containing four receptor genes. Proc Natl Acad Sci U S A 107:8660–8665

Grosse-Wilde E, Kuebler LS, Bucks S, Vogel H, Wicher D, Hansson BS et al (2011) Antennal transcriptome of Manduca sexta. Proc Natl Acad Sci 108:7449–7454. https://doi.org/10.1073/pnas.1017963108. PMID:21498690

Gulia-Nuss M, Nuss AB, Meyer JM et al (2016) Genomic insights into the *Ixodes scapularis* tick vector of Lyme disease. Nat Commun 7:10507

Leal WS (2013) Odorant reception in insects: roles of receptors, binding proteins, and degrading enzymes. Annu Rev Entomol 58:373–391

Liénard MA, Wang HL, Lassance JM et al (2014) Sex pheromone biosynthetic pathways are conserved between moths and the butterfly Bicyclus anynana. Nat Commun 5:3957

Liu Y, Gu S, Zhang Y et al (2012) Candidate olfaction genes identified within the *Helicoverpa armigera* antennal transcriptome. PLoS One 7:e48260

Missbach C, Dweck HK, Vogel H et al (2014) Evolution of insect olfactory receptors. elife 3:e02115

Miura N, Nakagawa T, Tatsuki S et al (2009) A male-specific odorant receptor conserved through the evolution of sex pheromones in *Ostrinia* moth species. Int J Biol Sci 5:319–330

Nei M, Niimura Y, Nozawa M (2008) The evolution of animal chemosensory receptor gene repertoires: roles of chance and necessity. Nat Rev Genet 9:951–963

Nieberding CM, de Vos H, Schneider MV et al (2008) The male sex pheromone of the butterfly *Bicyclus anynana*: towards an evolutionary analysis. PLoS One 3(7):e2751

Regier JC, Mitter C, Zwick A et al (2013) A large-scale, higher-level, molecular phylogenetic study of the insect order Lepidoptera (moths and butterflies). PLoS One 8:e58568

Robertson HM (2015) The insect chemoreceptor superfamily is ancient in animals. Chem Senses 40:609–614

Robertson HM, Warr CG, Carlson JR (2003) Molecular evolution of the insect chemoreceptor superfamily in *Drosophila melanogaster*. Proc Natl Acad Sci U S A 100:173–180

Rytz R, Croset V, Benton R (2013) Ionotropic Receptors (IRs): chemosensory ionotropic glutamate receptors in *Drosophila* and beyond. Insect Biochem Mol Biol 43:888–897

Saina M, Busengdal H, Sinigaglia C et al (2015) A cnidarian homologue of an insect gustatory receptor functions in developmental body patterning. Nat Commun 6:6243

Sakurai T, Nakagawa T, Mitsuno H et al (2004) The sex pheromone receptor in the silkworm moth *Bombyx mori*. Proc Natl Acad Sci U S A 101:16653–16658

Sánchez-Gracia A, Vieira FG, Rozas J (2009) Molecular evolution of the major chemosensory gene families in insects. Heredity 103:208–216

SilkBase (2006) Univ. of Tokyo, Tokyo. http://silkbase.ab.a.u-tokyo.ac.jp/cgi-bin/index.cgi. Accessed Sept–Nov 2018

Sun M, Liu Y, Walker WB et al (2013) Identification and characterization of pheromone receptors and interplay between receptors and pheromone binding proteins in the diamondback moth, *P. xyllostella*. PLoS One 8:e62098

Tanaka K, Uda Y, Ono Y et al (2009) Highly selective tuning of a silkworm olfactory receptor to a key mulberry leaf volatile. Curr Biol 19:881–890

Tian Z, Sun L, Li Y et al (2018) Antennal transcriptome analysis of the chemosensory gene families in *Carposina sasakii* (Lepidoptera: Carposinidae). BMC Genomics 19:544

Vieira FG, Rozas J (2011) Comparative genomics of the odorant-binding and chemosensory protein gene families across the Arthropoda: origin and evolutionary history of the chemosensory system. Genome Biol Evol 3:476–490

Vogt RG, Grosse-Wilde E, Zhou JJ (2015) The Lepidoptera odorant binding protein gene family: gene gain and loss within the GOBP/PBP complex of moths and butterflies. Insect Biochem Mol Biol 62:142–153

Vosshall LB, Hansson BS (2011) A unified nomenclature system for the insect olfactory coreceptor. Chem Senses 36:497–498

Wanner KW, Nichols AS, Allen JE et al (2010) Sex pheromone receptor specificity in the European corn borer moth, *Ostrinia nubilalis*. PLoS One 5:e8685

Xu P, Garczynski SF, Atungulu E et al (2012) Moth sex pheromone receptors and deceitful parapheromones. PLoS One 7:e41653

Yang B, Ozaki K, Ishikawa Y et al (2015) Identification of candidate odorant receptors in Asian corn borer *Ostrinia furnacalis*. PLoS One 10:e0121261

Yang S, Cao D, Wang G et al (2017) Identification of genes involved in chemoreception in *Plutella xyllostella* by antennal transcriptome analysis. Sci Rep 7:11941

Yasukochi Y, Miura N, Nakano R et al (2011) A sex-linked pheromone receptor genes of the European corn borer, *Ostrinia nubilalis,* are in tandem arrays. PLOS ONE 6:e18843

Yasukochi Y, Ohno M, Shibata R et al (2016) A FISH-based chromosome map for the European corn borer yields insights into ancient chromosomal fusions in the silkworm. Heredity 116:75–83

Yasukochi Y, Yang B, Fujimoto T et al (2018) Conservation and lineage-specific rearrangements in the GOBP/PBP gene complex of distantly related ditrysian Lepidoptera. PLoS One 13:e0192762

Yuvaraj JK, Corcoran JA, Andersson MN et al (2017) Characterization of odorant receptors from a nonditrysian moth, *Eriocrania semipurpurella* sheds light on the origin of sex pheromone receptors in Lepidoptera. Mol Biol Evol 34:2733–2746

Yuvaraj JK, Andersson MN, Corcoran JA et al (2018a) Functional characterization of odorant receptors from *Lampronia capitella* suggests a non-ditrysian origin of the lepidopteran pheromone receptor clade. Insect Biochem Mol Biol 100:39–47

Yuvaraj JK, Andersson MN, Zhang D et al (2018b) Antennal transcriptome analysis of the chemosensory gene families from Trichoptera and basal Lepidoptera. Front Physiol 9:1365

Zhang S, Zhang Z, Wang H et al (2014) Antennal transcriptome analysis and comparison of olfactory genes in two sympatric defoliators, *Dendrolimus houi* and *Dendrolimus kikuchii* (Lepidoptera: Lasiocampidae). Insect Biochem Mol Biol 52:69–81

Chapter 11
Application of Olfactory Detection Systems in Sensing Technologies

Hidefumi Mitsuno, Takeshi Sakurai, and Ryohei Kanzaki

Abstract Insects are equipped with a sophisticated olfactory detection system that enables high sensitivity and selectivity detection of various types of environmental odorants. Recently, the olfactory mechanisms of insects have been better elucidated to the extent to which they can be implemented for sensing technologies. The sensing technologies reported in this chapter, which utilize living insects or their antennae, showed that moth-derived sex pheromones and plant-derived general odorants can be detected in the field. In addition, many olfactory receptors have been identified in various insect species, each exhibiting a different response profile to different odorants. With the recent advancement in genetic engineering technologies, we have been able to reconstruct these insect olfactory receptors in protein expression systems for use as sensing elements in odorant sensors. In this chapter, we briefly introduce the mechanisms of olfaction in insects and summarize their applications in sensing technologies.

Keywords Biosensor · Insect · Proboscis extension reflex · Electroantennogram · Olfactory receptor · Culture cells · Artificial lipid membrane · Transgenic silkmoth

11.1 Preface

Insects comprise at least 800,000 species that inhabit diverse environments ranging from an urban setting to forests, coasts, and arid regions. In the environment, insects use odorants for chemical communication, which is necessary for

H. Mitsuno (✉) · R. Kanzaki
Research Center for Advanced Science and Technology, The University of Tokyo, Tokyo, Japan
e-mail: mitsuno@brain.imi.i.u-tokyo.ac.jp; kanzaki@rcast.u-tokyo.ac.jp

T. Sakurai
Faculty of Agriculture, Tokyo University of Agriculture, Kanagawa, Japan
e-mail: ts206448@nodai.ac.jp

Y. Ishikawa (ed.), *Insect Sex Pheromone Research and Beyond*, Entomology Monographs, https://doi.org/10.1007/978-981-15-3082-1_11

many life activities, such as searching for food, attracting mates, and avoiding foreign enemies. It is difficult to detect and locate a specific odor in a natural environment, where there is an abundance of chemical substances; however, insect olfaction has been refined through evolution for approximately 480 million years. Thus, they can detect trace amounts of odorants in the environment. With advances in entomology research, the mechanism of insect olfaction is being revealed from behavioral, neurophysiological, and molecular levels. In recent years, the entity sensing odorants had been identified to be olfactory receptor proteins unique to insects. Furthermore, an artificial odorant sensor based on the olfactory receptor of insects has shown far higher performance than the existing odor sensors, which are made with engineering technology. In this chapter, we will first outline the mechanisms underlying odor detection in insects, introduce the excellent olfactory abilities of insects, and then introduce the development of odorant biosensor technology based on the insect olfactory system.

11.2 Insect Olfaction and Olfactory Receptors

11.2.1 Olfaction of Insects

In insects, the antenna is an olfactory organ equipped with the ability to detect various odorants with high sensitivity in real time. There are numerous hair structures, called olfactory sensilla, on the surface of the antennae, whereas the interiors of the antennae are filled with aqueous sensillum lymph (see Jacquin-Joly and Merlin 2004, Chap. 9 of this book). A few olfactory receptor neurons (ORNs) are located at the base of each sensillum, and the branches of the dendrites of these ORNs are extended to the sensillum lymph. The olfactory mechanism in insects is described as follows. First, odorant molecules enter the inside of the sensillum from olfactory pores with a diameter of approximately several tens of nanometers (Keil 1999; Steinbrecht 1999). Next, the odorant molecules dissolve in the sensillum lymph with the aid of odorant-binding proteins (OBPs), which are abundantly present in the lymph (Vogt and Riddiford 1981; Vogt et al. 1991). Subsequently, the dissolved odorant molecules are transported to the cell membrane of the dendrites of ORNs. On the dendrite membrane are olfactory receptors (ORs), which enable the detection of odorants with high sensitivity and selectivity. The interaction of solubilized odorant molecules with a specific OR evokes action potentials in ORNs, which are then transmitted to the antennal lobe, the first olfactory center of the insect brain, for information processing (Hildebrand and Shepherd 1997; Chap. 9 of this book).

11.2.2 Olfactory Receptors and Signal Transduction

An OR gene was first discovered in rats in 1991 by Linda Buck and Richard Axel, who won the 2004 Nobel Prize in Medicine and Physiology (Buck and Axel 1991). This research became a breakthrough in studies of ORs in living organisms. OR genes similar to those found in rats have been identified from various species, including humans, mice, fish, birds, and nematodes (Ngai et al. 1993; Ben-Arie et al. 1994; Chou et al. 1995; Nef et al. 1996). Meanwhile, the OR gene of insects was first discovered from the fruit fly *(Drosophila melanogaster)*, by analyzing its genomic sequence in 1999 (Clyne et al. 1999; Gao and Chess 1999; Vosshall et al. 1999). As genome sequencing progresses, OR genes have been identified from various insect species, for example, honeybee *(Apis mellifera)*, mosquito *(Anopheles gambiae)*, beetle *(Tribolium castaneum)*, silkmoth *(Bombyx mori)*, and human louse *(Pediculus humanus)* (Fox et al. 2001; Hill et al. 2002; Robertson and Wanner 2006; Wanner et al. 2007).

ORs in many organisms other than insects are G-protein-coupled receptors (GPCRs; Chap. 9 of this book). The GPCRs activate the intracellular metabolic pathway mediated by G protein upon binding odorants. Next, the produced metabolite, cyclic adenosine monophosphate (cAMP), acts on other ion channels to permeate ions into ORNs (Jones and Reed 1989; Bakalyar and Reed 1990; Takeuchi and Kurahashi 2003). On the other hand, insect ORs belong to a class of membrane proteins with 7-transmembrane (TM) domains and have different topology from that of mammals (Buck and Axel 1991; Lundin et al. 2007; Smart et al. 2008). The mechanism of signal transduction in insect ORs after odorant binding is also different from that in other organisms. Insect ORs are ligand-gated ion channels that open upon binding with odorant molecules (Sato et al. 2008, Nakagawa et al. 2012, Chap. 9 of this book). The ligand-gated ion channels in insects are a complex comprised of a conventional OR and a common olfactory receptor co-receptor (Orco), and they permeate ions directly into ORNs upon binding with odorant molecules (Benton et al. 2006; Sato et al. 2008). Interestingly, it has been reported that insect gustatory receptors (GRs) are also ligand-gated ion channels, whereas mammalian taste receptors belong to the GPCR family (Sato et al. 2011). In addition to OR and GR, insects have ionotropic receptors (IRs) that function mainly for detecting acids and amines; IRs are not found in vertebrates (Benton et al. 2009; Ai et al. 2010; Abuin et al. 2011). Based on these studies, it is believed that odor and taste receptors in insects have evolved independently from those in other animals.

Insect ORs are characterized by (1) the simplicity of the signal transduction mechanism, which enables quick induction of ion influx at a speed of several tens of milliseconds, (2) easiness of the reconstruction of the receptor proteins in heterologous expression systems, and (3) easiness of acquiring receptor responses by the electrophysiological or imaging method. A ligand repertoire of >100 ORs from various insect species, such as *D. melanogaster*, *A. gambiae*, and *B. mori*, has already been clarified (see Glatz and Bailey-Hill 2011). Similar to mammalian ORs, each insect OR has different degrees of responsiveness to different sets of odorants (Hallem and Carlson 2006; Carey et al. 2010; Wang et al. 2010). For example, the responsiveness of the individual olfactory receptors of *D. melanogaster* to various types of odorants is stored in a database (Database of Odorant Responses: DoOR; http://neuro.uni-konstanz.de/DoOR/default.html; Münch and Galizia 2016), and the ORs that respond to an odorants of interest can easily be found in that database.

As described above, insect ORs have the advantage of highly selective responsiveness to odorants, easy reconstruction, and fast response time. Furthermore, because a myriad of ORs with different ligand specificities are available, insect ORs are equipped with the characteristics suitable for use as sensing elements to detect diverse odorants.

11.3 Sensing Technology

Living organisms detect odorants using biomolecules, such as ORs and OBPs, in the olfactory organs. By using these biomolecules, we may be able to develop artificial odorant sensors with both higher sensitivity and selectivity than those of the existing odor sensors made using engineering technology. Based on this concept, various odorant-sensing technologies using mammalian ORs, tissues, and cells have been reported (see Du et al. 2013). The fundamental design principle for developing odorant biosensors consists of production of functional receptors as a detection element, immobilization of receptors to a transducer, and acquisition of signals, that is, responses to odorants, from the transducer. Functional detection elements are obtained either by expressing mammalian ORs in *Escherichia coli* or cultured cells, or by directly isolating tissues or cells from mammals. The detection element is immobilized to the electrode surface of a proper transducer (e.g., Field Effect Transistor (FET), Metal Oxide Semiconductor (MOS), and Quartz Crystal Microbalance (QCM)) by using immobilization techniques, such as physical adsorption, self-assembled membrane (SAM), and Au–S bond. Next, a signal derived from the interaction between the odorant and the detection element is acquired from the transducer, whereby an odorant is detected. Depending on the performance and appropriate combination of ORs and transducers, it has become possible to develop sensors that can detect odorants with higher selectivity and sensitivity than those of conventional odor sensors (see Kanzaki et al. 2016).

In contrast to those utilizing the mammalian olfactory system, studies on odorant-sensing technology utilizing the insect olfactory system are still limited. Such an odorant-sensing technology can be roughly divided into two categories: (1) detection technology utilizing a living insect body or isolated antennae and (2) detection technology utilizing biomolecules that interact with odorants, such as ORs and OBPs. With progresses in both genome analysis of various insect species and genetic engineering, it became possible to manipulate the genes of olfactory biomolecules; therefore, technologies for artificially creating an insect OR and using it as a detection element have been actively studied. Here, we focus on these two sensing technologies that exploit the olfactory function of insects.

11.3.1 Sensing Technologies Using Whole Insects and Their Antennae

11.3.1.1 Whole Insects

Insects have small bodies that are equipped with sophisticated mechanisms for recognizing various types of odorants with high sensitivity. Therefore, it is possible to utilize living insects directly as detection elements. For example, a technique to utilize the proboscis extension reflex (PER) of honeybee *(A. mellifera)* as a sensor signal has been proposed. When the antennae of honeybees come into contact with sugar water, they reflexively exhibit PER, which is the extension of their proboscis. Simultaneous sugar water and odor stimuli may be used to establish an associative learning between sugar water and the odor, so that only stimulation with that particular odor causes PER in the trained honeybees. By using a desired target odorant for stimulation, techniques for detecting specific odors by observation of PER as an index have been established. By applying this technology, it is possible to learn of various types of odorants as targets. For example, according to Bromenshenk et al. (2003, 2015), it is possible for honeybees to learn the odor of 2,4-dinitrotoluene (2,4-DNT), an explosive, and other general odorants. Moreover, it has been reported that trained honeybees can detect 2,4-DNT with sensitivity as high as several tens of parts per trillion (ppt). In addition to the honeybees, genetically modified silkmoth (*B. mori*) has also been developed as a detection element (see Sect. 11.3.2). In summary, utilization of the characteristic behaviors of insects as indicators enables us to develop a highly sensitive and selective odor-sensing technology.

11.3.1.2 Insect Antennae

Insect antennae can also be used as odorant-detecting elements. The method used to acquire the electrical signals of the antennae, namely the electroantennogram (EAG), has been studied for a long time to analyze the olfactory function of the antennae of various insect species (Schneider 1957). Because insect antennae exhibit electrical responses in the form of EAG upon receiving sex pheromone components or general odorants, EAG can be used as an indicator to detect specific odorant. Baker and Haynes (1989), for the first time, successfully recorded EAG in the field by applying their mobile EAG system on oriental fruit moth *(Grapholita molesta)*. Since then, the portable EAG system has been further developed (Sauer et al. 1992; Suckling et al. 1994; Rumbo et al. 1995; Milli et al. 1997). Furthermore, the EAG system can be used to measure the diffusion and concentration of moth sex pheromones in the field, where various crops are cultivated (Karg and Sauer 1995).

Measurements from the antennae of various insect species can be integrated into a compact EAG measuring device. For example, van der Pers and Minks (1998) developed an EAG-based portable small device using the male antennae of Noctuid moth *(Chrysodeixis chalcites)*, and showed that the device can measure pheromone concentrations diffused in a greenhouse. Another study showed that a detection technology using intact antennae can be used to detect potatoes infected with pathogenic bacteria. Schutz et al. (1999) developed a biosensor based on the intact antennae of Colorado potato beetle *(Leptinotarsa decemlineata)* and showed that this biosensor system was able to detect 2-ethyl-hexan-1-ol, a biomarker of *Phytophthora infestans* infection, even in a potato storage room (Schutz et al. 1999). Park et al. (2002) found that the antennae of five different insect species, *D. melanogaster*, *Heliothis virescens*, *Helicoverpa zea*, *Ostrinia nubilalis*, and *Microplitis croceipes,* exhibit different EAG response spectra to 20 volatile compounds. By using a four-antenna array (Quadro-probe; Fig. 11.1), a device that can simultaneously measure EAG responses of four antennae, the authors were able to distinguish different odors in real time by comparing relative EAG responses in a wind tunnel and in a field. Myrick et al. (2009) have shown that a multichannel insect-antenna-based chemosensor system can classify individual odorant/mixture strands at subsecond temporal resolution and a few tens of millimeters of spatial resolution. Furthermore, Myrick et al. (2011) integrated an EAG measurement technology with an instrument that included a Global Positioning System (GPS) equipped with a digital compass and ultrasonic 2D anemometer. By analyzing the data collected at several

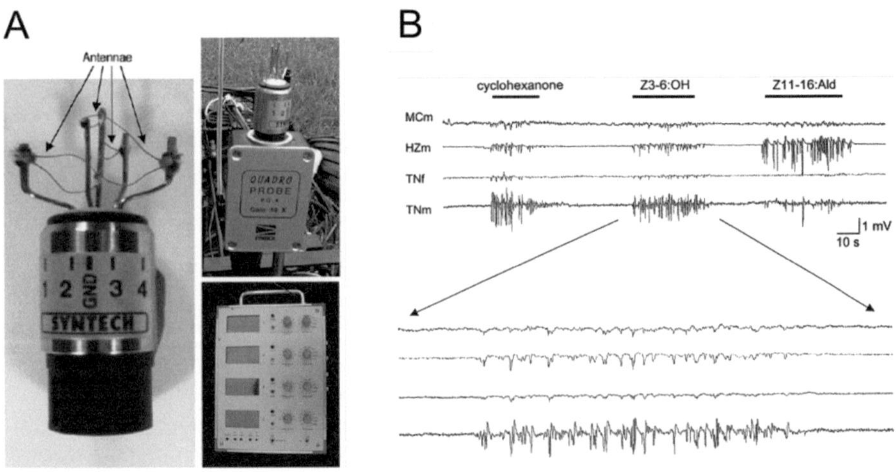

Fig. 11.1 A portable Quadro-probe EAG recording system with four different antennae. (**a**) Images of the Quadro-probe EAG stage (left), preamplifier (top right), and main amplifier (bottom right). (**b**) EAG responses of the four different antennae to odorants. (Reproduced from Park et al. 2002)

downwind locations, the authors succeeded in estimating the position of the odor source with an accuracy of <20 cm (Myrick et al. 2011).

EAG measurement technology can potentially be used as a sensor for odor source-searching robots. Kuwana and Shimoyama (1998) developed a mobile robot equipped with antennae isolated from male silkmoth. By incorporating a recurrent neural network that generates silkmoth-like behavior (zig-zag turn behavior), they demonstrated that the mobile robot with antennae was able to localize a pheromone source in a wind tunnel like a silkmoth. Furthermore, Kuwana et al. (1999) also developed a silkmoth-sized mobile robot with an improved antenna holder for EAG measurement (Fig. 11.2). As described above, EAG measurement technologies can be utilized not only for elucidating the olfactory function of insects but also as a sensor for detecting sex pheromones in the field and as a sensory device of a robot.

Field-effect transistors (FETs) incorporating various biological molecules are widely used as biosensors because they can acquire signals with high sensitivity and can be miniaturized and highly integrated into a system. FETs can detect a target substance by immobilizing biomolecules, such as an antibody, enzyme, or nucleic

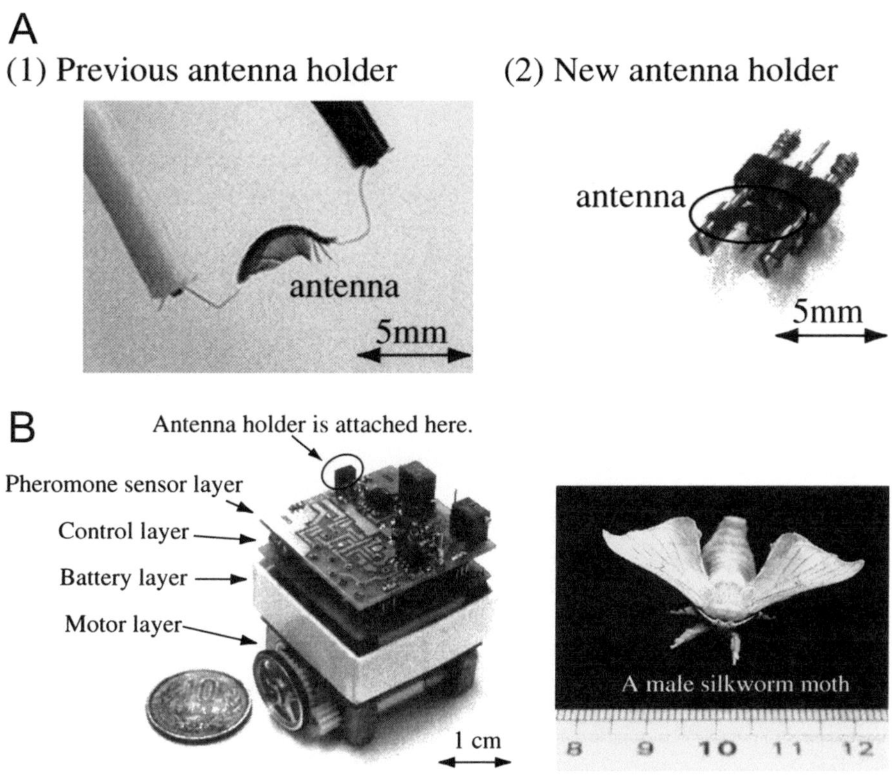

Fig. 11.2 Pheromone-guided mobile robot with a new antenna holder. (**a**) Comparison of the previous and new antenna holders for measuring EAG responses in silkmoth. (**b**) Images of the pheromone-guided mobile robot and male silkmoth. (Reproduced from Kuwana et al. 1999)

acid, on the surface of an insulating film at the gate of a transistor, which is a semiconductor element. An odorant-sensing technique involving the application of FET to the antenna of an insect, namely insect-based biosensor field-effect transistor (BioFET), has been proposed (Schutz et al. 1997). Schoning et al. (1998) established a technique to acquire the electrical response of the antennae of Colorado potato beetle, *L. decemlineata*, by connecting the tip of the antenna directly to the gate electrolyte of FET and placing the reference electrode on the body of the insect or on the base of the antennae (Fig. 11.3). The authors showed that insect-based BioFET was able to detect plant-derived *Z*-3-hexen-1-ol; the beetle responded to the substance at a concentration range of 0.01–100 ppm with a short response time of <1 s. Furthermore, they improved the BioFET with respect to the insulating film and size of the gate to detect *Z*-3-hexen-1-ol at a high sensitivity of 1 ppb (Schroth et al. 1999; Schutz et al. 2000). Thus, it is possible to develop an odorant sensor possessing the high sensitivity and selectivity of insects by combining isolated insects' antennae and an appropriate device to detect electrical signals.

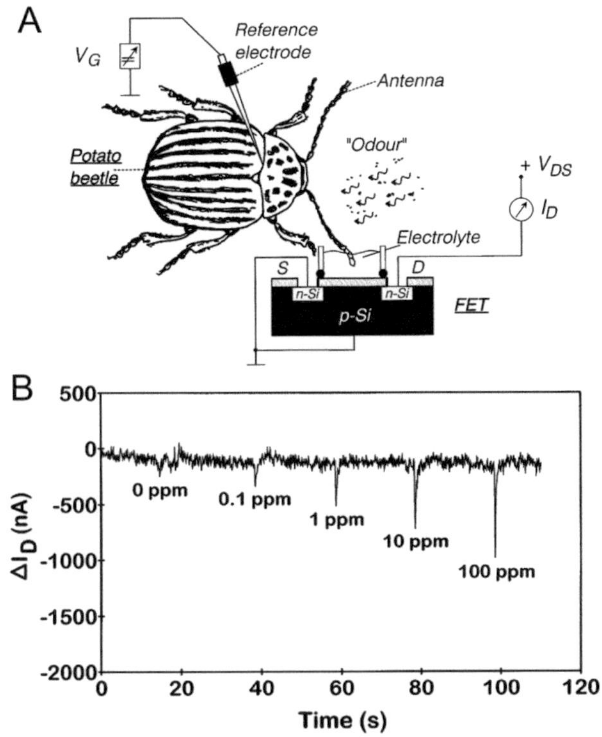

Fig. 11.3 Whole-beetle or isolated-antenna BioFET. (**a**) Schematic illustration of a biosensor using a whole beetle and a FET device. (**b**) Responses of the isolated-antenna BioFET to increasing concentrations of *Z*-3-hexen-1-ol. (Reproduced from Schoning et al. 1998)

11.3.2 Sensing Technologies Using Olfactory Receptors

11.3.2.1 Cell-Based Sensing Technology

Membrane proteins, such as ion channels and GPCRs, are functionally expressed in cultured cells, and they have been subjected to functional analyses for elucidating their response profiles to various ligands, as well as to three-dimensional (3D) structure analyses. Cells, such as oocytes derived from *Xenopus laevis* frog, Sf9 and Sf21 cells derived from insects, and human embryonic kidney 293 (HEK293) and HeLa cells derived from mammals, are mainly used in heterologous expression systems for functional analysis of membrane proteins (see Glatz and Bailey-Hill 2011). Among these, *Xenopus* oocytes have been commonly used for the expression of ion channels and receptors of various organisms, including mammals, insects, plants, and microorganisms. The functions of many insect ORs have been analyzed using the *Xenopus* oocyte expression system. Oocyte is a spherical cell with a diameter of approximately 1 mm, which is very large for a single cell. Therefore, the DNA or mRNA of a target OR gene can be easily microinjected into the oocyte using a glass capillary. A weak ion current across the oocyte membrane generated in response to a stimulus is then measured by using an electrophysiological method called two-electrode voltage clamping (TEVC), in which two glass electrodes are inserted into the oocyte. By using this method, Wetzel et al. (2001) conducted a functional analysis of an insect OR by co-expressing the G protein α subunit. They showed that oocytes co-expressing *D. melanogaster* Or43a and human Gα15 exhibit selective electrical responses to four kinds of odorants (cyclohexanol, cyclohexanone, benzaldehyde, and benzyl alcohol), indicating that *Xenopus* oocytes can be used as a heterologous functional expression system of insect ORs. Similarly, by co-expressing silkmoth BmOR1 with BmGαq in the oocyte, BmOR1 responded specifically to bombykol, the sex pheromone of silkmoth (Sakurai et al. 2004). Later, Nakagawa et al. (2005) reported that the co-expression of Orco with OR is important for sensitive detection of odorants using insect ORs. Finally, it became clear that insect ORs constitute a ligand-gated ion channel together with Orco (Sato et al. 2008).

An odorant sensor based on *Xenopus* oocytes co-expressing insect ORs and Orco has been proposed as an application for sensing technology. Misawa et al. (2010) developed a sensor chip capable of measuring ion current from oocytes in a fluidic channel based on the principle of TEVC (Fig. 11.4). The authors reported that by trapping oocytes co-expressing sex pheromone receptors or other ORs with Orco onto the sensor chip, sex pheromones or target odorants were detected with high sensitivity in the order of parts per billion (ppb). In addition, the sex pheromone of *B. mori* and its analogs (bombykol and bombykal, respectively) can be detected with one sensor chip by arraying two oocytes, one expressing BmOR1 and another expressing BmOR3, onto the chip. Furthermore, because this sensor chip is compact and portable, it can be used as the nose of a robot. It was shown

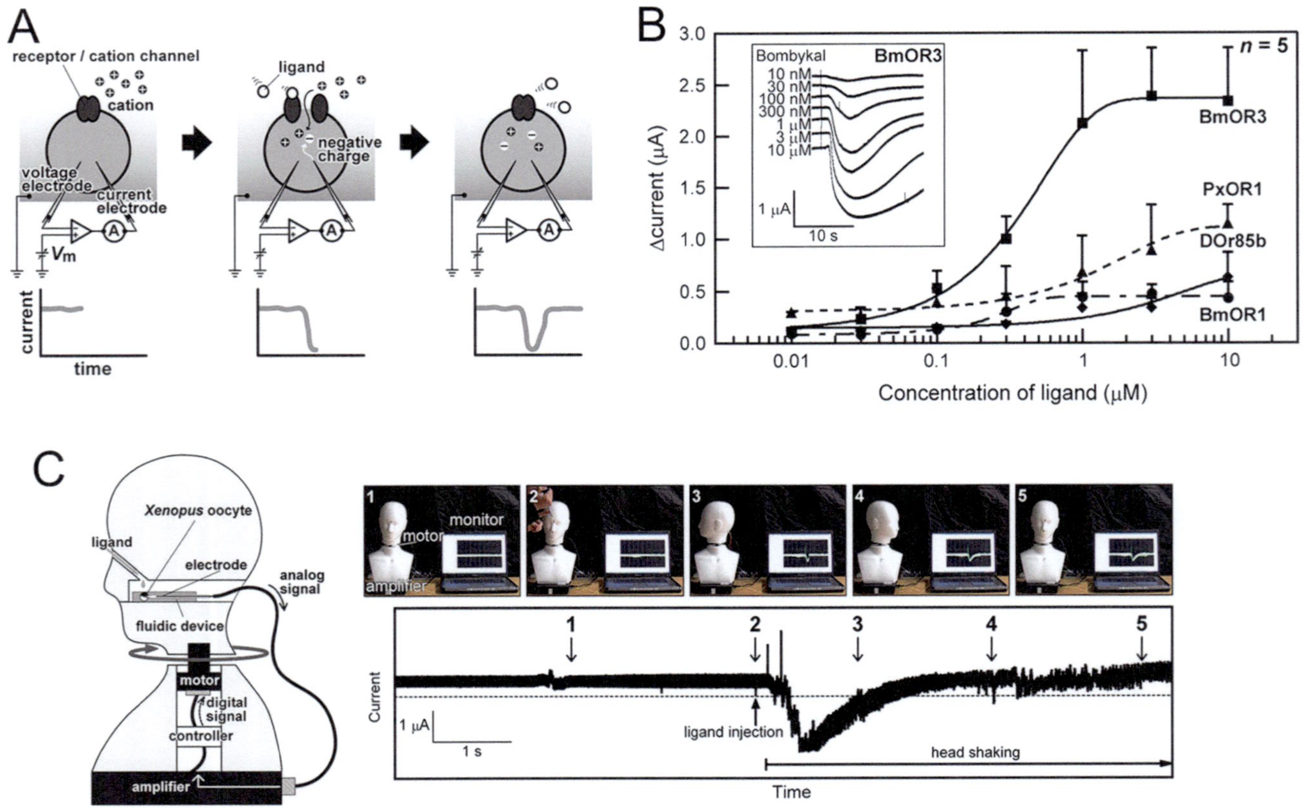

Fig. 11.4 An odorant sensor utilizing *Xenopus laevis* oocytes expressing insect olfactory receptors. (**a**) Image depicting the principle of two-electrode voltage clamp (TEVC) in measuring inward current from an oocyte expressing insect olfactory receptors. (**b**) Dose-dependent responses of the oocyte sensors to the target odorants. (OR; ligand) BmOR1; bombykol, BmOR3; bombykal, PxOR1; Z-11-hexadecenal (Z11-16:Ald), DOr85b; 2-heptanone. (**c**) Schematic illustration and sequential images of a head-shaking robot system integrating the oocyte sensor. (Reproduced from Misawa et al. 2010)

for the first time that it is possible to develop a highly sensitive and selective odorant sensor by using insect ORs.

In addition to *Xenopus* oocytes, cultured cells (HEK293 cell, HeLa cell, Sf9 cell, and Sf21 cell) derived from mammals and insects are widely used for the functional analysis of insect ORs. Regarding mammalian cultured cells, Neuhaus et al. (2005) reported that HEK293 cells co-expressing *D. melanogaster* OR and Orco exhibited higher fluorescence responses to odorants than those expressing OR alone. Later, Sato et al. (2008) constructed an ion channel composed of an insect OR and Orco in HeLa and HEK293 cells, and indicated that it is possible to examine the activity of ion channels by the patch-clamp method. Kiely et al. (2007) cultured insect cells and, by calcium imaging using Fluo-4, obtained the selective odorant responses of *D. melanogaster* Or22a functionally expressed in Sf9 cells. The functions of ORs from various insect species have been analyzed by using Sf9 cells (see Glatz and Bailey-Hill 2011).

For applications in sensing technology, Mitsuno et al. (2015) genetically introduced three genes encoding silkmoth pheromone receptor (BmOR1 or BmOR3), Orco, and calcium indicator protein (GCaMP3) into Sf21 cells to develop cell lines as odorant sensor elements that can detect target pheromone components by the increase in fluorescent intensity. They also reported that an odorant sensor chip can be developed by immobilizing cells onto a microfluidic glass chip prepared by photolithography technology. Similarly, two Sf21 cell lines that selectively detect geosmin and 1-octen-3-ol were developed by using *D. melanogaster* Or56a (for geosmin) and Or13a (for 1-octen-3-ol), DmelOrco, and GCaMP6s (Fig. 11.5; Termtanasombat et al. 2016, Mujiono et al. 2017). Moreover, Termtanasombat et al. (2016) separately arrayed four Sf21 cell lines expressing different ORs (Or56a, Or13a, BmOR1, and BmOR3) onto the glass surface of a chip by using a cell adhesion agent, oleyl-PEG-NHS, and showed that the chip enables us to distinguish four target odorants (geosmin, 1-octen-3-ol, bombykol, and bombykal) by the fluorescent patterns.

Biosensors using neuronal cultured cells have also been proposed. Tanada et al. (2012) transfected expression vectors containing the BmOR1 and BmOrco genes in dissociated primary neuronal cultured cells from rodent embryo. Calcium signals in response to bombykol stimulation were observed not only in the transfected cells but also in the nontransfected cells because of the neural network in the neuronal cultures. Furthermore, techniques for detecting chemical vapor using insect OR complexes have also been reported. Sato and Takeuchi (2014) produced spheroids from HEK293T cells that functionally expressed *A. gambiae* OR (GPROR2) and AgOrco (Fig. 11.6). Using an electrophysiological technique, they successfully measured the responses of the spheroid, which was embedded in a hydrogel microchamber system, to the headspace vapor of 2-methylphenol solution.

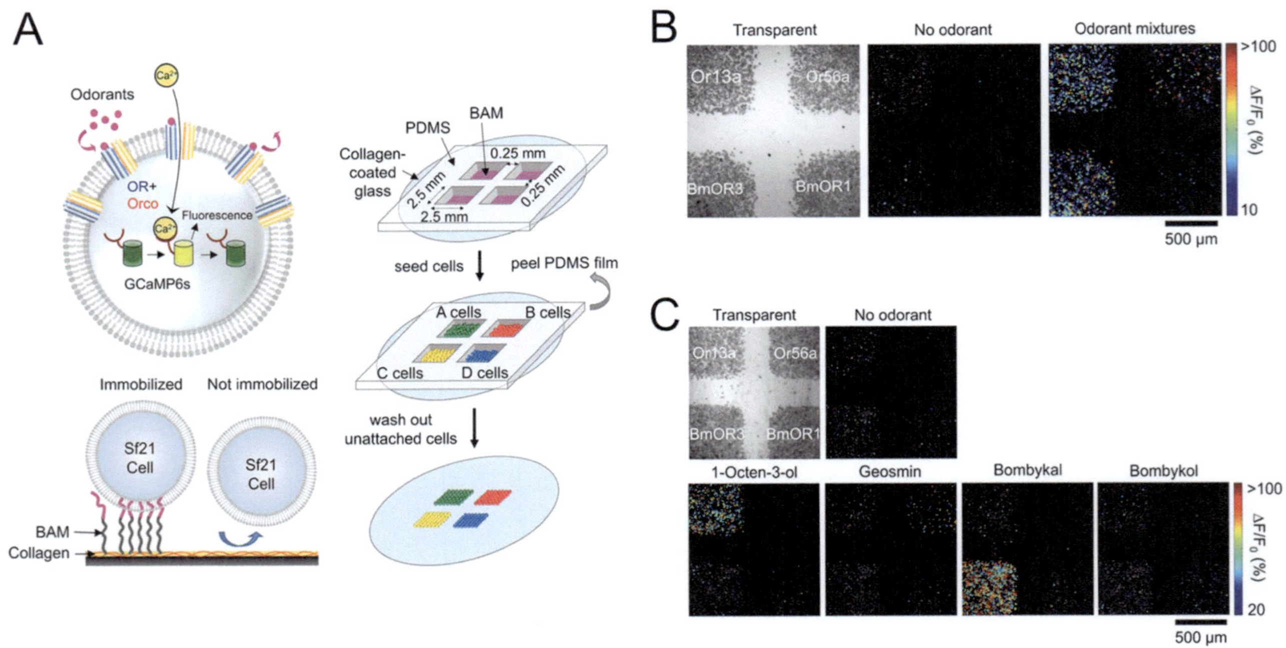

Fig. 11.5 Cell-based odorant sensor array. (**a**) Schematic illustration of the development of a cell array using four Sf21 cell lines expressing different insect olfactory receptors. (**b**) Response pattern of the cell-based sensor array to odorant mixtures. The odorant mixtures consist of four odorants (bombykol, bombykal, 1-octen-3-ol, and geosmin). (**c**) Response patterns of the cell-based sensor array to a single odorant. (OR; ligand) BmOR1; bombykol, BmOR3; bombykal, Or13a; 1-octen-3-ol, Or56a; geosmin. (Reproduced from Termtanasombat et al. 2016)

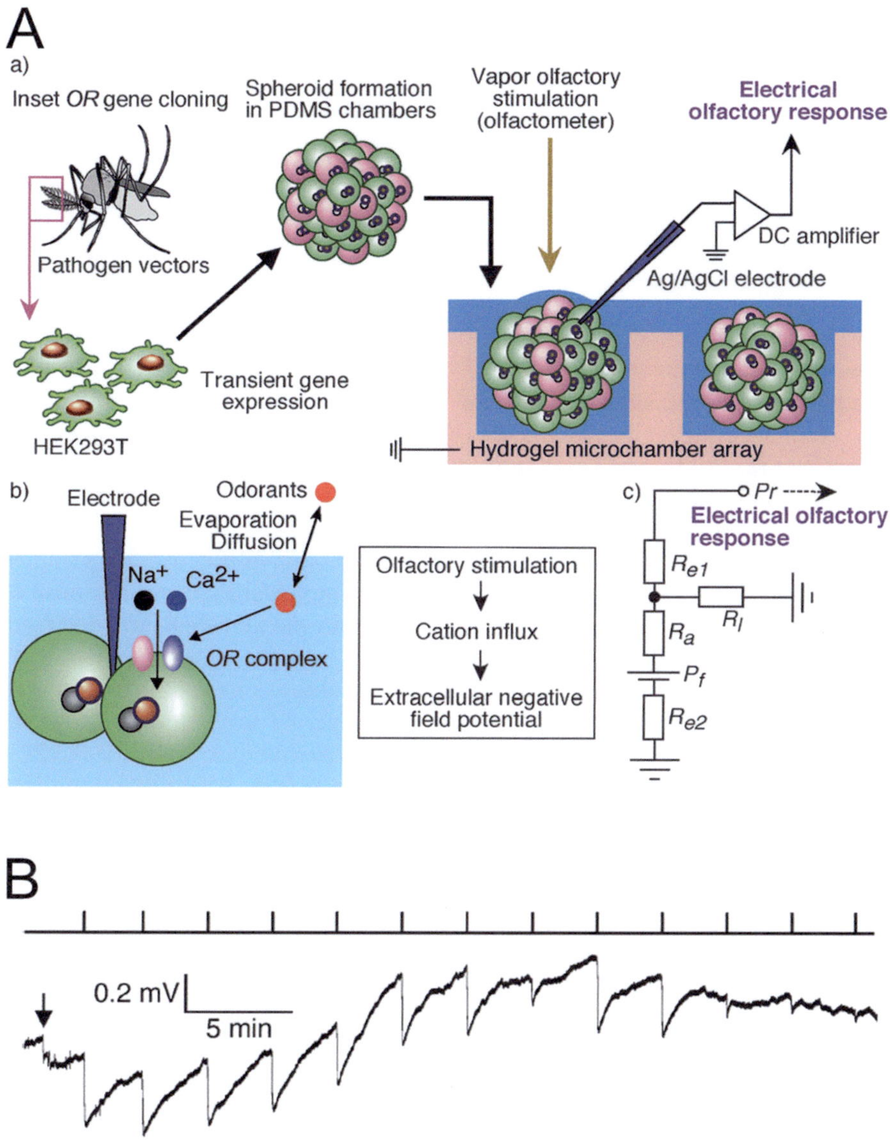

Fig. 11.6 Chemical vapor detection using OR-expressing spheroids. (**a**) Schematic illustration of the formation and field potential measurement of insect OR-expressing HEK293T cell spheroids. (**b**) Responses of the GPROR2-expressing spheroids to the headspace vapor of odorant (2-methylphenol) solution. (Reproduced from Sato and Takeuchi 2014)

11.3.2.2 Other Sensing Technologies

The functional expression of membrane proteins, including ORs and ion channels, can also be achieved by cell-free protein expression systems using cell extracts (see Sachse et al. 2014). Cell extracts contain abundant materials necessary for the functional expression of proteins. Therefore, it is possible to produce large amounts of functional proteins in vitro by mixing energy sources (adenosine triphosphate, ATP) and amino acids together with the target nucleic acid (mRNA). At present, extracts of *E. coli*, rabbit reticulocytes, human cells, wheat germs, and insect cells can be selected depending on the application. It has also been reported that a human OR protein, hOR17-4, was synthesized in vitro as an odorant sensor element by using such cell-free expression system. The purified OR was shown to respond to the ligand undecanal by surface plasmon resonance (SPR) measurements (Kaiser et al. 2008). In insects, functional reconstruction of *B. mori* sex pheromone receptor complexes in an artificial lipid membrane was reported. Hamada et al. (2014) encapsulated the mRNA of BmOR1 and BmOrco into giant vesicles (GV) together with insect cell-free protein synthesis reagent, and confirmed the expression of the receptor complex that was embedded onto the membrane of GV. The inward current response of the BmOR1/BmOrco complex to bombykol was for the first time recorded from the GV by using the patch-clamp recording.

More recently, it has been reported that heterologous insect ORs expressed in Sf9 cells were integrated into the lipid bilayer of an artificial liposome. Khadka et al. (2019) expressed and purified *D. melanogaster* Or10a, Or22a, and Or71a using a baculovirus-based expression system, integrated them into artificial nanoliposomes, and immobilized them onto the gold electrode in an Electrochemical Impedance Spectroscopy (EIS) system (Fig. 11.7). This method using insect ORs enabled the detection of methyl salicylate (Or10a), methyl hexanoate (Or22a), and 4-ethylguaiacol (Or71a) with ultrahigh sensitivity in the order of femtomolar. These studies represented the possibility of next-generation sensing technology utilizing artificial lipid membranes incorporating insect ORs.

In some insects, it is possible to express target genes in desired tissues and cells by genetic modification (Tamura et al. 2000). One of the insects that can be genetically modified is *B. mori* silkmoth. Males of this moth sensitively and selectively detect bombykol, which is the sex pheromone released by the female moths, and use this information to find and localize females (Kramer 1975). This pheromone source-searching behavior is triggered by the action of sex pheromone-sensitive ORNs expressing BmOR1, which specifically responds to bombykol (Sakurai et al. 2004). Therefore, by expressing the OR of the target odorants in the sex pheromone-sensitive ORNs, it is possible to create a male silkmoth that searches for the target odorants. Sakurai et al. (2011) created a genetically modified silkmoth whose sex pheromone-sensitive ORNs are expressing PxOR1, the sex pheromone receptor of diamondback moth *(Plutella xylostella)* (Chap. 9 of this book). The transgenic silkmoths

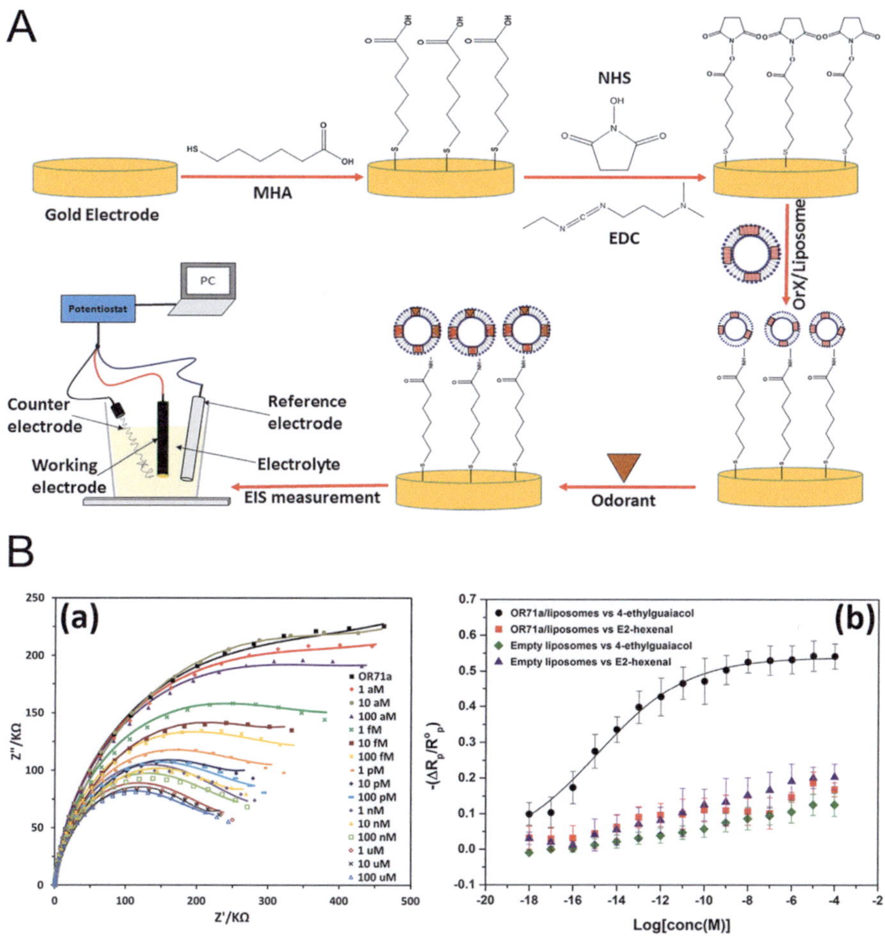

Fig. 11.7 Biosensor using nanoliposome integrating insect olfactory receptors. (**a**) Schematic illustration of immobilization of ORx/liposomes into the gold electrode of an EIS system. (**b**) Odorant detection using an EIS sensor with Or71a/liposome. (Reproduced from Khadka et al. 2019)

exhibited full pheromone-source searching behavior after stimulation with Z11-16:Ald, a ligand of PxOR1. It was shown that the transgenic male silkmoths successfully localized a female of diamondback moth in a wind tunnel experiment. Knockout of BmOR1 has also been used to establish a technique to produce genetically modified silkmoths that do not respond to bombykol, the sex pheromone of this moth species (Sakurai et al. 2015). Consequently, by expressing the desired OR in the sex pheromone-sensitive ORNs of BmOR1-knockout male silkmoth, it is possible to generate transgenic silkmoths (i.e., "sensor insect") that can detect the target odorant and search for the odor source upon detection.

11.4 Summary

The olfactory function of insects has drawn the attention of researchers, leading to the development of odorant-sensing technology using insect ORs. The studies introduced in this chapter do not represent the whole progress in this research field. High selectivity and sensitivity detection of a wide variety of odorants, which could not be achieved with existing odor sensors, may be possible by utilizing the olfactory function of insects. In fact, a sensing technology using EAG can detect the sex pheromones emitted by moths and the odorants released by infected crops in the field. In addition, a smell-sensing technology utilizing the PER of honeybees is being considered for application to the field. On the other hand, there are still problems in terms of stability in biosensors utilizing biological elements. For this reason, the technologies introduced in this chapter have not been widely used. Stabilization of biological elements is essential to bring this technology into practice in the future.

The odorant-sensing technologies shown in this chapter can be applied in a large number (over 800,000) of insect species. Targeting various insect species and their ORs can help us develop a sensing technology capable of detecting a wide variety of odorants and chemicals. On the other hand, the detection of odorants by this sensing technology depends on the characteristics of the insect ORs, such as sensitivity, selectivity, and fast response. Therefore, to detect a target odorant, it is essential to clarify the response characteristics of the target insect species and their ORs to various types of odorants.

To date, the functions of more than 100 ORs have been clarified. Functional analyses of *D. melanogaster* ORs have been conducted under various experimental conditions by researchers worldwide, and the normalized odorant response data have been collected in the DoOR database (Galizia et al. 2010; Münch and Galizia 2016). This database contains detailed information on *D. melanogaster* ORs, such as expression stages, expressing tissues, and projection areas of ORNs in the brain, as well as the response characteristics of *D. melanogaster* ORs to various odorants. By compiling such databases for various kinds of insect species, a system that enables easy and quick finding of the ORs suitable for detecting a target odorant can be constructed. Furthermore, by combining such detailed information with the sensing technologies introduced in this chapter, it will be possible to develop an odorant sensor capable of high-sensitivity detection of desired odorants ranging from those encountered in daily life to those related to safety crisis management.

References

Abuin L, Bargeton B, Ulbrich MH, Isacoff EY, Kellenberger S, Benton R (2011) Functional architecture of olfactory ionotropic glutamate receptors. Neuron 69:44–60

Ai M, Min S, Grosjean Y, Leblanc C, Bell R, Benton R, Suh GS (2010) Acid sensing by the *Drosophila* olfactory system. Nature 468:691–695

Bakalyar HA, Reed RR (1990) Identification of a specialized adenylyl cyclase that may mediate odorant detection. Science 250:1403–1406

Baker TC, Haynes KF (1989) Field and laboratory electroantennographic measurements of pheromone plume structure correlated with oriental fruit moth behavior. Physiol Entomol 14:1–12

Ben-Arie N, Lancet D, Taylor C, Khen M, Walker N, Ledbetter DH, Carrozzo R, Patel K, Sheer D, Lehrach H (1994) Olfactory receptor gene cluster on human chromosome 17: possible duplication of an ancestral receptor repertoire. Hum Mol Genet 3:229–235

Benton R, Sachse S, Michnick SW, Vosshall LB (2006) Atypical membrane topology and heteromeric function of *Drosophila* odorant receptors in vivo. PLoS Biol 4:e20

Benton R, Vannice KS, Gomez-Diaz C, Vosshall LB (2009) Variant ionotropic glutamate receptors as chemosensory receptors in *Drosophila*. Cell 136:149–162

Bromenshenk JJ, Henderson CB, Smith GC (2003) Biological system (paper II), pp 273–283

Bromenshenk JJ, Henderson CB, Seccomb RA, Welch PM, Debnam SE, Firth DR (2015) Bees as biosensors: chemosensory ability, honey bee monitoring systems, and emergent sensor technologies derived from the pollinator syndrome. Biosensors 5:678–711

Buck LB, Axel R (1991) A novel multigene family may encode odorant receptors: a molecular basis for odor recognition. Cell 65:175–187

Carey AF, Wang G, Su CY, Zwiebel LJ, Carlson JR (2010) Odorant reception in the malaria mosquito *Anopheles gambiae*. Nature 464:66–71

Clyne PJ, Warr CG, Freeman MR, Lessing D, Kim J, Carlson JR (1999) A novel family of divergent seven-transmembrane proteins: candidate odorant receptors in *Drosophila*. Neuron 22:327–338

Du L, Wu C, Liu Q, Huang L, Wang P (2013) Recent advances in olfactory receptor-based biosensors. Biosens Bioelectron 42:570–580

Fox AN, Pitts RJ, Robertson HM, Carlson JR, Zwiebel LJ (2001) Candidate olfactory receptors from the malaria vector mosquito *Anopheles gambiae* and evidence of down-regulation in response to blood feeding. Proc Natl Acad Sci U S A 98:14693–14697

Galizia CG, Münch D, Strauch M, Nissler A, Ma S (2010) Integrating heterogeneous odor response data into a common response model: a DoOR to the complete olfactome. Chem Senses 35:551–563

Gao Q, Chess A (1999) Identification of candidate *Drosophila* olfactory receptors from genomic DNA sequence. Genomics 60:31–33

Glatz R, Bailey-Hill K (2011) Mimicking nature's noses: from receptor deorphaning to olfactory biosensing. Prog Neurobiol 93:270–296

Hallem EA, Carlson JR (2006) Coding of odors by a receptor repertoire. Cell 125:143–160

Hamada S, Tabuchi M, Toyota T, Sakurai T, Hosoi T, Nomoto T, Nakatani K, Fujinami M, Kanzaki R (2014) Giant vesicles functionally expressing membrane receptors for an insect pheromone. Chem Commun 50:2958–2961

Hildebrand JG, Shepherd GM (1997) Mechanisms of olfactory discrimination: converging evidence for common principles across phyla. Annu Rev Neurosci 20:595–631

Hill CA, Fox AN, Pitts RJ, Kent LB, Tan PL, Chrystal MA, Cravchik A, Collins FH, Robertson HM, Zwiebel LJ (2002) G protein-coupled receptors in *Anopheles gambiae*. Science 298:176–178

Jacquin-Joly E, Merlin C (2004) Insect olfactory receptors: contributions of molecular biology to chemical ecology. J Chem Ecol 30:2359–2397

Jones DT, Reed RR (1989) Golf: an olfactory neuron specific-G protein involved in odorant signal transduction. Science 244:790–795

Kaiser L, Graveland-Bikker J, Steuerwald D, Vanberghem M, Herlihy K, Zhang SG (2008) Efficient cell-free production of olfactory receptors: detergent optimization, structure, and ligand binding analyses. Proc Natl Acad Sci U S A 105:15726–15731

Kanzaki R, Nakatani K, Sakurai T, Misawa N, Mitsuno H (2016) Physiology of chemical sense and its biosensor application. In: Nakamoto T (ed) Essentials of machine olfaction and taste. Wiley, New York

Karg G, Sauer AE (1995) Spatial distribution of pheromone in vineyards treated for mating disruption of the grape vine moth *Lobesia botrana* measured with electroantennograms. J Chem Ecol 21:1299–1314

Keil TA (1999) In: Hansson BS (ed) Morphology and development of the peripheral olfactory organs, Insect olfaction, Springer, pp 5–48

Khadka R, Aydemir N, Carraher C, Hamiaux C, Colbert D, Cheema J, Malmström J, Kralicek A, Travas-Sejdic J (2019) An ultrasensitive electrochemical impedance-based biosensor using insect odorant receptors to detect odorants. Biosens Bioelectron 126:207–213

Kiely A, Authier A, Kralicek AV, Warr CG, Newcomb RD (2007) Functional analysis of a *Drosophila melanogaster* olfactory receptor expressed in Sf9 cells. J Neurosci Methods 159:189–194

Kramer E (1975) Orientation of the male silkmoth to the sex attractant bombykol. In: Denton DA, Coghlan JP (eds) Olfaction and taste, vol 5. Academic Press, New York, pp 329–335

Kuwana Y, Shimoyama I (1998) A pheromone-guided mobile robot that behaves like a silkworm moth with living antennae as pheromone sensors. Int J Robot Res 17:924–933

Kuwana Y, Nagasawa S, Shimoyama I, Kanzaki R (1999) Synthesis of the pheromone-orientated behaviour of silkworm moths by a mobile robot with moth antennae as pheromone sensors. Biosens Bioelectron 14:195–202

Lundin C, Kall L, Kreher SA, Kapp K, Sonnhammer EL, Carlson JR, Heijine G, Nilsson I (2007) Membrane topology of the Drosophila OR83b odorant receptor. FEBS Lett 581:5601–5604

Milli R, Koch UT, de Kramer JJ (1997) EAG measurement of pheromone distribution in apple orchards treated for mating disruption of *Cydia pomonella*. Entomol Exp Appl 82:289–297

Misawa N, Mitsuno H, Kanzaki R, Takeuchi S (2010) Highly sensitive and selective odorant sensor using living cells expressing insect olfactory receptors. Proc Natl Acad Sci U S A 107:15340–15344

Mitsuno H, Sakurai T, Namiki S, Mitsuhashi H, Kanzaki R (2015) Novel cell-based odorant sensor elements based on insect odorant receptors. Biosens Bioelectron 65C:287–294

Mujiono T, Sukekawa Y, Nakamoto T, Mitsuno H, Termtanasombat M, Kanzaki R, Misawa N (2017) Sensitivity improvement by applying lock-in technique to fluorescent instrumentation for cell-based odor sensor. Sens Mater 29:65–76

Münch D, Galizia CG (2016) DoOR 2.0--comprehensive mapping of Drosophila melanogaster odorant responses. Sci Rep 6:21841

Myrick AJ, Park KC, Hetling JR, Baker TC (2009) Detection and discrimination of mixed odor strands in overlapping plumes using an insect-antenna-based chemosensor system. J Chem Ecol 35:118–130

Myrick AJ, Park KC, Hetling JR, Baker TC (2011) Locating a compact odor source using a four-channel insect electroantennogram sensor. Bioinspir Biomim 6:016002

Nakagawa T, Sakurai T, Nishioka T, Touhara K (2005) Insect sex pheromone signals mediated by specific combinations of olfactory receptors. Science 307:1638–1642

Nakagawa T, Pellegrino M, Sato K, Vosshall LB, Touhara K (2012) Amino acid residues contributing to function of the Heteromeric insect olfactory receptor complex. PLoS One 7:e32372

Nef S, Allaman I, Fiumelli H, De Castro E, Nef P (1996) Olfaction in bird: differential embryonic expression of nine putative odorant receptor genes in the avian olfactory system. Mech Dev 55:65–77

Neuhaus EM, Gisselmann G, Zhang W, Dooley R, Stortkuhl K, Hatt H (2005) Odorant receptor heterodimerization in the olfactory system of *Drosophila melanogaster*. Nat Neurosci 8:15–17

Ngai J, Dowling MM, Buck L, Axel R, Chess A (1993) The family of genes encoding odorant receptors in the channel catfish. Cell 72:657–666

Park KC, Ochieng SA, Zhu J, Baker TC (2002) Odor discrimination using insect electroantennogram responses from an insect antennal array. Chem Senses 27:343–352

Robertson HM, Wanner KW (2006) The chemoreceptor superfamily in the honey bee, Apis mellifera: expansion of the odorant, but not gustatory, receptor family. Genome Res 16:1395–1403

Rumbo ER, Suckling DM, Karg G (1995) Measurement of airborne pheromone concentrations using electroantennograms: interactions between environmental volatiles and pheromone. J Insect Physiol 41:465–471

Sachse R, Dondapati SK, Fenz SF, Schmidt T, Kubick S (2014) Membrane protein synthesis in cell-free systems: from bio-mimetic systems to bio-membranes. FEBS Lett 588:2774–2781

Sakurai T, Nakagawa T, Mitsuno H, Mori H, Endo Y, Tanoue S, Yasukochi Y, Touhara K, Nishioka T (2004) Identification and functional characterization of a sex pheromone receptor in the silkmoth *Bombyx mori*. Proc Natl Acad Sci U S A 101:16653–16658

Sakurai T, Mitsuno H, Haupt SS, Uchino K, Yokohari F, Nishioka T, Kobayashi I, Sezutsu H, Tamura T, Kanzaki R (2011) A single sex pheromone receptor determines chemical response specificity of sexual behavior in the silkmoth *Bombyx mori*. PLoS Genet 7:e1002115

Sakurai T, Mitsuno H, Mikami A, Uchino K, Tabuchi M, Zhang F, Sezutsu H, Kanzaki R (2015) Targeted disruption of a single sex pheromone receptor gene completely abolishes *in vivo* pheromone response in the silkmoth. Sci Rep 5:11001

Sato K, Takeuchi S (2014) Chemical vapor detection using a reconstituted insect olfactory receptor complex. Angew Chem Int Ed 53:11798–11802

Sato K, Pellegrino M, Nakagawa T, Nakagawa T, Vosshall LB, Touhara K (2008) Insect olfactory receptors are heteromeric ligand-gated ion channels. Nature 452:1002–1006

Sato K, Tanaka K, Touhara K (2011) Sugar-regulated cation channel formed by an insect gustatory receptor. Proc Natl Acad Sci U S A 108:11680–11685

Sauer AE, Karg G, Koch UT, de Kramer JJ, Milli R (1992) A portable EAG system for the measurement of pheromone concentrations in the field. Chem Senses 17:543–553

Schneider D (1957) Elektrophysiologische Untersuchungen von Chemound Mechanoreceptoren de Antenne des Seidenspinners *Bombyx mori* L. Z Vergl Physiol 40:8–41

Schoning MJ, Schutz S, Schroth P, Weißbecker B, Steffen A, Kordos P, Hummel HE, Luth H (1998) A BioFET on the basis of intact insect antennae. Sensors Actuators B 47:234–237

Schroth P, Schoning MJ, Kordos P, Luth H, Schutz S, Weißbecker B, Hummel HE (1999) Insect-based BioFets with improved signal characteristics. Biosens Bioelectron 14:303–308

Schutz S, Schoning MJ, Riemer A, Weißbecker B, Kordos P, Luth H, Hummel HE (1997) Field effect transistor–insect antenna junction. Naturwissenschaften 84:86–88

Schutz S, Weißbecker B, Koch UT, Hummel HE (1999) Detection of volatiles released by diseased potato tubers using a biosensor on the basis of intact insect antennae. Biosens Bioelectron 14:221–228

Schutz S, Schoning MJ, Schroth P, Malkoc U, Weißbecker B, Kordos P, Luth H, Hummel HE (2000) An insect-based BioFET as a bioelectronic nose. Sensors Actuators B 65:291–295

Smart R, Kiely A, Beale M, Vargas E, Carraher C, Kralicek AV, Christie DL, Chen C, Newcomb RD, Warr CG (2008) Drosophila odorant receptors are novel seven transmembrane domain proteins that can signal independently of heterotrimeric G proteins. Insect Biochem Mol Biol 38:770–780

Steinbrecht RA (1999) Olfactory receptors. In: Eguchi E, Tominaga Y (eds) Atlas of arthropod sensory receptors-dynamic morphology in relation to function. Springer, Tokyo, pp 155–176

Suckling DM, Karg G, Bradley SJ, Howard CR (1994) Field electroantennogram and behavioral responses of Epiphyas postvittana (Lepidoptera: Tortricidae) under low pheromone and inhibitor concentrations. J Econ Entomol 87:1477–1487

Takeuchi H, Kurahashi T (2003) Identification of second messenger mediating signal transduction in the olfactory receptor cell. J Gen Physiol 122:557–567

Tamura T, Thibert C, Royer C, Kanda T, Abraham E, Kamba M, Komoto N, Thomas JL, Mauchamp B, Chavancy G, Shirk P, Fraser M, Prudhomme JC, Couble P (2000) Germline transformation of the silkworm *Bombyx mori* L. using a piggyBac transposon-derived vector. Nat Biotechnol 18:81–84

Tanada N, Sakurai T, Mitsuno H, Bakkum DJ, Kanzaki R, Takahashi H (2012) Dissociated neuronal culture expressing ionotropic odorant receptors as a hybrid odorant biosensor—proof-of-concept study. Analyst 137:3452–3458

Termtanasombat M, Mitsuno H, Misawa N, Yamahira S, Sakurai T, Yamaguchi S, Nagamune T, Kanzaki R (2016) Cell-based odorant sensor Array for odor discrimination based on insect odorant receptors. J Chem Ecol 42:716–724

Troemel ER, Chou JH, Dwyer ND, Colbert HA, Bargmann CI (1995) Divergent seven transmembrane receptors are candidate chemosensory receptors in *C. elegans*. Cell 83:207–218
van der Pers JNC, Minks AK (1998) A portable electroantennogram sensor for routine measurements of pheromone concentrations in greenhouses. Entomol Exp Appl 87: 209–215
Vogt RG, Riddiford LM (1981) Pheromone binding and inactivation by moth antennae. Nature 293:161–163
Vogt RG, Prestwich GD, Lerner MR (1991) Odorant-binding-protein subfamilies associate with distinct classes of olfactory receptor neurons in insects. J Neurobiol 22:74–84
Vosshall LB, Amrein H, Morozov PS, Rzhetsky A, Axel R (1999) A spatial map of olfactory receptor expression in the *Drosophila* antennae. Cell 96:725–736
Wang G, Carey AF, Carlson JR, Zwiebel LJ (2010) Molecular basis of odor coding in the malaria vector mosquito *Anopheles gambiae*. Proc Natl Acad Sci U S A 107:4418–4423
Wanner KW, Anderson AR, Trowell SC, Theilmann DA, Robertson HM, Newcomb RD (2007) Female-biased expression of odourant receptor genes in the adult antennae of the silkworm, *Bombyx mori*. Insect Mol Biol 16:107–119
Wetzel CH, Behrendt HJ, Gisselmann G, Stortkuhl KF, Hovemann B, Hatt H (2001) Functional expression and characterization of a *Drosophila* odorant receptor in a heterologous cell system. Proc Natl Acad Sci U S A 98:9377–9380

Part IV
Mechanisms Controlling Behavior and Its Application to Robotics

Chapter 12
Brain Premotor Centers for Pheromone Orientation Behavior

Shigehiro Namiki and Ryohei Kanzaki

Abstract Moths use sophisticated olfactory navigation tactics for resource localization. We herein discuss the neuronal mechanisms that generate locomotor command for pheromone-source localization in the silkmoth *Bombyx mori.* Flip-flop neural signal, a state-dependent persistent firing activity, which is observed in the area of the insect brain called the lateral accessory lobe, correlates with turning behavior during pheromone orientation. We describe the morphology and physiology of basic neuronal components in the circuit, and provide a comparative view of its organization. We also describe neurons in the posterior slope, which contains a population of neurons descending toward the body ganglia. Mechanisms that generate flip-flop signals are also discussed.

Keywords Bombykol · Intracellular recording · Neuron · Olfaction

12.1 Pheromone Orientation

Moths use sophisticated olfactory navigation tactics for resource localization. The behavioral tactics are composed of two different modules: surge and casting. When male moths detect pheromone molecules, they immediately begin making surge in their movement. Moths exhibit surge frequently, along with zigzag upwind locomotion in the pheromone plume. The movement is often in a straight-line trajectory, which depends on the pheromone plume structure (Mafra-Neto and Cardé 1995). Casting is observed following the loss of the pheromone plume, with spontaneous turning, and often, an increased turn duration (Vickers and Baker 1994). The zigzag movement, which is the successive counterturning toward the left and right sides, is one of the characteristic movements in the casting behavior. This movement increases the probability to recontact the sensory cue, that is, pheromone molecules

S. Namiki (✉) · R. Kanzaki
Research Center for Advanced Science and Technology, The University of Tokyo, Tokyo, Japan
e-mail: namiki@rcast.u-tokyo.ac.jp

Y. Ishikawa (ed.), *Insect Sex Pheromone Research and Beyond*, Entomology Monographs, https://doi.org/10.1007/978-981-15-3082-1_12

(Bell 1990), and this tactic has been well studied using the moth pheromone communication system (Cardé and Willis 2008). However, the neuronal mechanisms that generate the locomotor command are unclear.

Here, we describe the brain region generating locomotor commands in the insect brain with a focus on the silkmoth *Bombyx mori*, which is domesticated and has lost the ability to fly. Male silkmoths orient toward conspecific females by walking. The presence of the sex pheromone alone is sufficient to trigger pheromone orientation behavior. The behavioral pattern they exhibit while walking is similar to that of their flying counterparts. They exhibit straight walking (i.e., surge) in response to pheromone input. After the loss of plume contact, moths demonstrate casting behavior, which is zigzag motion without the translational component (Kramer 1975). They repeat clockwise and counterclockwise turning, which is often called the mating dance (Obara 1979). Pheromone-triggered behavior occurs in the laboratory, and has therefore been used as one of the model systems to study odor-source-oriented navigation. Males placed on a ‘sled’ after having their legs removed moved upwind in a pheromone plume along zigzagging tracks using power produced by wing beating alone (Kanzaki 1998). The orientation behavior is thought to be organized by the same principle as in flying moth species.

12.2 Flip-Flop Signal

Olberg (1983) performs extracellular recording of neural activity from the neck connective of the silkmoth, which can monitor population signal of descending neurons (DNs) arising from the brain. This study identifies a group of axons exhibiting state-dependent spiking activity that correlated with antennal positioning (Fig. 12.1a). The firing activity is long lasting, and switched between ON and OFF either spontaneously or triggered by external stimulus. This activity is termed ‘flip-flop’ after the toggle flip-flop in electronics. As the antennal position correlates with the turning direction of the body (Fig. 12.1b), flip-flop signal is thought to underlie the control of locomotion during pheromone-triggered orientation behavior.

A brain region called the lateral accessory lobe (LAL) is later identified as the area involved in controlling motor function for pheromone orientation (Kanzaki et al. 1994). Extracellular recording with suction electrodes, similar to the previous study, but with dye-filling after recording, is performed (Fig. 12.1c). The flip-flop activity is observed in both sides of the connectives. The activity phase is reversed between the hemispheres. When one side demonstrated persistent firing activity (‘UP’-state), the other was silent (‘DOWN’-state) (Fig. 12.1d). DNs exhibiting the flip-flop neural signal innervate the LAL and several cell clusters for the DNs have been identified. The state of neuronal activity can be switched by sensory stimuli at any time point during the activity cycle (Olberg 1983; Kanzaki et al. 1994). The resetting mechanism is a key feature of the zigzag (Baker 1990; Kanzaki et al. 1992), which distinguishes it from other simple oscillatory behaviors.

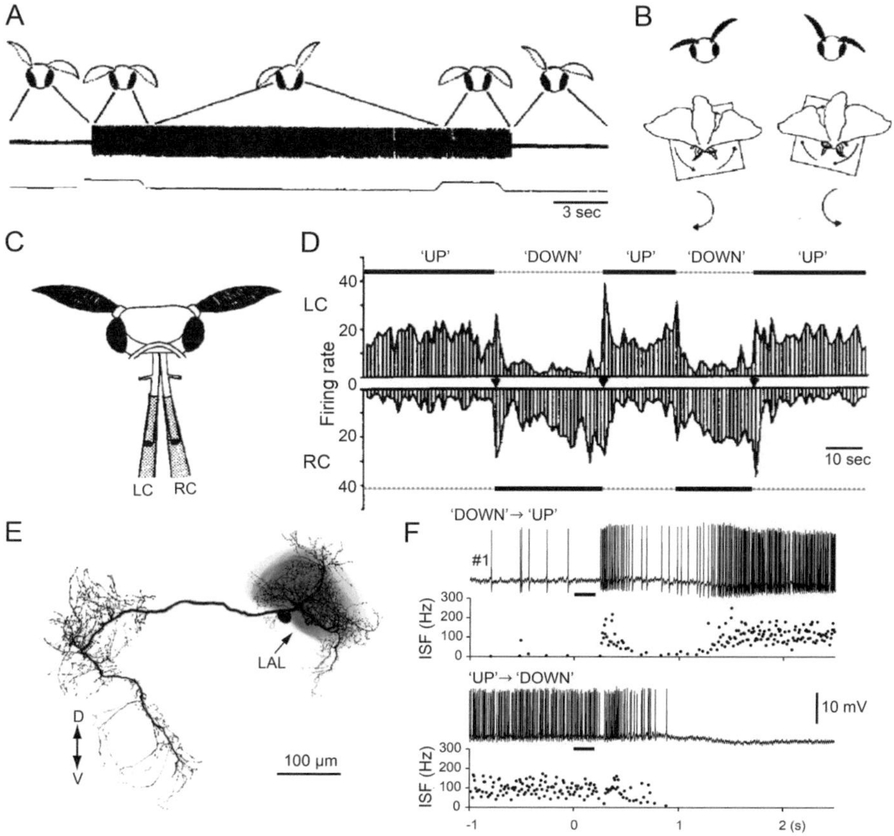

Fig. 12.1 State-dependent persistent firing activity in the silkmoth
(**a**) The correlation between antennal positions and neuronal activity recorded from the neck connective. The drawing shows the antennal position (top), neural activity (middle), and stimulus marker (bottom). The figure is modified from Olberg (1983). (**b**)**.** The correlation between antennal positions (top) and direction of turning (bottom). The figure is modified from Olberg (1983). (**c**) Preparation for extracellular recording with two suction pipettes. The electrodes were applied to the left (LC) and right connectives (RC). The figure is modified from Kanzaki et al. (1994). (**d**) Firing rate of the RC and LC. The activity on one side follows an alternating 'UP' and 'DOWN' state, which is negatively correlated between the two sides. Arrows indicate the timing of pheromone exposure. The figure is modified from Kanzaki et al. (1994). (**e**) Morphology of a flip-flopping descending neuron. The maximum intensity projection of confocal stacks is shown. The neuron innervates a wide area within the LAL. The volume of the LAL is shown in gray. (**f**) Physiology of a descending neuron exhibiting flip-flop responses to exposure to the sex pheromone. Changes in the membrane potential (top) and instantaneous firing frequency (bottom) associated with the state transition from 'DOWN' to 'UP' and 'UP' to 'DOWN' are shown. The black bar indicates the time period of pheromone exposure (500 ms). The figure is modified from Namiki et al. (2014)

Follow-up studies using intracellular recording with sharp glass microelectrodes characterize specific types of DNs, which exhibited state transition in their firing activity. The morphological features of individual DNs are described below (Sect. 12.4). DNs showed rapid state transition and identical sensory stimulation toggles the firing state (Namiki et al. 2014) (Fig. 12.1e, f). State transition occurs in response to the applied sensory stimuli (Mishima and Kanzaki 1999; Wada and Kanzaki 2005; Namiki et al. 2014) or may simply be spontaneous (Kanzaki et al. 1991a).

Behavioral studies are conducted to clarify the relationship between walking and wing movement during pheromone orientation (Ariyoshi and Kanzaki 1996; Kanzaki 1998). The timing of wing retraction toward the turning side is synchronized with the change in the direction of walking. The high correlation between the kinematics of wing motion and the pattern of zigzag walking suggests that a common motor program underlies pheromone-modulated flight and walking. This assumption is further supported by the observation that the flip-flop neural signal is present in a strain of *B. mori* that can fly (R. Kanzaki, unpublished observation). Furthermore, neurons with a similar morphology to the flip-flop DNs have been identified in flying moth species (Kanzaki et al. 1991a), suggesting that the mechanism is shared between walking and flying moths.

The neck angle is also synchronized with body direction during turning (Mishima and Kanzaki 1996, 1998). The head direction is controlled by neck motor neurons that innervate the cervical ventral muscles and the ventral muscles through the second cervical nerve. The cervical ventral 1 neck motor neuron (cv1), a member of this population, exhibits flip-flop responses that are synchronized with those of DNs innervating the LAL (Mishima and Kanzaki 1999). Double-labeling of DNs and cv1 has revealed the potential connectivity between these neurons. Some DNs innervating the LAL send information to the neck motor neurons (Wada and Kanzaki 2005), suggesting that the command signals for turning the body direction and head angle are conserved in the moth nervous system.

Recently, study using whole-brain imaging reports the flip-flop-like neuronal activity in the protocerebral bridge in *Drosophila* (Aimon et al. 2018). This area has dense connections with the LAL (Shih et al. 2015) and the interaction of flip-flop between these regions is of interest.

12.3 Lateral Accessory Lobe

The LAL is a paired midline neuropil located on the lateral side of the central complex (CX) in insects (Williams 1975; Ito et al. 2014) (Fig. 12.2). The CX and LAL are present in all insect species reported thus far and they have a common set of subdomains, suggesting the ground pattern of LAL organization (Namiki and Kanzaki 2016a, b). Due to the deep location and the lack of the clear anatomical boundaries, few studies have focused on the LAL. Thus, the function of

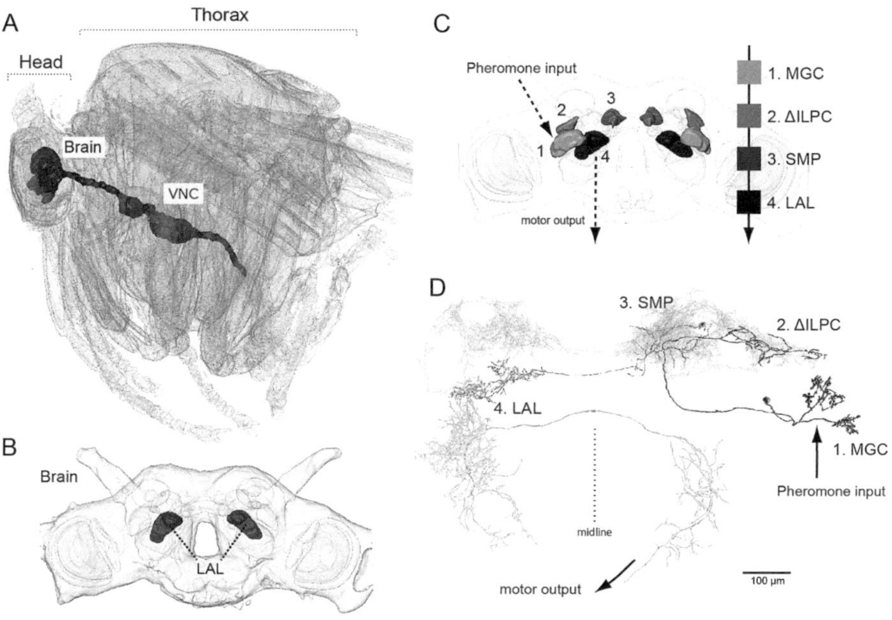

Fig. 12.2 The brain and neural circuits for pheromone information processing in the silkmoth *Bombyx mori*
(**a**) The location of the central nervous system in *Bombyx*. The volume of the nervous system was segmented from the image data obtained by X-ray micro computed tomography. The brain and ventral nerve cord (VNC) are shown in black. (**b**) The location of the lateral accessory lobe (LAL) in the brain of the silkmoth (black). The LAL was segmented from the data obtained by confocal microscopy. (**c**) Brain regions involved in sex pheromone information processing. The number represents the order from the sensory input. (**d**) Representative examples of neurons that constitute the pheromone information processing pathway. The original data for reconstruction of the neurons are taken from Namiki et al. (2014). *MGC* macroglomerular complex, *SMP* superior medial protocerebrum, *ΔILPC* the delta area of the inferior lateral protocerebrum

the LAL remains to be investigated. Several hypotheses about its function have been proposed such as playing a role in commanding behavior by selecting the activity of the descending output (Wolff and Strausfeld 2015).

A variety of response properties have been reported in neurons innervating the LAL in different species, and these studies suggest the involvement of the LAL in locomotion such as steering toward left or right, and moving forward or backward. For example, a DN innervating the LAL in crickets responds to the courtship song. The activation of this neuron elicits walking and steering on the contralateral side, and hyperpolarization causes the cessation of walking (Zorović and Hedwig 2013). Moreover, LAL lesions cause abnormal turning behavior in cockroaches (Harley and Ritzmann 2010). Although DNs innervate the LAL on one side, specific DNs in *Drosophila* innervate the LAL on both sides. The activation of these neurons causes backward locomotion (Bidaye et al. 2014). Furthermore, these DNs are present in the larval stage, and also induce backward crawling when activated using genetic manipulation techniques (Carreira-Rosario et al. 2018). Their involvement in both walking and crawling suggests that DNs function in command information irrespective of the gait condition.

Information flow of the neural circuits for sex pheromone processing has been characterized in the silkmoth (Fig. 12.2c, d) (Namiki et al. 2014). Pheromones are received at the receptor neurons on the antennae and the information is conveyed to the brain. Projection neurons, which are second-order olfactory neurons, receive input from the antennal lobe (AL) and convey information to the higher order center in the inferior lateral protocerebrum (Fig. 12.2c, ΔILPC) (Seki et al. 2005), where the terminals for projection neurons for pheromonal and non-pheromonal information are segregated. The information is then transmitted to the LAL via the superior medial protocerebrum (Fig. 12.2c, SMP) (Seki et al. 2005; Namiki et al. 2014). As all 'flip-flopping' neurons identified thus far innervate the LAL, the signal is thought to be generated at this stage. Projection neurons arising from earlier stages, such as the AL, ΔILPC, and SMP, respond to the pheromones, but do not show long-lasting firing activity.

The area surrounding the CX, including the LAL, is called the lateral complex (Ito et al. 2014) or the CX accessory regions (Lin et al. 2013). Comparative neuroanatomy for the organization of the lateral complex is currently being examined. The LAL and the bulb, a small satellite neuropil, are classified as members of the lateral complex in *Drosophila* (Ito et al. 2014). GABA- and serotonin-like immunoreactivity in the LAL and surrounding area in the silkmoth is shown in Fig. 12.3. There is a small satellite neuropil located dorsal to the LAL, termed the median olive (Iwano et al. 2010), which exhibits dense GABA immunoreactivity (Fig. 12.3, termed the bulb). This area is rarely connected with the LAL. Based on its position and immunoreactivity, this region may correspond to the lateral triangle in the monarch butterfly (Heinze and Reppert 2012), the bulb in *Drosophila* (Ito et al. 2014), the median olive (medial bulb), the lateral triangle (lateral bulb) in the locust (Träger et al. 2008; Heinze and Homberg 2008; el Jundi et al. 2014), and the lateral complex in the ant (Schmitt et al. 2015).

Homologous structures of the LAL have been reported in arthropods other than insects, including species in Malacostraca, Cephalocarida, Remipedia, Branchiopoda, and Chilopoda (Supplemental Table in Namiki and Kanzaki, 2016a). Although their basic layout is similar, several differences in the volume and connections are present, which may reflect evolutional and ecological contexts. For example, the relative volume of the LAL in non-insect arthropods is smaller than that in insects, although it was estimated by visual inspection (Stegner and Richter 2011; Kenning and Harzsch 2013; Krieger et al. 2015; Stegner et al. 2015). This observation may be related to their ability to use appendages. Many non-insect arthropods live under the water and do not walk well compared with insects.

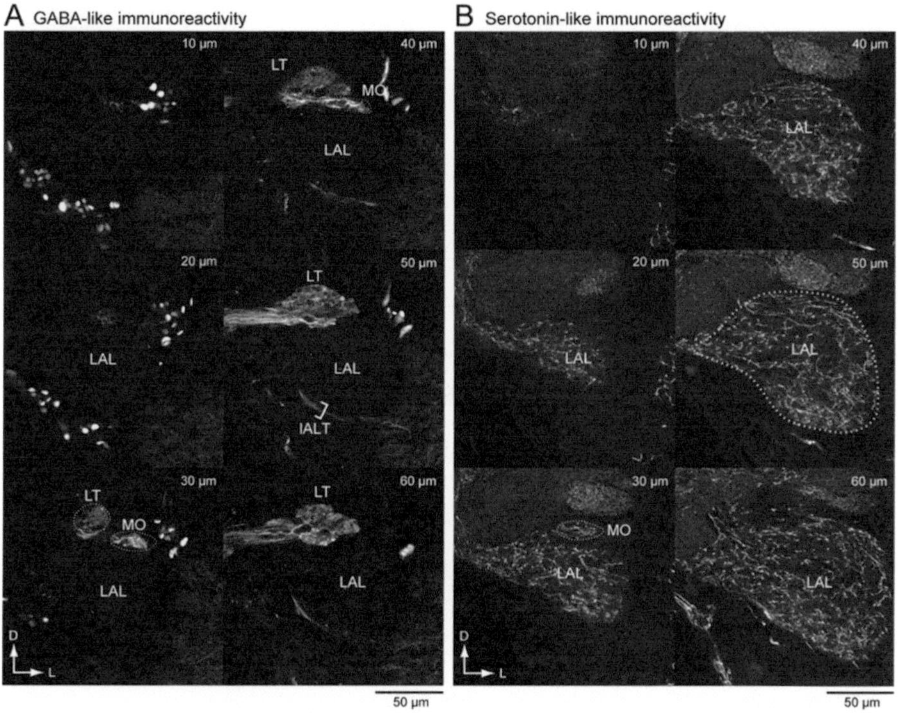

Fig. 12.3 GABA- and serotonin-like immunoreactivity in the lateral accessory lobe and surrounding areas in the silkmoth
Confocal stacks at different depths for GABA-like (**a**) and serotonin-like (**b**) immunoreactivity in the LAL on the right side. Anti-GABA antibody and anti-serotonin antibody were used for immunostaining. The depth from the anterior surface is shown in the top right. The region of the bulb is evident by GABA-immunolabeling. The mediolateral antennal lobe tract (ml-ALT) is labeled, which contains GABAergic projection neurons from the antennal lobe. The labeling pattern of serotonin is consistent with the shape of the LAL. The median olive (MO) is labeled by both antibodies. *D* dorsal, *L* lateral, *LT* lateral triangle

12.4 Descending Neurons

Descending neurons (DNs) run through the neck connective, and connect the brain and thoracic ganglia. Neuronal innervation by DNs and ascending neurons in the brain is presented in Fig. 12.4a–d. A fluorescent dye is injected from the cut end of the neck connective on both sides to label neurites of both DNs and ascending neurons. They are separable to an extent because the neurites of DNs stain well due to their larger axon diameter and because the cell bodies are located within the brain. DNs from the cerebral ganglion (a.k.a. supraesophageal ganglion) are anatomically classified based on their cell body locations. Group-I DNs have their soma on the anterior ventral surface of the brain. Group-II DNs have their soma on the anterior

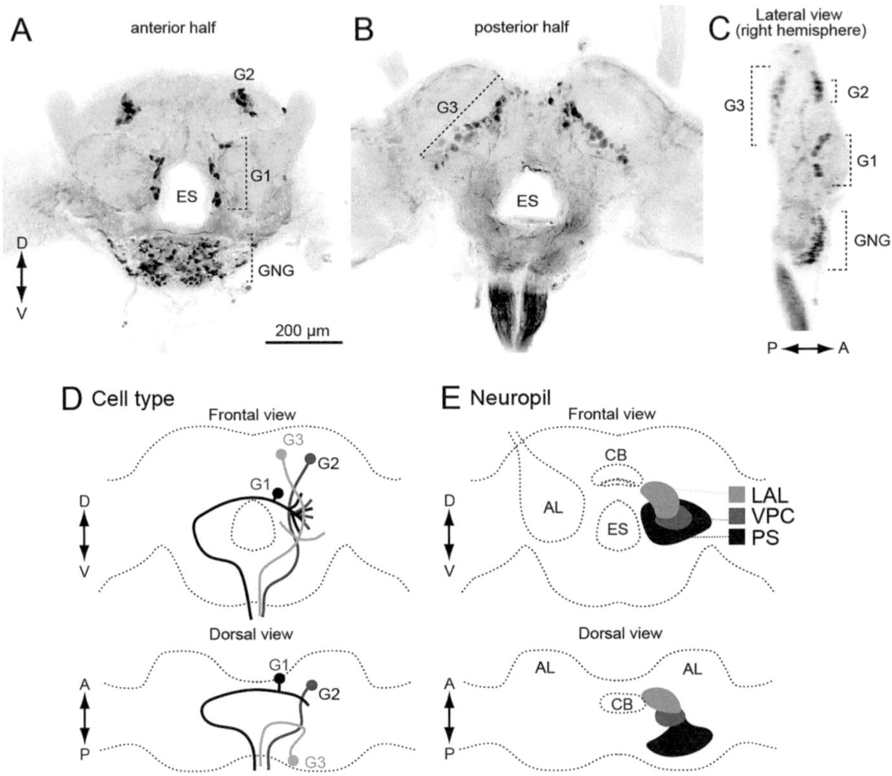

Fig. 12.4 Distribution of descending neurons in the brain
(**a–c**) Maximum intensity projection of confocal stacks for the sample that underwent backfill staining from the neck connective. Anterior (**a**), posterior (**b**), and lateral (**c**) views. Descending neurons (DNs) from the cerebral ganglion are classified based on their cell body locations: Group-I DNs (G1) have their soma on the anterior ventral surface of the brain; Group-II DNs (G2) have their soma on the anterior dorsal surface of the brain; Group-III DNs (G3) have their soma on the posterior surface of the brain. In addition, other DNs have their soma on the gnathal ganglia (GNG). (**d**) Schematics of DNs from the cerebral ganglion. Frontal and dorsal views are shown. (**e**) Schematics of brain neuropils, which contain the dendrites of a large number of DNs. The shapes of the brain and neuropils are indicated by broken lines. Frontal and dorsal views are shown. The original data are taken from Namiki et al. (2018b). *AL* antennal lobe, *CB* central body, *ES* esophagus

dorsal surface of the brain. The LAL, ventral protocerebrum (VPC), and posterior slope (PS) are well innervated by these DNs (Fig. 12.4e). The majority of DNs in the cerebral ganglion belong to group III, which have their soma on the posterior surface of the brain. In addition, a large number of DNs have their soma on the gnathal ganglia. They often have small cell bodies and a thin neurite diameter.

Pheromone information is processed by serial neural circuits in the silkmoth brain and the pheromone-evoked persistent firing activity is only observed in the

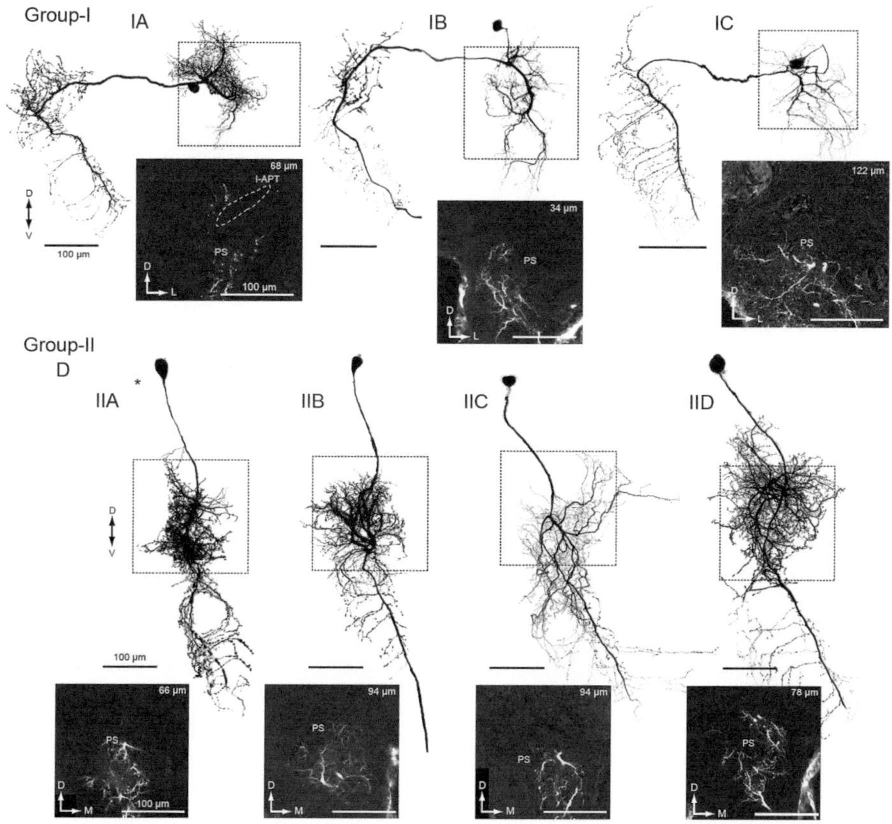

Fig. 12.5 Morphology of group-I and group-II descending neurons
Maximum intensity projections of brain innervation in the frontal view are shown for group-I (top) and group-II DNs (bottom). The inset shows a confocal stack for neurite innervation of the PS. The area of the section corresponds to the box indicated by broken lines. The depth from the posterior brain surface is shown in the top right. *l-APT* lateral antennal lobe tract, *PS* posterior slope

neurons innervating the LAL (Namiki et al. 2014), suggesting that the LAL is where the flip-flop neural signals are produced. Three types of DNs that exhibit flip-flop neural signals have been identified thus far (Mishima and Kanzaki 1999; Wada and Kanzaki 2005) (Fig. 12.5). These DNs are either group-I or group-II DNs. Group-IA DNs have smooth processes in the ipsilateral LAL and varicose processes in the contralateral LAL (Mishima and Kanzaki 1999), which are the indicators for the postsynaptic and presynaptic terminals (Cardona et al. 2010). Group-IIA and group-IID DNs have smooth processes in the ipsilateral LAL and descend to the ipsilateral neck connective (Mishima and Kanzaki 1999; Wada and Kanzaki 2005). Axonal projection in the ventral nervous system is unknown in the

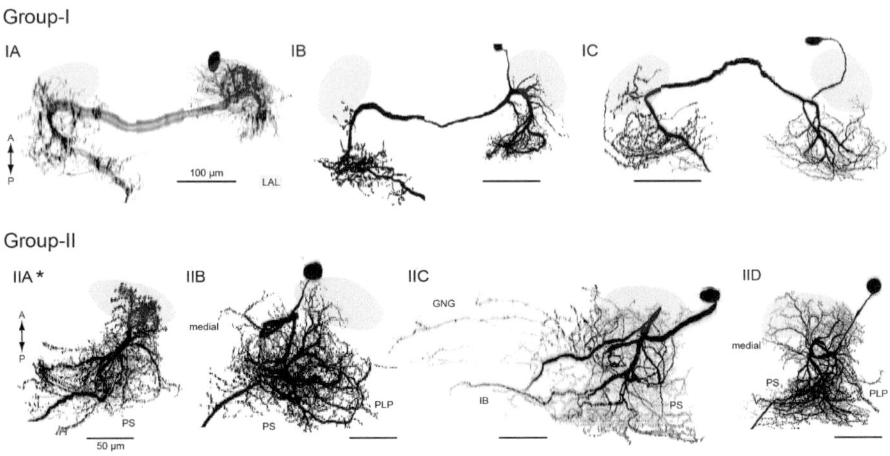

Fig. 12.6 Morphology of group-I and group-II descending neurons in the dorsal view
Maximum intensity projections of brain innervation in the dorsal view are shown for group-I (top) and group-II DNs (bottom). The volume of the LAL is shown in gray. DNs have additional innervation medial to the LAL ('medial'). *GNG* gnathal ganglion, *IB* inferior bridge, *PLP* posterior lateral protocerebrum, *PS* posterior slope

silkmoth. Although flip-flop DNs innervate the LAL, the medial part of the PS is innervated by all types of group-I/II DNs (Namiki et al. 2018b). The dorsal views of these DNs are shown in Fig. 12.6. The majority of the neurites are also located outside the LAL (labeled in gray).

Of note, neurons with a similar morphological profile are present in other insects. For example, a neuroanatomical study in *Drosophila* reports the overall neuron morphology, including projection to the thoracic ganglia by a group of ipsilateral DNs (termed anterior DNs) (Namiki et al. 2018a). The neurite innervation, running pathway, and the cell body location are similar to group-II DNs in the silkmoth. As in *Bombyx* group-II DNs, *Drosophila* anterior DNs mostly share neurite innervation in the medial side of the PS. Some anterior DNs have dendritic innervation within the LAL. Neuron types with axonal projection to both wing and leg motor regions are present, suggesting that the LAL is involved in both types of locomotion.

Examples of the morphology of group-III DNs are presented in Fig. 12.7. The function of group-III DNs in sex-pheromone information processing is unclear. Recently, we found that some group-III DNs are responsive to the sex pheromone with brief excitation (Fig. 12.7; Namiki et al. 2018b). The dendrites of these DNs are mostly confined within the PS and do not innervate the LAL, suggesting that the PS also processes sex-pheromone information. However, the flip-flop response has never been observed in group-III DNs. From these observations, it is possible that the group-III DNs are involved in controlling the surge phase but not the casting phase.

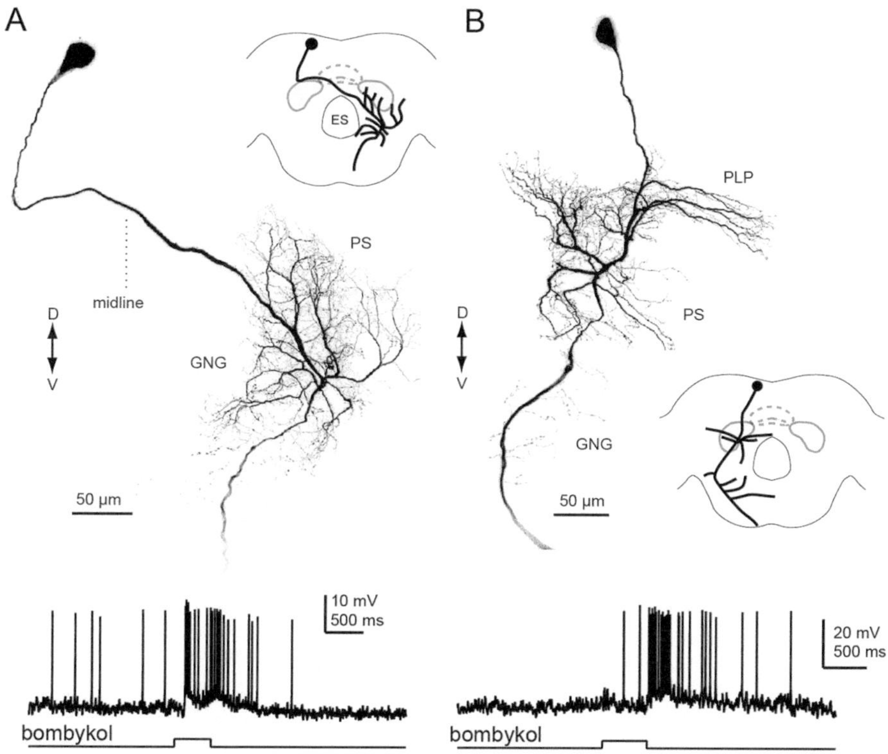

Fig. 12.7 Morphology and physiology of group-III descending neurons innervating the posterior slope
(**a**) Morphology of a group-III descending neuron (DN) that has a smooth process in the posterior slope (PS) and the gnathal ganglion (GNG). The neuron exhibited brief excitation in response to exposure to bombykol, the sex pheromone. The inset shows the schematics of the neuronal innervation. (**b**) Morphology of the group-III DN that has a smooth process in the PS and posterior lateral protocerebrum (PLP). The neuron exhibited brief excitation in response to exposure to bombykol. These neurons did not respond to a puff of clean air (data not shown). *ES* esophagus

12.5 Lateral Accessory Lobe Interneurons

One prominent feature in the circuit organization of the LAL is the dense connection between both hemispheres. The LAL bilateral neurons (BNs) connect both sides of the LAL (Homberg et al. 1987; Homberg and Hildebrand 1989; Breidbach 1990; Müller et al. 1997; Dacks et al. 2006). This neuronal population is thought to play an essential role in generating locomotor commands (Kanzaki 1997). The single-neuron morphology of the LAL BNs has been reported in moths, including *Heliothis virescens* (Pfuhl and Berg 2007), *Agrotis segetum* (Lei et al. 2001), and *Manduca sexta* (Kanzaki et al. 1991b), as well as in fruit flies (Hanesch et al. 1989), crickets (Zorović and Hedwig 2011), and locusts

(Müller et al. 1997; Heinze et al. 2009). A population of LAL BNs was identified by clustering analysis of the FlyCircuit Database in *Drosophila* (Clusters 31, 215 and 1037 in supercluster XII) (Chiang et al. 2011; Costa et al. 2016).

In the silkmoth, approximately 60 fibers run through the LAL commissure, a bundle of BNs connecting the LALs on both sides. Among these, many neurons exhibit GABA-like immunoreactivity (Iwano et al. 2010), which is thought to be necessary for flip-flop activity. Additionally, there are two pairs of BNs in the LAL with serotonin-like immunoreactivity, which are present also in other species that belong to orders such as Lepidoptera, Coleoptera, and Diptera (Dacks et al. 2006). A population of the LAL BNs with morphological heterogeneity is shown in Fig. 12.8. LAL BNs also connect the regions other than the LAL, including the ventral protocerebrum (VPC), superior medial protocerebrum, posterior slope, and posterior lateral protocerebrum. The VPC

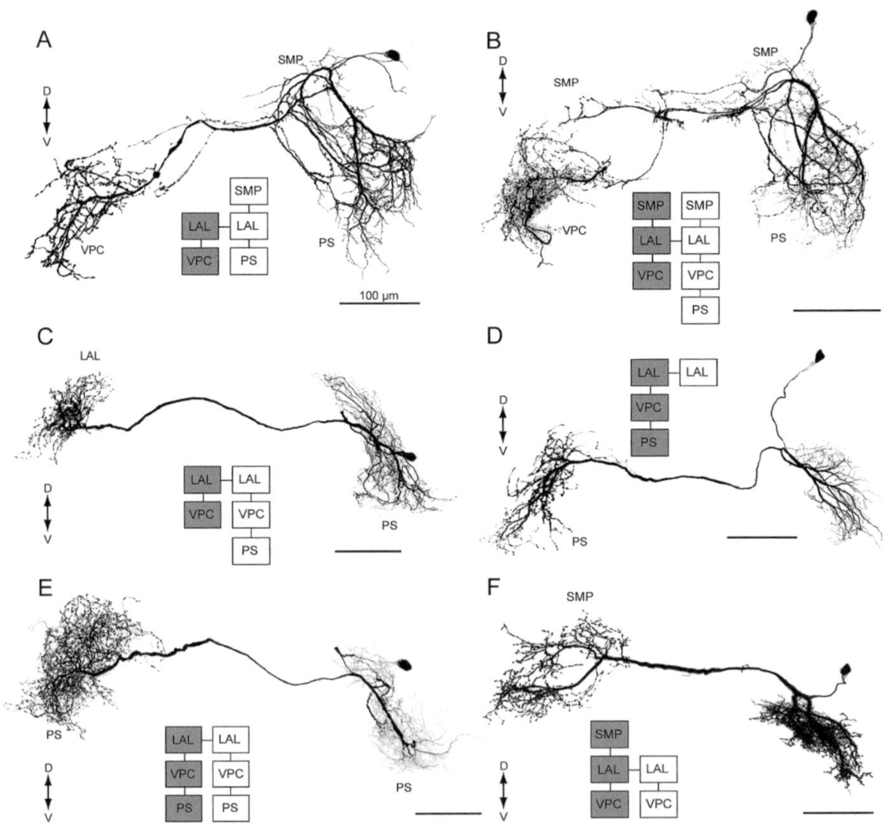

Fig. 12.8 Morphology of lateral accessory lobe bilateral neurons
Maximum intensity projection of confocal stacks for six different types of bilateral neurons (BNs). The inset shows the schematics representing the area of innervation. White and gray boxes indicate the innervation with smooth and varicose processes, respectively. *PS* posterior slope, *SMP* superior medial protocerebrum, *VPC* ventral protocerebrum

is an unstructured neuropil located posterior to the LAL. The LAL and VPC often share neuronal innervation, and there is no clear anatomical boundary. These regions are often referred to as the LAL-VPC unit (Iwano et al. 2010). The lateral antennal lobe tract (lALT), which is also known as the outer antenno-cerebral tract (Homberg et al. 1988), is used to define the anterior border with the LAL. The lALT is a bundle connecting the antennal lobe to the protocerebrum, and thus, does not fully define the boundary surface between the VPC and LAL. Although interneurons with innervation confined within the LAL are present, those with innervation confined within the VPC have not been identified (Iwano et al. 2010). The anatomical correspondence of the VPC in other insects is unclear.

12.6 Posterior Slope

We previously found that approximately half of LAL BNs innervate the posterior slope (PS) (Fig. 12.9). The PS is the inferior part of the posterior brain, where extensive arborization of descending and ascending neurons is observed (Strausfeld 1976; Ito et al. 2014). In all of the species studied thus far, the posterior-ventral part of the brain is densely labeled by backfilling from the neck connective (Cardona et al. 2009; Hsu et al. 2016; Kanzaki et al. 1994; Okada et al. 2003; Staudacher 1998). Lobula plate neurons supply this region, and are therefore considered to be involved in the processing of motion cues (Strausfeld and Bassemir 1985; Borst 2014).

The morphology of the LAL unilateral neurons, whose innervation is confined to one side of the hemisphere, is shown in Fig. 12.10. As for BNs, approximately half of unilateral interneurons innervate the PS. Although the connectivity between the LAL and PS remains unclear, neurons connecting the LAL and the posterior protocerebrum have been reported in the locust (Heinze and Homberg 2009) and butterfly (Heinze and Reppert 2011).

Although the number is small, there is a population of interneurons connecting the LAL and PS with opposite polarity. The morphology of a projection neuron that has smooth process in the PS of the ipsilateral hemisphere (the side of somata) and varicose process in the LAL of the contralateral hemisphere is shown in Fig. 12.11. Based on the terminal morphology, such neurons are considered to provide feedback from the PS to LAL. This example suggests that the LAL and PS are interconnected.

Group-I DNs, all of which exhibit bilateral innervation, have varicose terminal processes in the contralateral PS, and some group-II DNs studied thus far have varicose terminals in the ipsilateral PS (Mishima and Kanzaki 1999; Namiki et al. 2014). The neurons putatively homologous to the *Bombyx* group-I and group-II DNs in *Drosophila* have similar morphological features. They also additionally innervate the PS (e.g., VGlut-F-500726, VGlut-F-000150, and fru-F-100073; FlyCircuit Database) (Chiang et al. 2011). Group-I DNs have smooth

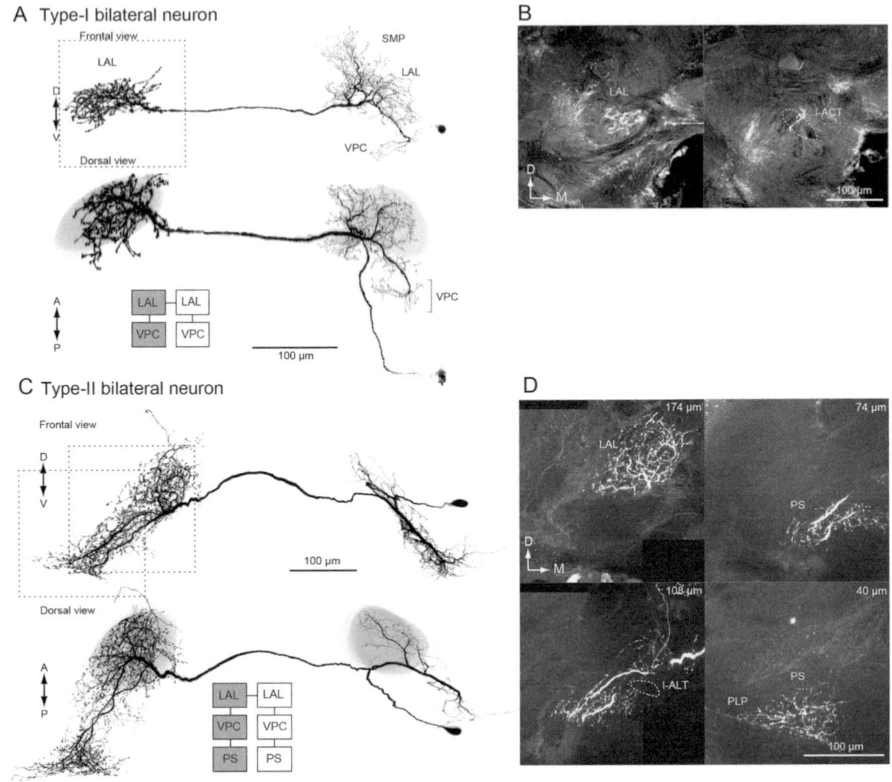

Fig. 12.9 Morphology of two morphological subtypes of LAL bilateral neurons
(**a**) Maximum intensity projection of confocal stacks for a type-I LAL BN. The varicose process in contralateral hemisphere is confined within the anterior brain. Frontal and dorsal views are shown. (**b**) Confocal stacks of the neuronal innervation. The depth from the posterior surface is shown in the top right. (**c**) Maximum intensity projection of confocal stacks for a type-II LAL BN. The innervation in the contralateral hemisphere extends to the posterior slope (PS). (**d**) Confocal stacks of the neuronal innervation. The neuron has a varicose process in the PS in addition to the LAL. *l-ALT* lateral antennal lobe tract, *PLP* posterior lateral protocerebrum, *SMP* superior medial protocerebrum, *VPC* ventral lateral protocerebrum

processes in the LAL on the ipsilateral side and varicose processes on the contralateral side. This pathway may mediate the information flow from the PS to the LAL (Fig. 12.4). These anatomical connections suggest a large degree of interplay between the LAL and PS.

However, the function of the PS in pheromone information processing is unclear. We hypothesize that the PS connects the LAL with the major population of DNs and then transmits the information to the thoracic motor centers. We assume that the PS is downstream of the LAL for the following two reasons: (i) the presence of interneurons connecting the LAL and PS (i.e., type-II LAL BNs), (ii) PS neurons are responsive to the sex pheromone and do not have connections to the circuits for pheromone

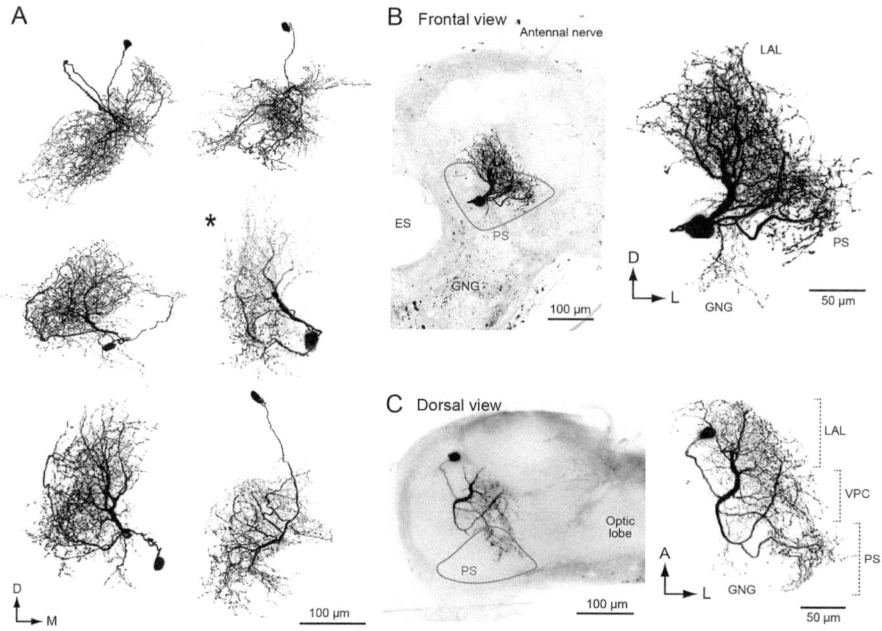

Fig. 12.10 Morphology of lateral accessory lobe unilateral neurons
(**a**) Morphology of LAL unilateral neurons (UNs). Maximum intensity projections of six different UNs are shown. One image is flipped for comparison (asterisk). Frontal (**b**) and dorsal (**c**) views of UNs innervating the posterior slope (PS). Low (10×, left) and high (40×, right) magnification images are shown. The shape of the PS is indicated with a gray line. *ES* esophagus, *GNG* gnathal ganglion, *VPC* ventral protocerebrum

information processing pathways except the LAL (Namiki et al. 2014). These observations suggest that the PS receives pheromone information from the LAL.

12.7 Candidate Neural Mechanisms

The flip-flop motif is widely observed in biological systems (Milo et al. 2002) and the underlying mechanisms have been analyzed. For example, the transition between the sleep and awake states has been proposed to be controlled by a mutual inhibitory network. The ventrolateral preoptic area, a sleep promoting area, inhibits ascending arousal pathways in the hypothalamus and brainstem, which in turn inhibit the sleep-promoting area (Saper et al. 2010). Furthermore, the cellular mechanisms, such as ionic conductance and integration properties of dendrite, are proposed for the flip-flop like neuronal activity (Egorov et al. 2002; Tahvildari et al. 2007, 2008; Idoux et al. 2008). We next consider neuronal mechanisms in the LAL with reference to other model systems.

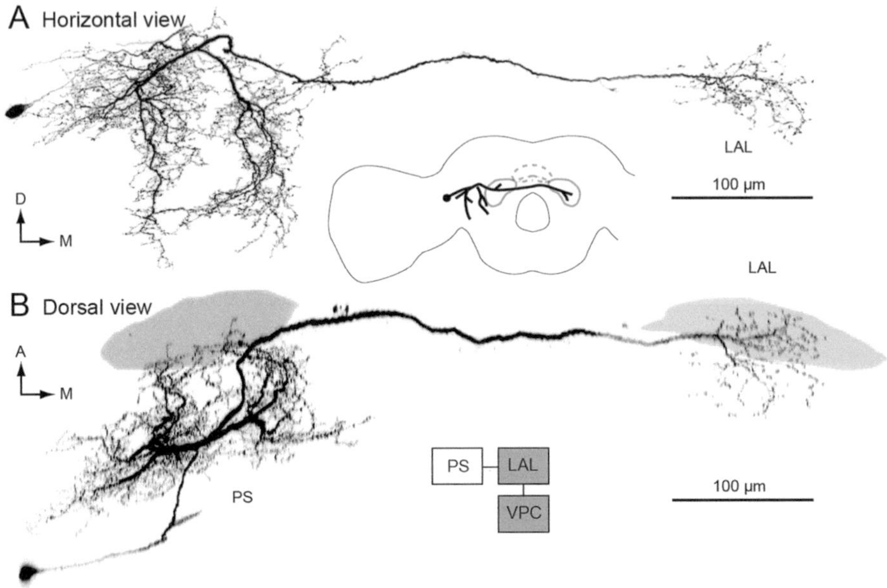

Fig. 12.11 Morphology of a projection neuron arising from the posterior slope and extending to lateral accessory lobe
Morphology of a putative feedback neuron extending from the posterior slope (PS) to the LAL. Horizontal (**a**) and dorsal (**b**) views of the maximum intensity projection of the neuron. The inset in (**a**) is a schematic illustration of the neuronal innervation in the brain. The inset in (**b**) is a schematic representing the connectivity. White and black boxes represent innervation with smooth and varicose processes, respectively. The original data are taken from Namiki and Kanzaki (2018)

12.7.1 Persistent Activity

Persistent neural activity is observed in a variety of nervous systems (Major and Tank 2004). Although neural mechanisms for persistent activity are mostly unknown in insects, the underlying mechanisms have been studied in vertebrate models. Persistent firing activity can be implemented by either cellular or synaptic mechanisms. Because persistent activity ceases after the blocking of synaptic transmission, circuit properties underlie the activity in the ferret cerebral cortex (Shu et al. 2003). An essential role of recurrent excitatory synapses in persistent firing has been reported in zebrafish descending interneurons (Li et al. 2006) and lamprey reticulospinal neurons (Antri et al. 2009). As the activity phase of persistent firing correlates among several DNs (Kanzaki et al. 1994), the involvement of a network mechanism is expected in the silkmoth. Recently, a recurrent synaptic circuit is identified in *Drosophila* and the connection implements persistent firing activity in the ellipsoid body of the central complex (Turner-Evans et al. 2017; Green et al. 2017). Such a reciprocal excitatory connection may enable persistent firing in the LAL. Iwano et al. (2010) propose that the reciprocal connection within the LAL-VPC unit underlies the persistent activity. Another candidate for the recurrent circuit is the connection between DNs. Group-I DNs receive input from the LAL in the ipsilateral hemisphere and send axonal projections to the contralateral hemisphere. Axonal projections of group-I DNs overlap with the dendritic innervation of group-II DNs (Namiki et al. 2018b). Thus, this potential connectivity between group-I and group-II DNs may form an excitatory loop.

Persistent firing activity, which is dependent on the cellular mechanism, has been reported in vertebrates (Egorov et al. 2002; Loewenstein et al. 2005; Tahvildari et al. 2007). Layer III principal neurons in the lateral entorhinal cortex demonstrate toggled flip-flop responses to current injection or synaptic activation (Tahvildari et al. 2007). Pharmacological experiments reveal that the ON-OFF transition of persistent firing is dependent on calcium-activated nonselective cationic (CAN) current (Tahvildari et al. 2008). The CAN current is known to mediate plateau potentials in other cortical regions (Schwindt et al. 1988; Andrade 1991). Investigation of whether LAL DNs, which exhibit flip-flop response, utilize this current in the silkmoth is of interest.

12.7.2 Mutual Inhibition

Another important aspect of the flip-flop activity is the antiphasic relationship between activities of the left and right hemispheres. Alternating turning toward the left and right sides corresponds to the state transition of flip-flop activity of DNs in the silkmoth (Olberg 1983; Kanzaki et al. 1994). Continuous turning behavior corresponds to the persistent firing of DNs on one side and concurrent silence on the opposite side. The activity of the left and right LALs is connected in a winner-take-all manner, with only one side being active at a time (Olberg 1983). The neuronal mechanism for this activity is still unknown in insects.

In a study using an integrated-fire neuron model, two recurrently connected loops of excitatory neurons with mutual inhibition, which enable flip-flop logic gate operation, are identified (Vogels and Abbott 2005). Several other modeling studies implement flip-flop operation using a similar circuit configuration (Wang 2002; Seely and Chow 2011; Zagha et al. 2015). The mutual inhibition contributes to the winner-take-all dynamics (Rutishauser et al. 2011). Neuronal components for mutual inhibition have been observed in nervous systems utilizing flip-flop activity: the feeding circuit in worms (Li et al. 2012), switching between REM and non-REM sleep in mice (Lu et al. 2006), and the neocortical UP-state in ferrets (Shu et al. 2003).

In the silkmoth, approximately 60 fibers run through the LAL commissure, a bundle of bilateral neurons connecting the LAL on both sides. Some neurons demonstrate GABA-like immunoreactivity, suggesting mutual inhibition between the left and right sides (Iwano et al. 2010). When the LAL commissure is ablated by microsurgery, moths exhibit straight-line walking or one-way looping, but zigzag motion is not observed (Kanzaki, unpublished observation). This suggests the importance of GABAergic connections for the state transition of the neuronal activity corresponding to left-right alternation. Additionally, there may be neuronal components for recurrent excitatory loops coupled with mutual inhibition. A population of interneurons innervating the LAL on one side is a candidate for this connection (Fig. 12.10) (Iwano et al. 2010).

One scenario for nervous system evolution is that all insects share the basic components of ancient circuits and numerous behaviors observed by modern insects share a single origin (Strausfeld 2012; Dickinson 2014). We summarized the current knowledge on the neural circuit of the LAL in the silkmoth, but the function of the LAL may not be limited to olfactory navigation. Indeed, we consider this mechanism to be applicable to a wide behavioral repertoire, such as wasp visual search and honeybee waggle dance (Namiki and Kanzaki 2016a). Comparative studies focusing on the circuit will contribute our understanding of the basic design of the insect nervous system.

References

Aimon S, Katsuki T, Jia T et al (2018) Fast near-whole brain imaging in adult *Drosophila* during responses to stimuli and behavior . bioRxiv 033803. https://doi.org/10.1101/033803

Andrade R (1991) Cell excitation enhances muscarinic cholinergic responses in rat association cortex. Brain Res 548:81–93. https://doi.org/10.1016/0006-8993(91)91109-E

Antri M, Fénelon K, Dubuc R (2009) The contribution of synaptic inputs to sustained depolarizations in reticulospinal neurons. J Neurosci 29:1140–1151. https://doi.org/10.1523/JNEUROSCI.3073-08.2009

Ariyoshi T, Kanzaki R (1996) Pheromone-triggered electromyograms of flight muscles in a male ***silkworm*** moth, Bombyx mori. Int J Wild Silkmoth Silk 2:21–25

Baker T (1990) Upwind flight and casting flight: complementary and tonic systems used for location of sex pheromone sources by male moths. In: Døving K (ed) Proceedings of the tenth international symposium on olfaction and taste. Oslo, Norway, pp 18–25

Bell WJ (1990) Searching behavior patterns in insects. Annu Rev Entomol 35:447–467. https://doi.org/10.1146/annurev.en.35.010190.002311

Bidaye SS, Machacek C, Wu Y, Dickson BJ (2014) Neuronal control of Drosophila walking direction. Science 344:97–101. https://doi.org/10.1126/science.1249964

Borst A (2014) Fly visual course control: behaviour, algorithms and circuits. Nat Rev Neurosci 15:590–599. https://doi.org/10.1038/nrn3799

Breidbach O (1990) Serotonin-immunoreactive brain interneurons persist during metamorphosis of an insect: a developmental study of the brain of *Tenebrio molitor* L. (Coleoptera). Cell Tissue Res 259:345–360. https://doi.org/10.1007/BF00318458

Cardé RT, Willis MA (2008) Navigational strategies used by insects to find distant, wind-borne sources of odor. J Chem Ecol 34:854–866. https://doi.org/10.1007/s10886-008-9484-5

Cardona A, Larsen C, Hartenstein V (2009) Neuronal fiber tracts connecting the brain and ventral nerve cord of the early *Drosophila* larva. J Comp Neurol 515:427–440. https://doi.org/10.1002/cne.22086

Cardona A, Saalfeld S, Preibisch S et al (2010) An integrated micro- and macroarchitectural analysis of the *Drosophila* brain by computer-assisted serial section electron microscopy. PLoS Biol. https://doi.org/10.1371/journal.pbio.1000502

Carreira-Rosario A, Zarin AA, Clark MQ et al (2018) MDN brain descending neurons coordinately activate backward and inhibit forward locomotion. elife 7:e38554. https://doi.org/10.7554/eLife.38554

Chiang AS, Lin CY, Chuang CC et al (2011) Three-dimensional reconstruction of brain-wide wiring networks in *Drosophila* at single-cell resolution. Curr Biol 21:1–11. https://doi.org/10.1016/j.cub.2010.11.056

Costa M, Manton JD, Ostrovsky AD et al (2016) NBLAST: rapid, sensitive comparison of neuronal structure and construction of neuron family databases. Neuron 91:293–311. https://doi.org/10.1016/j.neuron.2016.06.012

Dacks AM, Christensen TA, Hildebrand JG (2006) Phylogeny of a serotonin-immunoreactive neuron in the primary olfactory center of the insect brain. J Comp Neurol 498:727–746. https://doi.org/10.1002/cne.21076

Dickinson MH (2014) Death Valley, *Drosophila*, and the Devonian toolkit. Annu Rev Entomol 59:51–72. https://doi.org/10.1146/annurev-ento-011613-162041

Egorov AV, Hamam BN, Fransén E et al (2002) Graded persistent activity in entorhinal cortex neurons. Nature 420:173–178. https://doi.org/10.1038/nature01171

el Jundi B, Pfeiffer K, Heinze S, Homberg U (2014) Integration of polarization and chromatic cues in the insect sky compass. J Comp Physiol A Neuroethol Sens Neural Behav Physiol 200:575–589. https://doi.org/10.1007/s00359-014-0890-6

Green J, Adachi A, Shah KK et al (2017) A neural circuit architecture for angular integration in Drosophila. Nature 546:101–106. https://doi.org/10.1038/nature22343

Hanesch U, Fischbach KF, Heisenberg M (1989) Neuronal architecture of the central complex in Drosophila melanogaster. Cell Tissue Res 257:343–366. https://doi.org/10.1007/BF00261838

Harley CM, Ritzmann RE (2010) Electrolytic lesions within central complex neuropils of the cockroach brain affect negotiation of barriers. J Exp Biol 213:2851–2864. https://doi.org/10.1242/jeb.042499

Heinze S, Homberg U (2008) Neuroarchitecture of the central complex of the desert locust: intrinsic and columnar neurons. J Comp Neurol 511:454–478. https://doi.org/10.1002/cne.21842

Heinze S, Homberg U (2009) Linking the input to the output: new sets of neurons complement the polarization vision network in the locust central complex. J Neurosci 29:4911–4921. https://doi.org/10.1523/JNEUROSCI.0332-09.2009

Heinze S, Reppert SM (2011) Sun compass integration of skylight cues in migratory monarch butterflies. Neuron 69:345–358. https://doi.org/10.1016/j.neuron.2010.12.025

Heinze S, Reppert SM (2012) Anatomical basis of sun compass navigation I: the general layout of the monarch butterfly brain. J Comp Neurol 520:1599–1628. https://doi.org/10.1002/cne.23054

Heinze S, Gotthardt S, Homberg U (2009) Transformation of polarized light information in the central complex of the locust. J Neurosci 29:11783–11793. https://doi.org/10.1523/JNEUROSCI.1870-09.2009

Homberg U, Hildebrand JG (1989) Serotonin-immunoreactive neurons in the median protocerebrum and suboesophageal ganglion of the sphinx moth *Manduca sexta*. Cell Tissue Res 258:1–24. https://doi.org/10.1007/BF00223139

Homberg U, Kingan TG, Hildebrand JG (1987) Immunocytochemistry of GABA in the brain and suboesophageal ganglion of *Manduca sexta*. Cell Tissue Res 248:1–24. https://doi.org/10.1007/BF01239957

Homberg U, Montague et al (1988) Anatomy of antenno-cerebral pathways in the brain of the sphinx moth *Manduca sexta*. Cell Tissue Res 254:255–281. https://doi.org/10.1007/BF00225800

Idoux E, Eugène D, Chambaz A et al (2008) Control of neuronal persistent activity by voltage-dependent dendritic properties. J Neurophysiol 100:1278–1286. https://doi.org/10.1152/jn.90559.2008

Ito K, Shinomiya K, Ito M et al (2014) A systematic nomenclature for the insect brain. Neuron 81:755–765. https://doi.org/10.1016/j.neuron.2013.12.017

Iwano M, Hill ES, Mori A et al (2010) Neurons associated with the flip-flop activity in the lateral accessory lobe and ventral protocerebrum of the silkworm moth brain. J Comp Neurol 518:366–388. https://doi.org/10.1002/cne.22224

Kanzaki R (1997) Pheromone processing in the lateral accessory lobes of the moth brain: flip-flopping signals related to zigzagging upwind walking. In: Insect pheromone research. Springer, Boston, pp 291–303. https://doi.org/10.1007/978-1-4615-6371-6_26

Kanzaki R (1998) Coordination of wing motion and walking suggests common control of zigzag motor program in a male silkworm moth. J Comp Physiol A Sensory, Neural, Behav Physiol 182:267–276. https://doi.org/10.1007/s003590050177

Kanzaki R, Arbas EA, Hildebrand JG (1991a) Physiology and morphology of descending neurons in pheromone-processing olfactory pathways in the male moth *Manduca sexta*. J Comp Physiol A 169:1–14. https://doi.org/10.1007/BF00198168

Kanzaki R, Arbas EA, Hildebrand JG (1991b) Physiology and morphology of protocerebral olfactory neurons in the male moth *Manduca sexta*. J Comp Physiol A 168:281–298. https://doi.org/10.1007/BF00198348

Kanzaki R, Sugi N, Shibuya T (1992) Self-generated zigzag turning of *Bombyx mori* males during pheromone-mediated upwind walking. Zool Sci 9:515–527

Kanzaki R, Ikeda et al (1994) Morphological and physiological properties of pheromone-triggered flipflopping descending interneurons of the male silkworm moth, *Bombyx mori*. J Comp Physiol A Sensory, Neural, Behav Physiol 175:1–14. https://doi.org/10.1007/BF00217431

Kenning M, Harzsch S (2013) Brain anatomy of the marine isopod *Saduria entomon* Linnaeus, 1758 (Valvifera, Isopoda) with special emphasis on the olfactory pathway. Front Neuroanat 7:32. https://doi.org/10.3389/fnana.2013.00032

Kramer E (1975) Orientation of the male silkmoth to the sex attractant bombykol. In: Olfaction and Taste. Academic Press, New York, pp 329–335

Krieger J, Braun P, Rivera NT et al (2015) Comparative analyses of olfactory systems in terrestrial crabs (Brachyura): evidence for aerial olfaction? PeerJ 3:e1433. https://doi.org/10.7717/peerj.1433

Lei H, Anton S, Hansson BS (2001) Olfactory protocerebral pathways processing sex pheromone and plant odor information in the male moth *Agrotis segetum*. J Comp Neurol 432:356–370. https://doi.org/10.1002/cne.1108

Li W-C, Soffe SR, Wolf E, Roberts A (2006) Persistent responses to brief stimuli: feedback excitation among brainstem neurons. J Neurosci 26:4026–4035. https://doi.org/10.1523/JNEUROSCI.4727-05.2006

Li Z, Li Y, Yi Y et al (2012) Dissecting a central flip-flop circuit that integrates contradictory sensory cues in C. elegans feeding regulation. Nat Commun 3:776. https://doi.org/10.1038/ncomms1780

Lin CY, Chuang CC, Hua TE et al (2013) A comprehensive wiring diagram of the protocerebral bridge for visual information processing in the *Drosophila* brain. Cell Rep 3:1739–1753. https://doi.org/10.1016/j.celrep.2013.04.022

Loewenstein Y, Mahon S, Chadderton P et al (2005) Bistability of cerebellar Purkinje cells modulated by sensory stimulation. Nat Neurosci 8:202–211. https://doi.org/10.1038/nn1393

Lu J, Sherman D, Devor M, Saper CB (2006) A putative flip-flop switch for control of REM sleep. Nature 441:589–594. https://doi.org/10.1038/nature04767

Mafra-Neto A, Cardé RT (1995) Influence of plume structure and pheromone concentration on upwind flight of *Cadra cautella* males. Physiol Entomol 20:117–133. https://doi.org/10.1111/j.1365-3032.1995.tb00808.x

Major G, Tank D (2004) Persistent neural activity: prevalence and mechanisms. Curr Opin Neurobiol 14:675–684. https://doi.org/10.1016/j.conb.2004.10.017

Milo R, Shen-Orr S, Itzkovitz S et al (2002) Network motifs: simple building blocks of complex networks. Science 298:824–827. https://doi.org/10.1126/science.298.5594.824

Mishima T, Kanzaki R (1996) Pheromone-triggered 'flipflopping' neural signals correlate with activities of neck motor neurons of a male moth, *Bombyx mori*. Zool Sci 13:79–87

Mishima T, Kanzaki R (1998) Coordination of flipflopping neural signals and head turning during pheromone-mediated walking in a male silkworm moth *Bombyx mori*. J Comp Physiol A Sensory, Neural, Behav Physiol 183:273–282. https://doi.org/10.1007/s003590050255

Mishima T, Kanzaki R (1999) Physiological and morphological characterization of olfactory descending interneurons of the male silkworm moth, *Bombyx mori*. J Comp Physiol A Sensory, Neural, Behav Physiol 184:143–160. https://doi.org/10.1007/s003590050314

Müller M, Homberg U, Kühn A (1997) Neuroarchitecture of the lower division of the central body in the brain of the locust (*Schistocerca gregaria*). Cell Tissue Res 288:159–176. https://doi.org/10.1007/s0044100508

Namiki S, Kanzaki R (2016a) The neurobiological basis of orientation in insects: insights from the silkmoth mating dance. Curr Opin Insect Sci 15:16–26. https://doi.org/10.1016/j.cois.2016.02.009

Namiki S, Kanzaki R (2016b) Comparative neuroanatomy of the lateral accessory lobe in the insect brain. Front Physiol 7:244. https://doi.org/10.3389/fphys.2016.00244

Namiki S, Kanzaki R (2018) Morphology of visual projection neurons supplying premotor area in the brain of the silkmoth *Bombyx mori*. Cell Tissue Res 3:497–515. https://doi.org/10.1007/s00441-018-2892-0

Namiki S, Iwabuchi S, Pansopha Kono P, Kanzaki R (2014) Information flow through neural circuits for pheromone orientation. Nat Commun 5:5919. https://doi.org/10.1038/ncomms6919

Namiki S, Dickinson MH, Wong AM et al (2018a) The functional organization of descending sensory-motor pathways in *Drosophila*. elife 7:e34272. https://doi.org/10.7554/eLife.34272

Namiki S, Wada S, Kanzaki R (2018b) Descending neurons from the lateral accessory lobe and posterior slope in the brain of the silkmoth *Bombyx mori*. Sci Rep 8:9663. https://doi.org/10.1038/s41598-018-27954-5

Obara Y (1979) *Bombyx mori* mationg dance: an essential in locationg the female. Appl Entomol Zool 14:130–132

Okada R, Sakura M, Mizunami M (2003) Distribution of dendrites of descending neurons and its implications for the basic organization of the cockroach brain. J Comp Neurol 458:158–174. https://doi.org/10.1002/cne.10580

Olberg R (1983) Pheromone-triggered flip-flopping interneurons in the ventral nerve cord of the silkworm moth, *Bombyx mori*. J Comp Physiol A 152:297–307. https://doi.org/10.1007/BF00606236

Pfuhl G, Berg G (2007) Morphological and functional characterisation of the lateral accessory lobe in moth. In: The 8th congress of the international society for neuroethology, Vancouver

Rutishauser U, Douglas RJ, Slotine J-J (2011) Collective stability of networks of winner-take-all circuits. Neural Comput 23:735–773. https://doi.org/10.1162/NECO_a_00091

Saper CB, Fuller PM, Pedersen NP et al (2010) Sleep state switching. Neuron 68:1023–1042. https://doi.org/10.1016/j.neuron.2010.11.032

Schmitt F, Stieb SM, Wehner R, Rössler W (2015) Experience-related reorganization of giant synapses in the lateral complex: potential role in plasticity of the sky-compass pathway in the desert ant *Cataglyphis fortis*. Dev Neurobiol. https://doi.org/10.1002/dneu.22322

Schwindt PC, Spain WJ, Foehring RC et al (1988) Slow conductances in neurons from cat sensorimotor cortex in vitro and their role in slow excitability changes. J Neurophysiol 59:450–467. https://doi.org/10.1152/jn.1988.59.2.450

Seely J, Chow CC (2011) Role of mutual inhibition in binocular rivalry. J Neurophysiol 106:2136–2150. https://doi.org/10.1152/jn.00228.2011

Seki Y, Aonuma H, Kanzaki R (2005) Pheromone processing center in the protocerebrum of *Bombyx mori* revealed by nitric oxide-induced anti-cGMP immunocytochemistry. J Comp Neurol 481:340–351. https://doi.org/10.1002/cne.20392

Shih CT, Sporns O, Yuan SL et al (2015) Connectomics-based analysis of information flow in the *Drosophila* brain. Curr Biol 25:1249–1258. https://doi.org/10.1016/j.cub.2015.03.021

Shu Y, Hasenstaub A, McCormick DA (2003) Turning on and off recurrent balanced cortical activity. Nature 423:288–293. https://doi.org/10.1038/nature01616

Staudacher E (1998) Distribution and morphology of descending brain neurons in the cricket *Gryllus bimaculatus*. Cell Tissue Res 294:187–202. https://doi.org/10.1007/s004410051169

Stegner MEJ, Richter S (2011) Morphology of the brain in *Hutchinsoniella macracantha* (Cephalocarida, Crustacea). Arthropod Struct Dev 40:221–243. https://doi.org/10.1016/j.asd.2011.04.001
Stegner ME, Stemme T, Iliffe TM et al (2015) The brain in three crustaceans from cavernous darkness. BMC Neurosci 16:19. https://doi.org/10.1186/s12868-015-0138-6
Strausfeld NJ (1976) Atlas of an insect brain. Springer, Berlin
Strausfeld NJ (2012) Arthropod brains: evolution, functional elegance, and historical significance. Harvard University Press, Cambridge, MA
Strausfeld NJ, Bassemir UK (1985) Lobula plate and ocellar interneurons converge onto a cluster of descending neurons leading to neck and leg motor neuropil in *Calliphora erythrocephala*. Cell Tissue Res 240:617–640. https://doi.org/10.1007/BF00216351
Tahvildari B, Fransén E, Alonso AA, Hasselmo ME (2007) Switching between "on" and "off" states of persistent activity in lateral entorhinal layer III neurons. Hippocampus 17:257–263. https://doi.org/10.1002/hipo.20270
Tahvildari B, Alonso AA, Bourque CW (2008) Ionic basis of ON and OFF persistent activity in layer III lateral entorhinal cortical principal neurons. J Neurophysiol 99:2006–2011. https://doi.org/10.1152/jn.00911.2007
Träger U, Wagner R, Bausenwein B, Homberg U (2008) A novel type of microglomerular synaptic complex in the polarization vision pathway of the locust brain. J Comp Neurol 506:288–300. https://doi.org/10.1002/cne.21512
Turner-Evans D, Wegener S, Rouault H et al (2017) Angular velocity integration in a fly heading circuit. elife 6:e23496. https://doi.org/10.7554/eLife.23496
Vickers NJ, Baker TC (1994) Reiterative responses to single strands of odor promote sustained upwind flight and odor source location by moths. Proc Natl Acad Sci U S A 91:5756–5760. https://doi.org/10.1073/pnas.91.13.5756
Vogels TP, Abbott LF (2005) Signal propagation and logic gating in networks of integrate-and-fire neurons. J Neurosci 25:10786–10795. https://doi.org/10.1523/JNEUROSCI.3508-05.2005
Wada S, Kanzaki R (2005) Neural control mechanisms of the pheromone-triggered programmed behavior in male silkmoths revealed by double-labeling of descending interneurons and a motor neuron. J Comp Neurol 484:168–182. https://doi.org/10.1002/cne.20452
Wang XJ (2002) Probabilistic decision making by slow reverberation in cortical circuits. Neuron 36:955–968
Williams JLD (1975) Anatomical studies of the insect central nervous system: a ground-plan of the midbrain and an introduction to the central complex in the locust, *Schistocerca gregaria* (Orthoptera). J Zool 176:67–86. https://doi.org/10.1111/j.1469-7998.1975.tb03188.x
Wolff G, Strausfeld NJ (2015) The insect brain: a commentated primer. In: Structure and evolution of invertebrate nervous systems. Oxford University Press. https://doi.org/10.1093/acprof:oso/9780199682201.003.0047
Zagha E, Ge X, McCormick DA (2015) Competing neural ensembles in motor cortex gate goal-directed motor output. Neuron 88:565–577. https://doi.org/10.1016/j.neuron.2015.09.044
Zorović M, Hedwig B (2011) Processing of species-specific auditory patterns in the cricket brain by ascending, local, and descending neurons during standing and walking. J Neurophysiol 105:2181–2194. https://doi.org/10.1152/jn.00416.2010
Zorović M, Hedwig B (2013) Descending brain neurons in the cricket *Gryllus bimaculatus* (de Geer): auditory responses and impact on walking. J Comp Physiol A Neuroethol Sens Neural Behav Physiol 199:25–34. https://doi.org/10.1007/s00359-012-0765-7

Chapter 13
Coding and Evolution of Pheromone Preference in Moths

Teun Dekker and Zsolt Kárpáti

Abstract Olfaction is of overriding importance in defining the ecological niche of most insect species. Understanding the inner workings of olfaction can thus provide new ways for sustainable control of pests. However, getting at the neurogenetic blueprint of 'attraction' has been a slow process, owing to the immensely diverse odor environments and the *ditto* multidimensional complexity of the sense that 'navigates' in these. Pheromone preference coding in male moths offers several important advantages here: an often-binary signal, mirrored by a 'simple' peripheral detection system, a strongly enlarged brain center devoted to its processing, and fast and robust behaviors. Additionally, to avoid mating with heterospecific females, preference is often disjunct between closely related species, offering a perfect platform for comparative and evolutionary studies on preference codes. Here we mine through the moth pheromone research database to surface correlates of pheromone preference in male moths, from peripheral detection by olfactory receptors and sensory neurons, to their processing in the antennal lobes, and discuss the significance in the context of general odor preference, evolution, and application.

Keywords Odor coding · Pheromone preference · Evolution

13.1 Moth Pheromone Preference: A Primer for Understanding General Odor Coding

Insects rely heavily on innate odor preferences to effectively navigate to resources during their short lifespan. These odors have a great potential for use in novel and sustainable plant protection technologies and can, with minimal or no environmental impact, reduce the impact of pest insects by confusing their orientation or luring

T. Dekker
Swedish University of Agricultural Sciences, Uppsala, Sweden

Z. Kárpáti (✉)
CAR HAS Plant Protection Institute, Budapest, Hungary
e-mail: karpati.zsolt@agrar.mta.hu

Y. Ishikawa (ed.), *Insect Sex Pheromone Research and Beyond*, Entomology Monographs, https://doi.org/10.1007/978-981-15-3082-1_13

them out using traps or attract-and-kill (Cardé and Minks 1995; Cork 2016; Evenden 2016; Gregg et al. 2018). In the light of the global commitment to the Agenda 2030 for sustainable development, such innovations, backed by regulatory incentives, are sorely needed to replace the traditional environmentally disruptive techniques and contribute to securing food for a growing world population (www.un.org/sustainabledevelopment/sustainable-development-goals).

To facilitate odor-based innovations for sustainable pest control, a good understanding of the 'logic' of insect olfaction would be extremely helpful. However, currently the impact of basic olfactory research findings on application is negligible, largely because the 'blueprint' for attraction is still elusive. Coding rules that determine a species' olfactory affinity from genes to behavior are not clear yet. Even for the olfactory model *par excellence*, *Drosophila melanogaster*, it is largely unclear what constitutes 'banana' for its olfactory circuit. This is largely due to the enormously diverse odorous compounds in nature, the multidimensionality of odor circuits that detect these, and the ambiguous behavioral correlates that display an apparent built-in slack that can cope with signal variability and physiological state. Indeed, most insect olfactory circuits are highly multidimensional with many tens of differently tuned sensory unit types providing an olfactory interface to the external odor world (Su et al. 2009), making it a formidable task to extract elemental aspects of this code.

Thus, to get at the olfactory coding principles, there is a need for a simpler system, less multidimensional, less variable, and consisting of robust signal–response relationships. It is here that the *Lepidoptera* pheromone system offers unrivalled advantages. Species within this highly speciose order (over 160,000 species) typically rely on pheromones in mate finding. Pheromones from over 1000 species have been identified (www.pherobase.com) and show a narrow bandwidth in production, as well as in response, although the latter with possibly some genetic and environmental variability or bet-hedging (see Allison and Cardé 2016a, b, c for a review). Whether the narrow bandwidth is due to mutual stabilizing selection that frequently characterizes mate recognition, strong repercussions of heterospecific matings, the result of inherent restrictions within the underlying circuitry tuned to long-chain pheromone compounds, or a combination of factors is a matter of conjecture. How the production as well as the detection of pheromones can 'escape' from stabilizing selection and establish new pheromone blend optima is equally much debated.

Typically, females release blends consisting of simple binary or ternary signals of 10–18 aliphatic carbon esters, aldehydes, and alcohols, in ratios that are characteristic for a species. These disperse in wind to form plumes in which males orient upwind. To avoid heterospecific mating, both the blend composition and ratio, as well as the timing of release, are under tight control. The specificity of the blend may further be enhanced by the release of compounds that antagonize orientation of heterospecific males, sometimes without playing any apparent role in orientation of conspecific males themselves (e.g., Hillier and Baker 2016). In fact, pheromone glands may produce and selectively release many more compounds that are used in conspecific and heterospecific communication or do not serve an apparent function at all (www.pherobase.org). The genetic, molecular, and biochemical machinery that underlies pheromone production as well as the evolutionary variation and dynamics (Löfstedt 1993; Löfstedt et al. 2016; Groot et al. 2008; Hillier and Baker 2016) thereof is increasingly understood.

In contrast, the pheromone detection and response and its evolutionary genetic core is not well understood. Although the basic layout of the circuitry is well described in many species, the factors that determine preference are largely hypothetical. The responding sex, the male, has a set of sensory neurons that are tuned to the female-produced pheromone that matches the pheromone signal, and, in many cases, neurons that are dedicated to pheromone antagonism. Although the female may release several more compounds, males respond generally to only two to three compounds within these blends in a particular ratio. Of interest is further that where females release pheromones in a very narrow bandwidth, the male response system permits considerably more 'slack.' It is not clear if the imprecision in the response is the result of neural limitations to continually resolve blend quality during spatially and temporally highly dynamic sensory input or perhaps even an evolutionary strategy aimed at maximizing mating success and include occasional outliers, or both. The first idea has recently received strong support from research on *O. nubilalis* (Kárpáti et al. 2013). This study demonstrated that male moths are much more selective in initial stages of the orientation process (activation and wing fanning), while after taking flight males accepted a much broader range of blend ratios. This correlated with the peripheral circuitry not faithfully resolving blend ratios when continually challenged with puffs of pheromone blends. Regardless of the variation in response to a pheromone that is observed, it is very little compared to general odor circuits, which generally accepts large qualitative and quantitative variations in blends (Su et al. 2009). While already worth studying in their own right, studying moth pheromone coding may thus function as a 'rosetta stone' for deciphering olfactory coding in much more multidimensional and complex general odor circuits.

In the following sections we will describe the production of pheromones and their evolutionary and biochemical differences and outline the detection circuitry (Sects. 13.2.1 and 13.2.2). This is followed by a section describing how variation in production and preference may be linked to shifts in the layout in the pheromone detection and processing circuitry, and illustrate these with examples where available. Finally, the evolution of pheromone preference is also reconsidered in the context of other environmental cues which have shaped the evolutionary forces acting on the circuitry.

13.2 Moth Pheromone Dialects and Their Evolution

13.2.1 Pheromone Production and Types: Opportunities and Constraints

Sex pheromones from female moths are the most discovered pheromones in insects. The drive to identify moth pheromones is strongly incentivized by the fact that many Lepidoptera species are agricultural pests of economic importance. More than 1600 moth pheromones have been identified (www.pherobase.com). Most species synthesize their pheromone de novo in a specialized gland through enzymatic transformation of fatty acids, and generally consist of a relatively simple saturated or unsaturated hydrocarbon chain with an oxygenated functional group. Besides different chemical compositions, the precise ratio of the components is essential in reproductive isolation between closely related species. In ancestral Lepidoptera groups and in Trichoptera, a sister group of Lepidoptera, the pheromone production is located in the sternum V gland. In higher Lepidoptera species, whose pheromones are characterized by longer carbon chains, the site of production has shifted to the abdominal tip. How the shift in pheromone production site has come about, and the evolutionary forces that led to this, is unclear. In the biosynthetic pathways of pheromones in higher Lepidoptera, fatty acid enzymes play an important role to form the structure of the pheromone molecule. These enzymes have been a hotbed in the evolution of pheromone blends: duplications, AA substitutions, and differential expression, among others, have caused shifts in substrate affinity and led to the production of pheromones that are species-specific either in ratio and/or in composition. A limited number of enzymes are needed to change the chemical structure of the pheromone, of which desaturase genes have been found particularly important in the evolution of pheromone blends in numerous moth species. Minor changes in the genetic code of these enzymes cause robust changes in pheromone structure (Löfstedt 1993; Löfstedt et al. 2016).

At least four different types of moth pheromone production exist, which can be separated on the basis of distinct biochemical pathways, and are often linked to particular clades of Lepidoptera (Löfstedt et al. 2016). Type I pheromones are long-chain hydrocarbons (C10–18), having one or two double bonds in different positions and different functional groups such as alcohol, aldehyde, or acetate ester. They are by far the most common and present in all families with the exception of basal clades and butterflies. They are produced de novo in the pheromone gland (Jurenka 2004; Rafaeli and Jurenka 2003; Tillman et al. 1999). Type II pheromones consist of longer polyunsaturated hydrocarbons (C17–23) and their epoxide derivatives and they do not have functional groups at the terminal position. This type has been identified from highly evolved moth groups such as Geometroidea and Noctuoidea, which do not have the Type I pheromones (Lymantriidae, Arctiidae, Quadrifinae) (Ando et al. 2004). Type II pheromones are biosynthesized in the oenocytes and then transferred to the pheromone gland. Type III pheromones are saturated or unsaturated hydrocarbons with or without functional groups and contain one or more methyl branches. They are synthesized de novo in body parts other than the pheromone gland (Löfstedt et al. 2016). Type 0 pheromones are the aforementioned pheromones used by ancestral Eriocraniidae moths and caddisflies (Trichoptera) and consist of short-chain secondary alcohols and ketones, synthesized in the sternum V gland (Löfstedt et al. 2016).

Type I sex pheromone production is controlled by the pheromone biosynthesis activating neuropeptide (PBAN), which is produced by the subesophageal ganglion and released into the hemolymph (Jurenka 2004; Rafaeli and Jurenka 2003; Tillman et al. 1999). In the pheromone gland, PBAN activates a G protein-coupled receptor and the receptor–ligand complex opens the calcium channel and the Ca^{2+} influx initiates pheromone production via second messengers (Rafaeli et al. 2007).

In most cases, pheromones are produced and released by females. However, males are capable of producing similar compounds in analogue structures and in a few cases male-produced pheromones have been demonstrated to significantly contribute to mating success (e.g., male tiger moths, *Utetheisa ornatrix*, produce a courtship pheromone, hydroxydanaidal, and male sugarcane borers, *Eldana saccharina*, produce eldanolide, which can attract conspecific females, Zagatti 1981), although rarely involved in long-range attraction of females (Landolt and Heath 1989; Zagatti et al. 1987; Kunesch et al. 1981)

13.2.2 The Pheromone Detection Circuitry, Elements

Contrary to production, elements needed to shift pheromone preference are much less clear. Yet such factors are likely not only controlling the preference of pheromones, but are also critical in coding and evolution of attraction and repulsion in general odor circuits, and possibly even more widely of importance in neuronal coding in general. A handle on the molecular neurogenetic factors that determine pheromone preference will thus be of importance in informed, rational approaches to using odors in the control of insects. What are the basic elements of the olfactory circuitry on which evolution acts?

The pheromone detection and response circuitry can roughly be divided into three neuronal levels: the peripheral circuitry, the primary olfactory center (the antennal lobes, ALs), and the higher-order brain centers (mushroom body and lateral protocerebrum). Hydrophobic pheromone molecules interact with the 3D surface of the sensillum, in which olfactory sensory neurons (OSNs) are located. In moths, pheromone OSNs are almost exclusively localized in sensilla trichodea, which is covered by a lipid layer that adsorbs and passes pheromone molecules toward the pores into the sensillum lumen. The lipid coating of trichoid sensilla differentially adsorbs pheromones compared to other sensillum types (Kanaujia and Kaissling 1985; Maitani et al. 2010). In the sensillum lumen, which is filled with sensillar lymph, odorant-binding proteins (OBPs) bind the pheromone molecules and transport them to the dendrites of the OSNs. OBPs solubilize and transport the molecules to the pheromone receptor (PR), which sits in the membrane of the OSN dendrite. Pheromone-binding proteins (PBPs) are a subfamily of OBPs, which are to a more or lesser extent tuned to pheromone molecules and can enhance the specificity and sensitivity of the pheromone signal (Grosse-Wilde et al. 2006; Chang et al. 2015). In addition, to evoke a sensitive pheromone signal and rapid activation, the sensory neuron membrane protein 1 (SNMP1) appears essential in vivo but not in vitro (Benton 2007; Li et al. 2014; Pregitzer et al. 2014). This protein is located nearby the PR in the membrane of the OSN and is thought to bind and forward the pheromone molecules to the PR and involved in both activation and termination of binding, although the precise mechanisms and interactions are not known yet. Similarly, details of the interaction between PBP and PR are unclear (Gomez-Diaz et al. 2013). The PR genes are an atypical clade within the family of G-protein-coupled receptors (GPCRs), but are unusual in topology (inverted, with the N terminal inside the cell), and their requirement for heterodimerization with a coreceptor, Orco. Orco helps to localize the PRs in the dendritic membrane. Most PR genes have a male-biased expression, which is mirrored by a large portion of OSNs in males tuned to pheromone. Pheromone OSNs are often narrowly tuned and selectively bind based on carbon chain length, position of double bonds, and functional groups. This fits with the need for species-specific signaling, and its resulting stabilizing, purifying selection. However, more broad tuning has also been observed and associated with, for example, the large number of the PRs being co-expressed (Koutroumpa et al. 2014), although occasionally PRs themselves can also be more broadly tuned (van de Pers and Den Otter 1978). The PR being expressed is determined by binding motifs located up- or downstream from the receptors. These motifs bind transcription factors (TFs) (regulatory genes) that regulate (promote or suppress) the expression of a specific PR. In *D. melanogaster*, Jafari et al. (2012) identified seven TFs which are in different combinations responsible for the expression of a set of over 30 ORs in *Drosophila*. Evolutionary changes in the DNA binding motif of TFs can cause a new pattern of OR expression across the different sensory neuron types, which causes a shift in the perception of odors or pheromones (Jafari et al. 2012).

The OSNs send the odor information via the OSN axons to the moth brain's first olfactory center, the antennal lobe (AL). The AL contains glomeruli, which are spherical structures that receive axonal input from a single OSN type. Within the glomerulus, synaptic contacts are established between OSNs, projection neurons (PNs), and local interneurons (LNs). In the male moth AL, some enlarged glomeruli form a macroglomerular complex (MGC) at the entrance of the antennal nerve. The MGC receives information about the presence of female-produced sex pheromones (Bretschneider 1924; Koontz and Schneider 1987). The number of behaviorally relevant pheromone components correlates with the number of glomeruli in the MGC. Axons from the same pheromone component specific ORN arborize in one MGC glomerulus (Hansson et al. 1992; Ochieng et al. 1995; Berg et al. 1998). Different combinations of heterophilic cell adhesion proteins are important in correctly guiding OSNs to the right glomerulus. For instance, the defective proboscis response (Dpr) family proteins and their heterophilic binding partners – Dpr-Interacting Proteins (DIPs) – expressed in OSNs guide the axons of OSNs to specific areas in the developing antennal lobe (Barisch et al. 2018). Within glomeruli, OSNs synapse with projection neurons (PNs) and local interneurons (LNs). Local (and, e.g., centrifugal) interneurons are involved in temporal and spatial signal enhancement (Christensen et al. 1993; Sun et al. 1993) and modulate signal sensitivity using neuromodulators (such as neuroamines, and neuropeptides, e.g., Olsen and Wilson 2008; Ignell et al. 2009; and others), depending on the internal state of the insect (Saveer et al. 2012). Instead, PNs send the information to the higher brain center, the mushroom bodies, and the lateral horn of the protocerebrum from where axon collaterals are sent to the motor control center. Numerous reviews have provided schematic overviews of the insect olfactory circuits, to which we refer here for further details.

13.3 Determinants of Pheromone Preference in Moths

The olfactory circuitry offers a wide variety of potential targets for evolutionary changes that impact preference, from peripheral elements to targets located deep inside the central nervous system. The likelihood of each of these elements to de facto contribute to evolutionary changes in preferences, however, depends on a variety of factors, including whether these changes require single or multiple mutations in single or multiple genes, which roles genes serve in other contexts, or whether there are pleiotropic effects (on other elements) in the olfactory circuitry or beyond. As some elements may cause saltational shifts in preference whereas others only cause minor shifts, their importance and likely involvement may further differ depending on the constraints posed by evolution. For instance, in situations where intermediate preferences suffer a high fitness cost, those evolutionary changes that dramatically change preference and bridge intermediate stages may be favored. Conversely, in scenarios where intermediate preferences support survival, changes in circuitry elements that only gradually shift preference may be favored. Under strong stabilizing selection regimes and narrow bandwidth preferences, shifts in male moth pheromone preference are perhaps expected to predominantly fall in the first category, as incremental changes are likely to be selected against. The number of targets that can cause preference change may thus in reality be more limited than one imagines at first glance, and may depend on the evolutionary forces that underlie change.

Most, if not all, reported correlates of change in pheromone preference in male moths involve peripheral factors that then lead to perceptual change in the CNS of a moth. While this ‘peripheral bias’ may be due to the accessibility and extensive knowledge of the function of the peripheral circuitry compared to more central processes, it may also be the result of the periphery being relatively unconstrained and accepting changes with little spillover effects on other circuitry elements. In the section below, we provide an overview of the factors that could be involved in a change in preference, evaluate their potential in modulating change along the above listed constraints and opportunities (multiple functions, pleiotropic effects,

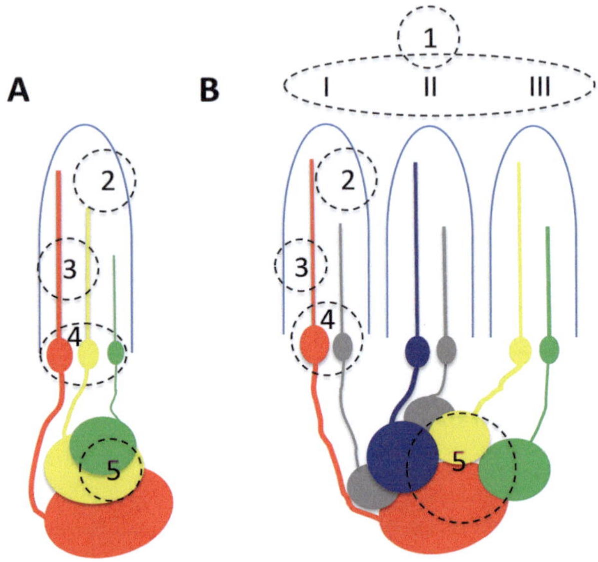

Fig. 13.1 Schematic of the elements in the pheromone circuitry that may be involved in the evolution of preference. (**a**) a layout resembling that of *Ostrinia nubilalis*, with only one sensillum type detecting all pheromone signals. (**b**) a layout resembling that of heliothine moths, with up to three different sensillum types underlying the detection of pheromone and antagonist signals. Correlates of pheromone evolution include (*1*) the ratio of sensillum types, typically most abundant for the one carrying the main pheromone receptor neuron (not in *Ostrinia* spp. which has only a single sensillum type); (*2*) perireceptor events such as sensillar structure and biochemistry, OBPs, ODEs, and CSPs; (*3*) pheromone receptors and their affinity for pheromone components; (*4*) OR expression patterns, axonal diameter, sensillar abundance, and preference; and finally (*5*) wiring in the central nervous system and the volume of glomeruli. Red neurons and glomeruli signify those sensitive to the main pheromone component, yellow those that respond to pheromone antagonists, and gray neurons and glomeruli are 'silent neurons', that is, neurons that do not appear to respond to any pheromone or odor (Berg et al. 2014)

constrainedness, their potential saltational effects), and provide, where available, examples from the moth pheromone literature to illustrate their involvement in preference changes (see for an overview Fig. 13.1).

13.3.1 *Peri-receptor Events: Sensillum Ultrastructure, OBPs, ODEs, and CSP*

Olfactory sensory neurons are housed in diverse cuticular protrusions, called sensilla. Those that detect pheromones in moths appear, with few exceptions (Ansebo et al. 2005), to be restricted to trichoid sensilla. This may reflect an evolutionary cul-de-sac, that is, an evolutionary trajectory, once established, continues, not necessarily because of canonical restrictions, but rather because of a post hoc locking of genes and function to the environment they have evolved in, or, alternatively, because it requires the least rearrangements in an already functional system. Alternatively (or in addition) it has a functional significance. Support for the latter comes from the fact that receptors for the *Drosophila* pheromone Z11-18:OAc do not function well in basiconic sensilla, but do in their cognate trichoid sensilla (Syed et al. 2010; Jin et al. 2008). The same holds true for *Bombyx mori* bomkykol receptors heterologously expressed in basiconic versus trichoid sensilla of *D. melanogaster* (Syed et al. 2010). Several factors may underlie this phenomenon, including the cuticular hydrocarbon coating of sensilla and how these may gate or block pheromones (see also Böröczky et al. 2013), the mosaic of odor-binding proteins, degrading enzymes, and chemosensory proteins expressed through support cells, as well as membrane-bound chemosensory proteins (sensory neuron membrane protein, Benton 2007; Li et al. 2014; Pregitzer et al. 2014), which appear to play a role in proper trichoid functioning.

Sensillum Ultrastructure Using atomic force microscopy and chemical force microscopy, Maitani et al. (2010) found that trichoid sensilla of *Helicoverpa zea*, the corn earworm, display ridges and interspaced regions of alternating relatively hydrophilic and hydrophobic hydrocarbons. Although functional evidence will have to be generated, these ridges may facilitate in diffusing aldehyde pheromones into the pores and extracellular lymph surrounding the dendrites of sensory neurons, thereby possibly decreasing the detection threshold and/or increase the temporal resolution. Of interest is that, in *Utetheisa ornatrix*, the ornate moth, trichoid sensilla, did not have such a heterogeneous trichoid surface. Instead the surface was more lipophilic, which may be better suited for the hydrocarbon (type II) pheromones this taxon uses. Apparently, the surface structures and composition may differ between trichoids of different species and possibly evolve to best fit the affinities of their pheromone types (see Sect. 13.2.1). Further studies should resolve whether the ultrastructures of trichoid sensilla diverge among subtypes or in comparison with other sensillum types, and verify whether that links to functional characteristics of the neurons housed in these sensilla. However, as sensilla generally house multiple sensory neurons, and cuticular changes would impact all neurons in that sensillum, it is difficult to envision scenarios where the sensillum ultrastructure has shaped the evolution of pheromone preferences and the speciation process in moths in more recent evolutionary history.

Sensillar Lymph Proteins: Odor-Binding and -Degrading Enzymes and Chemosensory Proteins Besides differences in the sensillar ultrastructure, several proteins expressed within the sensillar lymph may contribute to the functional divergence of the olfactory system between species. These include odor-binding proteins and -degrading enzymes as well as chemosensory proteins. Odor-binding proteins were first described from the giant silkmoth, *Antheraea polyphemus* (Saturnidae). Olfactory receptors described for this species insufficiently explained pheromone specificity. However, a set of three different pheromone-binding proteins was found to be expressed by support cells and released in the extracellular lymph and were hypothesized to contribute to the specificity for pheromones observed in vivo (Maida et al. 2003; Mohl et al. 2002; Bette et al. 2002), one of which, ApolPBP2, was subsequently found to play a role in rendering specificity and sensitivity of pheromone receptors expressed in HEK293 cells (Forstner et al. 2009). Similarly, pheromone-binding protein of *B. mori*, BmorPBP, is selectively involved in activation pheromone receptor BmOR1 upon stimulation with bombykol, but not of bombykal (Grosse Wilde et al. 2006). Today, OBPs appear in countless transcriptome studies, sometimes featuring functional characterization. For instance, *Agrotis ipsilon* expressed three PBPs (AipsPBP1-3) in the sensillum lymph of the male antennae, with each of them displaying selective sensitivity to either the two main pheromone components (AipsPBP1) or its synergists (AipsPBP2-3, Gu et al. 2013). In *Grapholita molesta*, four abundantly expressed OBPs showed partially overlapping divergent affinities with two of them binding mostly to pheromone-type compounds (Chen et al. 2018a, b), whereas a mosaic of four *Chilo suppressalis* OBPs enhanced sensitivity and specificity of pheromone receptors in a *Xenopus* oocyte expression system (Chang et al. 2015). The examples in literature illustrate that pheromone sensitivity and selectivity of neurons can be supported by OBPs. However, as odor-binding proteins are, as a rule, not very ligand-selective compared to olfactory receptors (Pelosi et al. 2014), are shared between neurons in the same sensillum, frequently expressed in multiple sensillum types (Sun et al. 2018), are sometimes abundantly expressed outside the olfactory system (Chen et al. 2018a), and can be evolutionarily constrained due to multiple roles they fulfill even outside chemoreception (e.g., scavenging proteins, immune response, general solubilizing agent: Pelosi et al. 2017), a generalized role in mediating specificity of sensory neurons seems unlikely. In the evolution of pheromone preference, OBPs may thus have played a role in selected cases only.

Less is known about other proteins that are present in the sensillar lymph, including odor-degrading enzymes (ODEs) and chemosensory proteins (CSPs). ODEs are thought to be important in termination of the sensory response by depleting the sensillar lymph from pheromone. In *Antheraea polyphemus*, the half-life of pheromones due to antenna-specific esterase ApolSE was an estimated 15 ms only (Vogt et al. 1985), although estimates in other studies and other moth species are less favorable (Ishida and Leal 2005; Merlin et al. 2007; He et al. 2014). Cytochrome P450s are also abundantly expressed in the sensillar lymph of certain species, and in *Mamestra brassica* appears to be tuned to pheromones (Maïbèche-Coisne et al. 2002). These and other ODE-candidate genes belong to gene families that serve other functions, such as scavenging, detoxification, immune responses, and xenobiotic degradation, and appear to have secondarily acquired a role in olfaction. While such molecules likely assure that the pheromone signal's extrinsic dynamics (odor plume) is conserved in the moth CNS (Vickers et al. 2001), their role in pheromone specificity of sensory neurons and thus in the evolution of pheromone preference is likely minor. On a similar note,

CSPs, which is a family of small soluble polypeptides, appear to have multiple functions judging from their expression pattern (Dani et al. 2011, e.g., Zhu et al. 2016) and broad binding affinity (Jacquin-Joly et al. 2001; Younas et al. 2018), and therefore are not a primary candidate in the evolution of pheromone preference.

13.3.2 Olfactory Receptors Detecting Pheromones (Pheromone Receptors, PRs)

Of all potential candidates that may be involved in the evolution of pheromone preference, olfactory receptors (ORs) are undoubtedly the most studied. Since the discovery of olfactory receptors at the turn of the century, much of our research focus has shifted to the identification of receptors across species. This has demonstrated that ORs evolve rapidly, with numerous taxa-specific clades and with relatively little conservation even in orthologue receptors. It is generally assumed that olfactory preference is encoded by the repertoire of ORs and the odors they encode. Since in male moth, as a rule, each pheromone component is detected by a single OSN type, and since each OSN expresses generally one OR, it would appear that the highly selective blend preference in male moth should be dictated by the ORs and their affinities.

Yet, how can male moths evolve new PR functions for genes that are under stabilizing selection? Zhang and Löfstedt (2013) looked at selection regimes of *Agrotis segetum* PR paralogues. *A. segetum* PRs from recent duplication events were under positive or relaxed selection and functionally diverged, whereas orthologues with low sequence similarity but high purifying selection tended to be functionally more conserved. Under such relaxed scenarios, PR paralogues that are not or lowly expressed may be a breeding ground for acquisition of new receptor–ligand affinities, which may, when shifts in pheromone blends arise, subsequently be put under strong positive selection. In *Ostrinia* species, relaxed purifying or neutral selection of one of seven receptors predated the evolution of a new pheromone optimum and possibly a new species, *O. furnacalis* (Leary et al. 2012). Unusually, the receptor has evolved from a specifically tuned receptor, only responding to E11-14:OAc to one that includes both components of *O. furnacalis*, and may have provided a preference bridge for *O. nubilalis* males. Indeed, cross-attraction between the species, although rare, can be found (Linn et al. 2003, 2007), males of which tend to display slightly altered OSN affinities (Domingues et al. 2007a, b)

Thus, as with ordinary olfactory receptors, pheromone receptors are highly dynamically evolving both in sequence and function, be it under a more directional and restrictive sexual selection regime (see, e.g., Leary et al. 2012; Zhang and Löfstedt 2015). Indeed, within clades of orthologous receptors shifts in affinities in position of the double bond (Leary et al. 2012; Albre et al. 2012), carbon chain length (Zhang and Löfstedt 2015), and functional group (Zhang and Löfstedt

2015) can be found, even with closely related receptors. This can readily be deduced by plotting PR affinities reported in literature onto trees of orthologous PRs (see, e.g., Zhang and Löfstedt 2013; Bastin-Heline et al. 2019).

13.3.3 OR Expression Patterns, Axonal Diameter, and Sensillar Abundance

In spite of PR evolution, commonly related moth species produce and detect the same pheromone-type components, and instead change the preference for ratios, or alternatively have altered the valence from pheromone to antagonist or vice versa. In such preference shifts, the PR repertoire is likely functionally identical, and PRs do not underlie the shifts preference. So, what does?

In the crambid moth *Ostrinia nubilalis*, two strains produce and prefer diametrically opposite ratios of a binary pheromone blend of Z11–14:OAc and E11–14:OAc (Roelofs et al. 1987; Hansson et al. 1987). This was paralleled by the relative volumes of the two glomeruli tuned to each pheromone component, with the largest glomerulus always tuned to the main pheromone component (Kárpáti et al. 2008). In hybrids of the two strains, which produce and prefer intermediate ratios of the two pheromone components, the relative volume of the two MGC glomeruli was also intermediate (Kárpáti et al. 2010). PR expression studies, electrophysiology, and anterograde backfills to the antennal lobes demonstrated that the two glomeruli are targeted by two antennal neurons cohoused in a single sensillum, one small and one large (Kárpáti et al. 2008, 2010; Koutroumpa et al. 2014). These two neurons are each tuned to one of the two pheromone components, but their affinity is reversed between the two strains, with the large neuron always responding to the major pheromone component and projecting to the large MGC glomerulus. Apparently the (relative) size of PR-expressing OSNs matters, as this translates into the (relative) volume of the corresponding glomerulus, which in turn regulates preference. The relationship between glomerular volume and preference seems not only to hold for pheromone preference, but also for general odor preference (Dekker et al. 2006, 2015; Ibba et al. 2010; Linz et al. 2013).

The link between preferred ratio of pheromone components and size is corroborated by the fact that in all moth species the large OSN in pheromone-responding sensilla is tuned to the major pheromone component (Hansson et al. 1987) as is the largest MGC glomerulus (Baker et al. 2004; Lee et al. 2006a, b) neurons arborizing in a glomerulus. Commonly, moth species express more than one subtype of trichoid that responds to pheromone components, and invariably, the trichoid subtype containing the OSN responding to the major pheromone component is most abundant in the antenna. For instance, in *Helicoverpa armigera*, 78% of the trichoid sensilla are tuned to the major pheromone component Z11–16:Ald, the remainder to the minor Z9–16:Ald, whereas the ratio between these trichoid subtypes is reversed in *H. assulta*, fitting with this species' opposite ratio of the two pheromone components (Cossé et al. 1998; Baker et al. 2004; Lee et al. 2006a, b; Wu et al. 2013; Berg et al. 2014). Similar patterns have been observed for Agrotis species and strains (Löfstedt et al. 1986; Wu et al. 1999).

Clearly, whereas the evolution of pheromone components is regulated by the PRs and their affinity, the evolution of ratio preference is regulated by the relative abundance and size of OSNs and its corresponding glomerular volume. Similarly, shifts from pheromone component to antagonist are regulated by factors other than PR affinity shifts, and typically involves rearranging affinities of OSN identity within and possibly between sensilla. Indeed, an extensive genetic mapping study in *O. nubilalis* demonstrated that the genetic factor responsible for the difference in pheromone preference between the two strains is located on the sex chromosome, and pointing toward a set of genes previously described to be involved in neurogenesis (Koutroumpa et al. 2016). Similarly, a *B. mori* strain, mutant for a sex-linked transcription factor, has apparently swapped the expression of the bombykol and bombykal PRs that are colocalized in the same sensillum, thereby changing its behavioral preference to bombykal instead of bombykol (Fuji et al. 2011). Apparently, the OSN identity can be swapped using neurogenesis genes or transcription factors, thereby changing the interpretation of the message, without changing the detection of each compound per se. In *D. melanogaster*, a combinatorial pattern of transcription factors (TFs) determines which OR is expressed in an OSN (Jafari et al. 2012). Changing TF binding motifs up- or downstream of an OR not only affect the expression of that OR, but also that of others, which may be expressed in its stead. One could envision that changing the binding affinity of the TF, through a mutation such as in *B. mori*, has similar effects. Also in *D. melanogaster* it was found that a series of TFs combinatorially determine sensillar (sub) type identity, with mutants overexpressing or suppressing certain subtypes (Li et al. 2013). Such a mechanism could underlie the aforementioned differences in sensillar subtype abundance between species (e.g., *H. assulta* and *H. armigera*). The aforementioned highlights the importance of other factors than PRs in the evolution of pheromone preference. These factors complement evolution of PR affinity in terms of evolution of signal valence and ratio specificity, and are only indirectly (expression patterns of PRs in different OSNs), or not at all (sensillar identity) linked to PRs. The relative importance for these different routes of evolution of moth pheromone preference depends on the selection pressure experienced on the pheromone blend and the pheromone receptors (purifying, relaxed, or positive) and things such as the number of genetic (amino acid) changes needed to create the saltational shift toward the putative new pheromone optimum.

13.3.4 Other Peripheral Mechanisms: Modulation of OSN Sensitivity

Besides changes in PR affinity, expression patterns of PRs in different OSN, and sensillum subtypes, and sensillar (sub)type abundance, preference shifts may also be mediated through shifts in relative sensitivity of sensory neurons. The mechanism, exemplified in particular by research on the cabbage looper, *Trichoplusia ni*, has not received much attention, but may be more common than perhaps assumed.

In the laboratory, a mutant cabbage looper moth, *Trichoplusia ni*, was found, which overproduced (Z)-9–14:OAc on the cost of (Z)-7–12:OAc compared to the wildtype blend. Only after more than 25 generations under strong positive selection did males respond to this new blend (Liu and Haynes 1994). Males of this selection line did however continue to respond to the old blend. Sensory recordings demonstrated that this was paralleled by a strongly reduced sensitivity of Z9–14:OAc responding neurons in the evolved strain, which likely allowed males to tolerate the excessive amount of Z9–14:OAc (Domingue et al. 2009). The mechanism behind the desensitization observed here is unknown, but could be linked to a change in PR affinity, in PR expression levels, or in the intracellular signaling cascade which contains many factors that can modulate sensitivity (Guo and Smith 2017; Wicher 2018). We are not aware of cases similar to *T. ni* nature, but this may well be due to the fact that such forms are likely to be transitional, evolving toward evolutionary stable optima.

13.3.5 Modulation in the Central Nervous System

Olfactory input is known to be modulated in the CNS, depending on other external stimuli as well as the internal state of the insect (Saveer et al. 2012; Ignell et al. 2009). This also holds for responses to pheromones, which appears to be modulated by experience, energy levels, and regulated through neuromodulators such as biogenic amines, neuropeptides, etc. (Ignell et al. 2009; Saveer et al. 2012). However, although pheromone blend *recognition* certainly involves some neuronal gauge that assesses if the quality of the incoming signal permits gating it onward, until now no other than peripheral factors appear to be involved in the *evolution* of pheromone preference. This also makes sense, as changing CNS neuronal connections that integrate external and internal signals would likely have strong side effects on the whole neural circuitry, and not only affect pheromone preference.

In contrast, the periphery offers more degrees of freedom for changes with little pleiotropic effects. Indeed, until now no CNS mechanism had to be invoked to explain patterns of evolution of pheromone preference. Although the periphery thus appears as the primary evolutionary playground for change in pheromone preference, it does not preclude central mechanisms to be directly involved in preference. In studies in which the imaginal antennal disks were transplanted between strains and species, in some cases the preference remained that of the moth receiving the heterospecific discs, indicating that CNS rather than antennal factors determine preference (Ochieng et al. 2003; Vickers et al. 2005; Lee et al. 2016; Linn et al. 1999). Koutroumpa et al.'s (2016) study pointed toward a factor that regulated neuronal guidance and growth, which again may indicate a potential involvement of the CNS in mediating shifts in preference. However, none of the above studies have conclusively determined the molecular mechanisms for preference shifts. Further research is needed to evaluate whether in these selected cases the CNS did indeed play a direct role in preference shifts, or whether the CNS 'read-out centers' simply follow the changes in the periphery.

13.3.6 Coevolution of Pheromone and Plant Odors in Specialist Moths

Traditionally, moth pheromone perception has been treated as a separate subcircuitry, devoted to the detection of and response to intra- and inter-specific olfactory messages. However, moth pheromones are rarely detected in isolation and in some moth species, environmental odors can play an important or even essential role in mate orientation (Barrozo and Dekker 2015). In at least one group of moths, Yponomeutidae, this has had strong consequences for the evolution of pheromone preference. Interestingly, pheromones alone provided insufficient separation between sympatric Yponomeuta spp. Yet, heterospecific attraction and mating was not observed. It was subsequently found that host plant odors synergized with pheromone: females perched on host plants increased calling rates and attracted more males (Hendrikse and Vos-Buennemeyer 1987). This fits the monophagous life style of these species well (Liénard and Löfstedt 2016), and may be a phenomenon that may be more common for specialist species (Den Otter and Thomas 1979). Rendezvous sites for males and females may have consequences for the selection pressure the pheromone circuitry is under. More specifically, one would expect relaxed purifying selection on and a much less narrow bandwidth tuning curve of OSNs. Indeed, different OSNs types tuned to pheromone components showed overlapping activation patterns (Van Der Pers and Den Otter 1978; Liénard and Löfstedt 2016).

Also, in other moth species plant odors have been observed to synergize attraction to pheromones. For instance, in the tortricid moth, *Cydia pomonella* (Torticidae), synergy was noted for a combination of host-emitted pear ester ((*E,Z*)-2,4-decadienoate) and the pheromone ((*E,E*)-8,10-dodecadienol), depending on the type of orchard (pear, walnut, or apple) in which the tests were performed (Kutinkova et al. 2005; Light and Knight 2005; Knight et al. 2010). These interactions were reflected in synergistic responses to combinations of pheromone and plant odor in the male MGC (Trona et al. 2013). Similarly, pheromone and grapevine volatiles act synergistically for the grapevine moth, *Lobesia botrana* (Tortricidae) (von Arx et al. 2012). It has not been investigated whether these synergistic interactions may have had an impact on the strength and direction of the selection regime for the pheromone circuitry of these less-specialized species.

13.4 Outlook

The summary given highlights some of the work and the current status of understanding of the evolution of pheromone preference in male moths. The discovery of olfactory receptors along with the recent expansion of the molecular genetic toolkit, and at increasingly reasonable prices, is dramatically accelerating our understanding of the innerworkings of pheromone preference. We are rapidly advancing in unraveling the peripheral circuitries in numerous species, and we will increasingly find correlates of the evolution of pheromone preference. This will be important in understanding how more multidimensional olfactory circuits tuned to general odors 'make sense' of complex olfactory input, such as host odors. A profound understanding of preference would provide tools for more rationally unraveling olfactory preference in pest insects, and designing species-selective blends based on that, which can be used in sustainable control.

References

Albre J, Liénard MA, Sirey TM, Schmidt S, Tooman LK et al (2012) Sex pheromone evolution is associated with differential regulation of the same desaturase gene in to genera of leafroller moths. PLoS Genet 8:e1002489. https://doi.org/10.1371/journal.pgen.1002489

Allison JD, Cardé RT (2016a) Pheromones: reproductive isolation and evolution in moths. In: Pheromone communication in moths: evolution, behavior, and application. University of California Press, Oakland, pp 11–23

Allison JD, Cardé RT (2016b) Pheromone communication in moths: evolution, behavior, and application. University of California Press, Oakland

Allison JD, Cardé RT (2016c) Variation in moth pheromones. In: Pheromone communication in moths: evolution, behavior, and application. University of California Press, Oakland

Ando T, Inomata SI, Yamamoto M (2004) Lepidopteran sex pheromones. In: Schulz S (ed) The chemistry of pheromones and other semiochemicals I. Springer, Berlin/Heidelberg, pp 51–96

Ansebo L, Ignell R, Löfqvist J, Hansson BS (2005) Responses to sex pheromone and plant odours by olfactory receptor neurons housed in sensilla auricillica of the codling moth, *Cydia pomonella* (Lepidoptera: Tortricidae). J Insect Physiol 51:1066–1074. https://doi.org/10.1016/j.jinsphys.2005.05.003

Baker TC, Ochieng SA, Cossé AA, Lee SG, Todd JL, Quero C, Vickers NJ (2004) A comparison of responses from olfactory receptor neurons of *Heliothis subflexa* and *Heliothis virescens* to components of their sex pheromone. J Comp Physiol A 190:155. https://doi.org/10.1007/s00359-003-0483-2

Barish S, Nuss S, Strunilin I, Bao S, Mukherjee S, Jones CD, Volkan PC (2018) Combinations of DIPs and Dprs control organization of olfactory receptor neuron terminals in *Drosophila*. PLoS Genet 14:e1007560. https://doi.org/10.1371/journal.pgen.1007560

Barrozo R, Dekker T (2015) Contextual modulation of moth pheromone perception by plant odors. In: Allison JD, Cardé RT (eds) Pheromone communication in moths. University of California Press, Berkely, pp 101–112

Bastin-Héline L, Fouchier A, Cao S, Koutroumpa F, Caballero-Vidal G, Robakiewicz S, Monsempes C, François M-C, Ribeyre T, Cian A, Walker WB, Wang G, Jacquin-Joly E, Montagné N (2019) A novel lineage of candidate pheromone receptors for sex communication in moths. eLife 8. https://doi.org/10.1101/707174

Benton R (2007) Sensitivity and specificity in *Drosophila* pheromone perception. Trends Neurosci 30:512–519. https://doi.org/10.1016/j.tins.2007.07.004

Berg BG, Almaas TJ, Bjaalie JG, Mustaparta H (1998) The macroglomerular complex of the antennal lobe in the tobacco budworm moth *Heliothis virescens*: specified subdivision in four compartments according to information about biologically significant compounds. J Comp Physiol A 183:669–682

Berg BG, Zhao X-C, Wang G (2014) Processing of pheromone information in related species of heliothine moth. Insects 5:742–761. https://doi.org/10.3390/insects5040742

Bette S, Breer H, Krieger J (2002) Probing a pheromone binding protein of the silkmoth *Antheraea polyphemus* by endogenous tryptophan fluorescence. Insect Biochem Mol Biol 32:241–246

Böröczky K, Wada-Katsumata A, Batchelor D, Zhukovskaya M, Schal C (2013) Insects groom their antennae to enhance olfactory acuity. Proc Natl Acad Sci U S A 110:3615–3620. https://doi.org/10.1073/pnas.1212466110

Bretschneider F (1924) Über die Gehirne des eichenspinners und des Seidenspinners (*Lasiocampa quercus* L. und *Bombyx mori* L.). Jena. Z Naturw 60:563–570

Carde RT, Minks AK (1995) Control of moth pests by mating disruption: successes and constraints. Annu Rev Entomol 40:559–585. https://doi.org/10.1146/annurev.en.40.010195.003015

Chang H, Liu Y, Yang T, Pelosi P, Dong S, Wang G (2015) Pheromone binding proteins enhance the sensitivity of olfactory receptors to sex pheromones in *Chilo suppressalis*. Sci Rep 5:13093. https://doi.org/10.1038/srep13093

Chen X-L, Su L, Li B-L, Li G-W, Wu J-X (2018a) Molecular and functional characterization of three odorant binding proteins from the oriental fruit moth *Grapholita molesta* (Busck) (Lepidoptera: Tortricide). Arch Insect Biochem Physiol 9:e2145698. https://doi.org/10.1002/arch.21456

Chen X-L, Li G-W, Xu X-L, Wu J-X (2018b) Molecular and functional characterization of odorant binding protein 7 from the oriental fruit moth *Grapholita molesta* (Busck) (Lepidoptera: Tortricidae). Front Physiol 9:1762. https://doi.org/10.3389/fphys.2018.01762

Christensen TA, Waldrop BR, Haroow ID, Hildebrand JG (1993) Local interneurons and information processing in the olfactory glomeruli of the moth *Manduca sexta*. J Comp Physiol A 173:385–399

Cork A (2016) Pheromones as management tools mass trapping and lure-and-kill. In: Allison JD, Cardé RT (eds) Pheromone communication in moths. University of California Press, Berkely

Cossé A, Todd J, Baker TJ (1998) Neurons discovered in male *Helicoverpa zea* antennae that correlate with pheromone-mediated attraction and interspecific antagonism. Comp Physiol A 182:585. https://doi.org/10.1007/s003590050205

Dani FR, Michelucci E, Francese S, Mastrobuoni G, Cappellozza S, Marca G, Niccolini A, Felicioli A, Moneti G, Pelosi P (2011) Odorant-binding proteins and chemosensory proteins in pheromone detection and release in the silkmoth *Bombyx mori*. Chem Senses 36:335–344. https://doi.org/10.1093/chemse/bjq137

Dekker T, Ibba I, Siju KP, Stensmyr MC, Hansson BS (2006) Olfactory shifts parallel superspecialism for toxic fruit in *Drosophila melanogaster* sibling, *D. sechellia*. Curr Biol 16:101–109

Dekker T, Revadi S, Mansourian S et al (2015) Loss of *Drosophila* pheromone reverses its role in sexual communication in Drosophila suzukii. Proc Biol Sci 282:20143018. https://doi.org/10.1098/rspb.2014.3018

Den Otter CJ, Thomas G (1979) Olfactory preference in insects: a synthesis of behaviour and electrophysiology. In: Kroeze GHA (ed) Preference behaviour and chemoreception. Information Retrieval Limited, London, pp 171–182

Domingue MJ, Musto CJ, Linn CE, Roelofs WL, Baker TC (2007a) Altered olfactory receptor neuron responsiveness in rare *Ostrinia nubilalis* males attracted to the *O. furnacalis* pheromone blend. J Insect Physiol 53:1063–1071. https://doi.org/10.1016/j.jinsphys.2007.05.013

Domingue MJ, Musto CJ, Linn CE, Roelofs WL, Baker TC (2007b) Evidence of olfactory antagonistic imposition as a facilitator of evolutionary shifts in pheromone blend usage in *Ostrinia* spp. (Lepidoptera: Crambidae). J Insect Physiol 53:488–496. https://doi.org/10.1016/j.jinsphys.2007.05.013

Domingue MJ, Haynes KF, Todd JL, Baker TC (2009) Altered olfactory receptor neuron responsiveness is correlated with a shift in behavioral response in an evolved colony of the cabbage looper loth, *Trichoplusia ni*. J Chem Ecol 35:405. https://doi.org/10.1007/s10886-009-9621-9

Evenden M (2016) Mating disruption of moth pests in integrated pest management a mechanistic approach. In: Allison JD, Cardé RT (eds) Pheromone communication in moths. University of California Press, Berkely

Forstner M, Breer H, Krieger J (2009) A receptor and binding protein interplay in the detection of a distinct pheromone component in the silkmoth *Antheraea polyphemus*. Int J Biol Sci 5:745–757. https://doi.org/10.7150/ijbs.5.745

Fujii T, Fujii T, Namiki S, Abe H, Sakurai T, Ohnuma A, Kanzaki R, Katsuma S, Ishikawa Y, Shimada T (2011) Sex-linked transcription factor involved in a shift of sex-pheromone preference in the silkmoth *Bombyx mori*. Proc Natl Acad Sci U S A 108:18038–18043

Gomez-Diaz C, Reina JH, Cambillau C, Benton R (2013) Ligands for pheromone-sensing neurons are not conformationally activated odorant binding proteins. PLoS Biol 11:e1001546. https://doi.org/10.1371/journal.pbio.1001546

Gregg PC, Del Socorro AP, Landolt PJ (2018) Advances in attract-and-kill for agricultural pests: beyond pheromones. Annu Rev Entomol 63:453–470. https://doi.org/10.1146/annurev-ento-031616-035040

Groot AT, Marr M, Schöfl G, Lorenz S, Svatos A, Heckel DG (2008) Host strain specific sex pheromone variation in *Spodoptera frugiperda*. Front Zool 5:20. https://doi.org/10.1186/1742-9994-5-20

Grosse-Wilde E, Svatos A, Krieger J (2006) A pheromone-binding protein mediates the bombykol-induced activation of a pheromone receptor in vitro. Chem Senses 31:547–555. https://doi.org/10.1093/chemse/bjj059

Gu S-H, Zhou J-J, Wang G-R, Zhang Y-J, Guo Y-Y (2013) Sex pheromone recognition and immunolocalization of three pheromone binding proteins in the black cutworm moth *Agrotis ipsilon*. Insect Biochem Mol Biol 43:237e251. https://doi.org/10.1016/j.ibmb.2012.12.009

Guo H, Smith DP (2017) Odorant receptor desensitization in insects. J Exp Neurosci 11:1179069517748600. https://doi.org/10.1177/1179069517748600

Hansson BS, Löfstedt C, Roelofs WL (1987) Inheritance of olfactory response to sex pheromone components in *Ostrinia nubilalis*. Naturwissenschaften 74:497. https://doi.org/10.1007/BF00447935

Hansson BS, Ljunberg H, Hallberg E, Löfstedt C (1992) Functional specialization of olfactory glomeruli in a moth. Science 256:1313–1315

He P, Zhang J, Li Z-Q, Zhang Y-N, Yang K, Dong S-L (2014) Functional characterization of an antennal esterase from the noctuid moth, *Spodoptera exigua*. Arch Insect Biochem Physiol 86:85–99. https://doi.org/10.1002/arch.21164

Hendrikse A, Vos-Buennemeyer E (1987) Role of host-plant stimuli in sexual behaviour of small ermine moths (Yponomeuta). Ecol Entomol 12:363–371

Hillier NK, Baker TC (2016) Pheromones of heliothine moths. In: Allison JD, Cardé RT (eds) Pheromone communication in moths. University of California Press, Berkely

Ibba I, Angioy AM, Hansson BS, Dekker T (2010) Macroglomeruli for fruit odors change blend preference in *Drosophila*. Naturwissenschaften 97:1059–1066. https://doi.org/10.1007/s00114-010-0727-2

Ignell R, Root CM, Birse RT, Wang JW, Nassel DR, Winther AME (2009) Presynaptic peptidergic modulation of olfactory receptor neurons in *Drosophila*. Proc Natl Acad Sci U S A 106:13070–13075

Ishida Y, Leal WS (2005) Rapid inactivation of a moth pheromone. Proc Natl Acad Sci U S A 102:14075–14079

Jacquin-Joly E, Vogt RG, François M-C, Meillour PN (2001) Functional and expression pattern analysis of chemosensory proteins expressed in antennae and pheromonal gland of *Mamestra brassicae*. Chem Senses 26:833–844. https://doi.org/10.1093/chemse/26.7.833

Jafari S, Alkhori L, Schleiffer A, Brochtrup A, Hummel T et al (2012) Combinatorial activation and repression by seven transcription factors specify *Drosophila* odorant receptor expression. PLoS Biol 10:e1001280. https://doi.org/10.1371/journal.pbio.1001280

Jin X, Ha TS, Smith DP (2008) SNMP is a signaling component required for pheromone sensitivity in *Drosophila*. Proc Natl Acad Sci U S A 105:10996–11001

Jurenka R (2004) Insect pheromone biosynthesis. In: Schulz S, 239 (eds) Chemistry of pheromones and other semiochemicals I. Springer, Berlin, pp 97–131
Kanaujia S, Kaissling K-E (1985) Interactions of pheromone with moth antennae: adsorption, desorption and transport. J Insect Physiol 31:71–81. https://doi.org/10.1016/0022-1910(85)90044-7
Kárpáti Z, Dekker T, Hansson BS (2008) Reversed functional topology in the antennal lobe of the male European corn borer. J Exp Biol 211:2841–2848. https://doi.org/10.1242/jeb.017319
Kárpáti Z, Olsson SB, Hansson BS, Dekker T (2010) Inheritance of central neuroanatomy and physiology related to pheromone preference in the male European corn borer. BMC Evol Biol 10:1–12
Kárpáti Z, Tasin M, Cardé RT, Dekker T (2013) Early quality assessment lessens pheromone specificity in a moth. Proc Natl Acad Sci U S A 110:7377–7382. https://doi.org/10.1073/pnas.1216145110
Knight A, Haworth J, Lingren B, Hebert V (2010) Combining pear ester with codlemone improves management of codling moth. IOBC/wprs Bull 2010:345–348
Koontz MA, Schneider D (1987) Sexual dimorphism in neuronal projections from the antennae of silk moths (*Bombyx mori, Antheraea polyphemus*) and the gypsy moth (*Lymantria dispar*). Cell Tissue Res 249:39–50. https://doi.org/10.1007/BF00215416
Koutroumpa FA, Kárpáti Z, Monsempes C, Hill SR, Hansson BS, Jacquin-Joly E, Krieger J, Dekker T (2014) Shifts in sensory neuron identity parallel differences in pheromone preference in the European corn borer. Front Ecol Evol 2:65. https://doi.org/10.3389/fevo.2014.00065
Koutroumpa FA, Monsempes C, François M-C, Cian A, Royer C, Concordet J-P, Jacquin-Joly E (2016) Heritable genome editing with CRISPR/Cas9 induces anosmia in a crop pest moth. Sci Rep 6:29620
Kunesch G, Zagatti P, Lallemand JY, Debal A, Vigneron JP (1981) Structure and synthesis of the wing gland pheromone of the male african sugar-cane borer: *Eldana saccharida* (wlk.) (Lepidoptera, pyralidae). Tetrahedron Lett 22:5270–5274. https://doi.org/10.1016/S0040-4039(01)92478-5
Kutinkova H, Subchev M, Light DM (2005) Interactive effects of ethyl (E,Z)-2, 4-decadienoate and sex pheromone lures to codling moth: apple orchard investigations in Bulgaria. J Plant Protect Res 45:49–52
Landolt PJ, Heath RR (1989) Attraction of female cabbage looper moths (Lepidoptera: Noctuidae) to male-produced sex pheromone. Ann Entomol Soc Am 82:520–525. https://doi.org/10.1093/aesa/82.4.520
Leary GP, Allena JE, Bungera PL, Luginbill JB, Linn CL, Macallisterd IE (2012) Single mutation to a sex pheromone receptor provides adaptive specificity between closely related moth species. Proc Natl Acad Sci U S A 109:14081–14086
Lee SG, Poole K, Linn CE Jr, Vickers NJ (2016) Transplant antennae and host brain interact to shape odor perceptual space in male moths. PLoS One 11:e0147906. https://doi.org/10.1371/journal.pone.0147906
Lee S-G, Vickers NJ, Baker TC (2006a) Glomerular targets of Heliothis subflexa male olfactory receptor neurons housed within long trichoid sensilla. Chem Senses 31(9):821–834. https://doi.org/10.1093/chemse/bjl025
Lee S-G, Carlsson MA, Hansson BS, Todd JL, Baker TC (2006b) Antennal lobe projection destinations of Helicoverpa zea male olfactory receptor neurons responsive to heliothine sex pheromone components. J Comp Physiol A 192:351. https://doi.org/10.1007/s00359-005-0071-8
Li Q, Ha TS, Okuwa S, Wang Y, Wang Q, Millard SS, Smith DP, Volkan PC (2013) Combinatorial rules of precursor specification underlying olfactory neuron diversity. Curr Biol 23:2481–2490
Li Z, Ni JD, Huang J, Montell C (2014) Requirement for Drosophila SNMP1 for rapid activation and termination of pheromone-induced activity. PLoS Genet 10:e1004600. https://doi.org/10.1371/journal.pgen.1004600

Liénard MA, Löfstedt C (2016) In: Allison JD, Cardé RT (eds) Small ermine moths: role of pheromones in reproductive isolation and speciation. University of California Press, Berkely, pp 211–224

Light DM, Knight AL (2005) Specificity of codling moth (Lepidoptera: Tortricidae) for the host plant kairomone, ethyl (E, Z)-2,4-decadienoate: field bioassays with pome fruit volatiles, analogue, and isomeric compounds. J Agric Food Chem 53:4046–4053

Linn C Jr, Poole K, Zhang A, Roelofs W (1999) Pheromone-blend discrimination by European corn borer moths with inter-race and inter-sex antennal transplants. J Comp Physiol A 184:273. https://doi.org/10.1007/s003590050325

Linn C Jr, O'Connor M, Roelofs W (2003) Silent genes and rare males: a fresh look at pheromone blend response specificity in the European corn borer moth, *Ostrinia nubilalis*. J Insect Sci 3(15):1

Linn C Jr, Musto CJ, Roelofs WL (2007) More rare males in ostrinia: response of Asian corn borer moths to the sex pheromone of the European corn borer. J Chem Ecol 33:99. https://doi.org/10.1007/s10886-006-9204-y

Linz J, Baschwitz A, Strutz A, Dweck HK, Sachse S, Hansson BS et al (2013) Host plant-driven sensory specialization in *Drosophila erecta*. Proc R Soc B 280:20130626

Liu YB, Haynes KFJ (1994) Evolution of behavioral responses to sex pheromone in mutant laboratory colonies of Trichoplusia ni. Chem Ecol 20:231. https://doi.org/10.1007/BF02064433

Löfstedt C (1993) Moth pheromone genetics and evolution. Philos Trans R Soc Lond B 340:167–177. https://doi.org/10.1098/rstb.1993.0055

Löfstedt C, Löfqvist J, Lanne BS, Van Der Pers JNC, Hansson BS (1986) Pheromone dialects in European turnip moths Agrotis segetum. Oikos 46(2):250–257. https://doi.org/10.2307/3565474. https://www.jstor.org/stable/3565474

Löfstedt C, Wahlberg N, Millar JG (2016) Evolutionary patterns of pheromone diversity in lepidoptera. In: Allison JD, Cardé RT (eds) Pheromone communication in moths. University of California Press, Berkely

Maïbèche-Coisne M, Jacquin-Joly E, Francois MC, Meillour PN (2002) cDNA cloning of biotransformation enzymes belonging to the cytochrome P450 family in the antennae of the noctuid moth *Mamestra brassicae*. Insect Mol Biol 11:273–281

Maida R, Ziegelberger G, Kaissling KE (2003) Ligand binding to six recombinant pheromone-binding proteins of *Antheraea polyphemus* and *Antheraea pernyi*. J Comp Physiol B 173:565–573

Maitani MM, Allara DL, Park KC, Lee SG, Baker TC (2010) Moth olfactory trichoid sensilla exhibit nanoscale-level heterogeneity in surface lipid properties. Arthropod Struct Dev 39:1–16. https://doi.org/10.1016/j.asd.2009.08.004

Merlin C, Rosell G, Carot-Sans G, Francois MC, Bozzolan F, Pelletier J, Jacquin-Joly E, Guerrero A, Maïbèche-Coisne M (2007) Antennal esterase cDNAs from two pest moths, *Spodoptera littoralis* and *Sesamia nonagrioides*, potentially involved in odorant degradation. Insect Mol Biol 16:73–81

Mohl C, Breer H, Krieger J (2002) Species-specific pheromonal compounds induce distinct conformational changes of pheromone binding protein subtypes from *Antheraea polyphemus*. Invertebr Neurosci 4:165–174

Ochieng SA, Anderson P, Hansson BS (1995) Antennal lobe projection patterns of olfactory receptor neurons involved in sex pheromone detection in *Spodoptera littoralis* (Lepidoptera: Noctuidae). Tissue Cell 27:221–232

Ochieng S, Poole K, Linn C, Vickers N, Roelofs W, Baker T (2003) Unusual pheromone receptor neuron responses in heliothine moth antennae derived from inter-species imaginal disc transplantation. J Comp Physiol A 189:19. https://doi.org/10.1007/s00359-002-0371-1

Olsen SR, Wilson RI (2008) Lateral presynaptic inhibition mediates gain control in an olfactory circuit. Nature 452:956–960

Pelosi P, Iovinella I, Felicioli A, Dani FR (2014) Soluble proteins of chemical communication: an overview across arthropods. Front Physiol 5:320

Pelosi P, Iovinella I, Zhu J, Wang G, Dani FR (2017) Beyond chemoreception: diverse tasks of soluble olfactory proteins in insects. Biol Rev 93:184–200. https://doi.org/10.1111/brv.12339

Pherobase: www.pherobase.com

Pregitzer P, Greschista M, Breer H, Krieger J (2014) The sensory neuron membrane protein SNMP1 contributes to the sensitivity of a pheromone detection system. Insect Mol Biol 23:733–742. https://doi.org/10.1111/imb.12119

Rafaeli A, Jurenka R (2003) PBAN regulation of pheromone biosynthesis in female moths. In: Blomquist GJ, Vogt R (eds) Pheromone biochemistry and molecular biology. Elsevier, London, pp 107–136

Rafaeli A, Bober R, Becker L, Choi MY, Fuers EJ, Jurenka R (2007) Spatial distribution and differential expression of the PBAN receptor in tissues of adult *Helicoverpa* spp. (Lepidoptera: Noctuidae). Insect Mol Biol 16:287–293

Roelofs WL, Glover TJ, Tang X-H, Robbins PS, Löfstedt C, Hansson BS, Bengtsson BO, Sreng I, Eckenrode CJ (1987) Sex pheromone production and perception in European corn borer moths is determined by both autosomal and sex-linked genes. Proc Natl Acad Sci U S A 84:7585–7589

Saveer AM, Kromann SH, Birgersson G, Bengtsson M, Lindblom T, Balkenius A, Hansson BS, Witzgall P, Becher PG, Ignell R (2012) Floral to green: mating switches moth olfactory coding and preference. Proc Biol Soc 279(1737):2314–2322. https://doi.org/10.1098/rspb.2011.2710

Su C-Y, Menuz K, Carlson JR (2009) Olfactory perception: receptors, cells, and circuits. Cell 139:45–59. https://doi.org/10.1016/j.cell.2009.09.015

Sun XJ, Fonta C, Masson C (1993) Odour quality processing by bee antennal lobe interneurons. Chem Senses 18:355–377

Sun JS, Xiao S, Carlson JR (2018) The diverse small proteins called odorant-binding proteins. Open Biol 8:180208. https://doi.org/10.1098/rsob.180208

Syed Z, Kopp A, Kimbrell DA, Leal WS (2010) Bombykol receptors in the silkworm moth and the fruit fly. Proc Natl Acad Sci U S A 107:9436–9439. https://doi.org/10.1073/pnas.1003881107

The UN Sustainable Development Goals. https://www.un.org/sustainabledevelopment/development-agenda/

Tillman JA, Seybold SJ, Jurenka RA, Blomquist GJ (1999) Insect pheromones – an overview of biosynthesis and endocrine regulation. Insect Biochem Mol Biol 29:481–514

Trona F, Anfora G, Balkenius A, Bengtsson M, Tasin M, Knight A, Janz N, Witzgall P, Ignell R (2013) Neural coding merges sex and habitat chemosensory signals in an insect herbivore. Proc R Soc B 280:2013–0267

Van der Pers JNC, Den Otter CJ (1978) Single cell responses from olfactory receptors of small ermine moths to sex-attractants. J Insect Physiol 24:337–343

Vickers NJ, Christensen TA, Baker TC, Hildebrand JG (2001) Odour-plume dynamics influence the brain's olfactory code. Nature 410:466–470

Vickers NJ, Poole K, Linn CE Jr (2005) Plasticity in central olfactory processing and pheromone blend discrimination following interspecies antennal imaginal disc transplantation. J Comp Neurol 491:141–156. https://doi.org/10.1002/cne.20725

Vogt RG, Riddiford LM, Prestwich GD (1985) Kinetic properties of a sex pheromone-degrading enzyme: the sensillar esterase of *Antheraea polyphemus*. Proc Natl Acad Sci U S A 82:8827–8831

Von Arx M, Schmidt-Büsser D, Guerin PM (2012) Plant volatiles enhance behavioral responses of grapevine moth males, Lobesia botrana to sex pheromone. J Chem Ecol 38:222–225

Wicher D (2018) Tuning insect odorant receptors. Front Cell Neurosci 12:94. https://doi.org/10.3389/fncel.2018.00094

Wu W, Cottrell CB, Hansson BS, Löfstedt C (1999) Comparative study of pheromone production and response in Swedish and Zimbabwean populations of turnip moth, Agrotis segetum. J Chem Ecol 25:177. https://doi.org/10.1023/A:1020849419193

Wu H, Hou C, Huang L-Q, Yan F-S, Wang C-Z (2013) Peripheral coding of sex pheromone blends with reverse ratios in two Helicoverpa species. PLoS One 8:e70078. https://doi.org/10.1371/journal.pone.0070078

Younas A, Waris MI, Qamar MT, Shaaban M, Prager SM, Wang M-Q (2018) Functional analysis of the chemosensory protein MsepCSP8 from the oriental armyworm *Mythimna separata*. Front Physiol 9:872. https://doi.org/10.3389/fphys.2018.00872

Zagatti P (1981) Comportement sexuel de la Pyrale de la Canne a sucre *Eldana saccharina* (Wlk.) lie a deux pheromones emises par le male. Behaviour 78:81–98

Zagatti P, Kunesch G, Ramiandrasoa IF, Malosse C, Hall DR, Lester R, Nesbitt BF (1987) Sex pheromonesof rice moth, *Corcyra cephalonica* Stainton I. Identification of male pheromone. J Chem Ecol 13:1561–1573. https://doi.org/10.1007/BF00980200

Zhang D-D, Löfstedt C (2013) Functional evolution of a multigene family: orthologous and paralogous pheromone receptor genes in the turnip moth, *Agrotis segetum*. PLoS One 8:e77345. https://doi.org/10.1371/journal.pone.0077345

Zhang D-D, Löfstedt C (2015) Moth pheromone receptors: gene sequences, function, and evolution. Front Ecol Evol. https://doi.org/10.3389/fevo.2015.00105

Zhu J, Iovinella I, Dani FR, Liu Y-L, Huang L-Q, Liu Y, Wang C-Z, Pelosi P, Wang G (2016) Conserved chemosensory proteins in the proboscis and eyes of Lepidoptera. Int J Biol Sci 12:1394–1404. https://doi.org/10.7150/ijbs.16517

Chapter 14
Toward Robotic Applications of Insect Sex Pheromone-Searching Behavior: Lessons from the Silkmoth

Noriyasu Ando and Poonsup Pansopha Kono

Abstract The sex pheromone-searching behavior of insects is innate and highly reproducible. Therefore, pheromone-searching behavior has been an excellent model for understanding the odor-searching strategies of animals. Furthermore, pheromone-searching behavior has attracted the attention of engineers, aiming at developing autonomous odor-searching robots. In this chapter, we will first explain the fundamental strategy of pheromone-searching behavior used by insects, and then review recent advances in knowledge about the searching strategy of our model insect, the silkmoth (*Bombyx mori*), in terms of visual and olfactory integration. We next introduce our recent studies on a silkmoth-driven hybrid robot and show how silkmoths adaptively behave in challenging circumstances. Lastly, we discuss the future direction of biomimetic approaches to robotic odor searching.

Keywords Pheromone-searching behavior · Odor-searching robot · *Bombyx mori*

14.1 Introduction

Tracking an odor plume and localizing its source is an essential ability for animals to find food, nests, and mating partners. On the other hand, robotic odor-searching has been one of the challenges in engineering. As odorants are intermittently distributed in the air, the odor concentration detected by animals changes abruptly (Murlis and Jones 1981). Therefore, it is difficult to find the odor source by continuously detecting and following the odor gradient. Odor-searching strategies of animals,

N. Ando (✉)
Research Center for Advanced Science and Technology, University of Tokyo, Tokyo, Japan

Department of Systems Life Engineering, Maebashi Institute of Technology, Gunma, Japan
e-mail: ando@brain.imi.i.u-tokyo.ac.jp; ando@maebashi-it.ac.jp

P. Pansopha Kono
Research Center for Advanced Science and Technology, University of Tokyo, Tokyo, Japan

Y. Ishikawa (ed.), *Insect Sex Pheromone Research and Beyond*, Entomology Monographs, https://doi.org/10.1007/978-981-15-3082-1_14

which have evolved over the long history of life, are thus attractive to researchers in robotics. Indeed, there have been many studies on robotic odor-searching in which robots behave based on the algorithms of animals (Naeem et al. 2007; Kowadlo and Russell 2008; Russell et al. 2003; Ishida et al. 2012). Among the odor-searching behaviors of animals, the sex pheromone-searching behaviors of insects are excellent models for the study of behavioral strategies and their neural bases because they are innate and highly reproducible.

The pheromone-searching behaviors of flying moths are the representative examples of odor-searching behaviors of insects, and have been investigated for many decades (Cardé 2016; Bau and Cardé 2015). Male moths have the ability to track a highly intermittent and fluctuating sex-pheromone plume, and successfully locate its source from relatively long distances (Fabre and Miall 1912; Wall and Perry 1987; Elkinton et al. 1987). The odor-searching strategy of moths consists of two fundamental behaviors: upwind surge and crosswind casting (Fig. 14.1a). The surge is a straight upwind flight, and is initiated shortly after the moth encounters the pheromone filaments (Vickers and Baker 1996; Baker et al. 1985). When tracking a pheromone plume with high-frequency dispersion, moths perform a reiterative upwind surge due to the high encounter rate of pheromone filaments, which results in a straighter flight (Mafra-Neto and Carde 1994; Vickers and Baker 1996). The function of the reiterative upwind surge is believed to be reorientation toward the pheromone source (Murlis et al. 1992). On the other hand, the moth initiates cross-

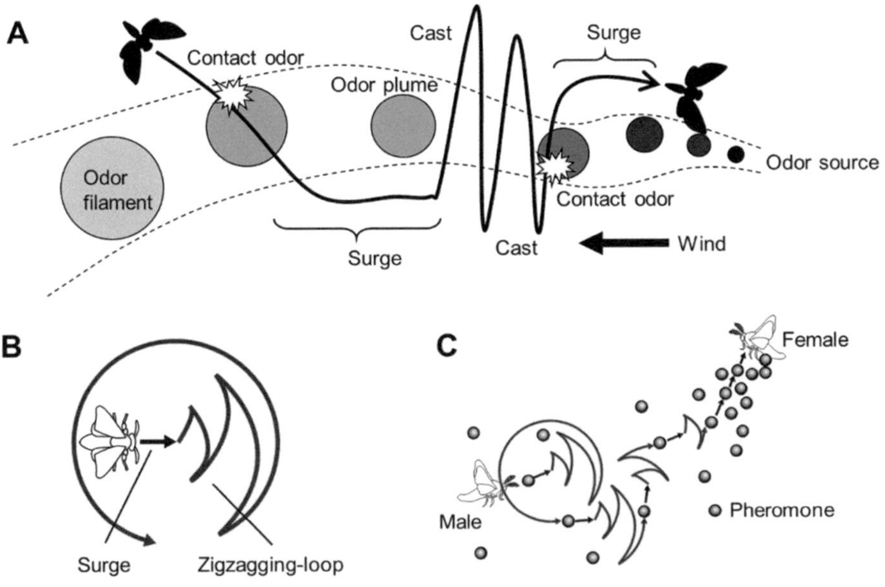

Fig. 14.1 Odor-searching strategies of moths. (**a**) A flying moth tracking an odor plume with surge and casting. (**b**) Pheromone-triggered sequential behavior of a male silkmoth. The turn angle during zigzagging monotonically increases by turns due to the increase in turn duration. Note that the actual behavior during zigzagging and looping consists of point turns. (**c**) Orientation of a male to a female silkmoth. The pheromone-triggered behavioral sequence is reset by additional pheromone reception

wind casting when it loses contact with the pheromone plume. The casting is a self-steering counterturning and is presumably an intrinsically programmed behavior (Mafra-Neto and Cardé 1994; Kennedy 1983), and its function is to recontact with the windborne odor. The effectiveness of the surge and casting strategy of flying moths has been evaluated by computer simulations (Kramer 1997; Belanger and Arbas 1998) or by using robotic platforms (Edwards et al. 2005; Willis 2008). Mobile robots are advantageous because we can evaluate an algorithm in a real odor plume. Furthermore, such mobile robots will be prototypes of the future biomimetic robots in engineering. The so-called 'moth-inspired' algorithm has been examined not only for testing biological hypotheses but also for aiming for practical use with various robotic platforms, such as land, underwater, and flying robots (Wei et al. 2006; Neumann et al. 2013; Pyk et al. 2006).

The fundamental strategy of flying moths can be observed in the odor-searching walk by the male silkmoths (*Bombyx mori*) when they respond to conspecific female sex pheromones. Their walking behavior (note that *B. mori* cannot fly) is a well-defined programmed behavior consisting of surge (straight-line walking) during pheromone reception, and zigzagging turns followed by looping after the loss of the pheromone (Fig. 14.1b; Kanzaki et al. 1992). The behavioral sequence is reset by an additional pheromone reception; therefore, their trajectories become straight when they repetitively encounter the pheromone (Fig. 14.1c). The male silkmoth is an excellent model for odor-searching behavior because its neural basis has been well documented in behavioral, genetic, neurophysiological, and robotic studies (Wada and Kanzaki 2005; Kanzaki 2007; Iwano et al. 2010; Namiki et al. 2014; Sakurai et al. 2014; see also Chap. 12). Furthermore, the behavioral algorithm and abstracted neural network models were implemented to miniaturized mobile robots to generate the behaviors (Fig. 14.2 and Table 14.1) (Kuwana et al. 1996; Nagasawa et al. 1999; Kanzaki et al. 2004; Kanzaki 2007; Katsumata et al. 2009). These studies were extensions of the biological analyses aiming at confirming the insect models based on experimental data. On the other hand, these studies are also expected to be the first step for the application of insect odor-searching for mobile robots for practical use.

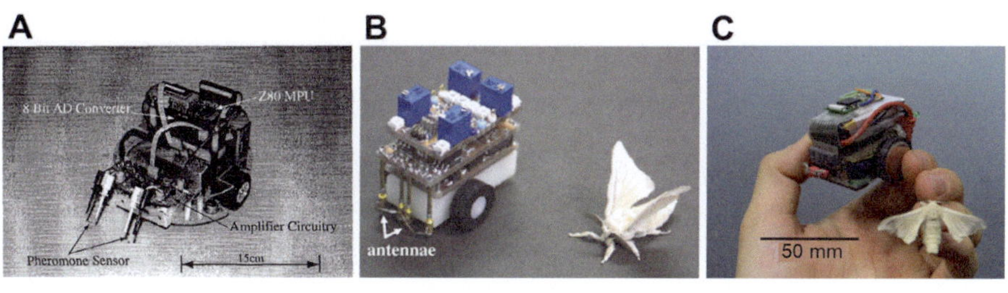

Fig. 14.2 Silkmoth robots. (**a**) A mobile robot with silkmoth antennae as pheromone sensors. The zigzagging behavior was generated by a recurrent neural network with bilateral reciprocal inhibition. (© 1996 IEEE. Reprinted from Kuwana et al. (1996) with permission.) (**b**) A silkmoth-sized mobile robot with the antennae. A brain model abstracted from neurophysiological data was implemented. (Reprinted from Kanzaki 2007, with permission from Elsevier.) (**c**) A small mobile robot with atmospheric ion detectors (Katsumata et al. 2009). A behavior-based model with osmotropotaxis was implemented. An ionizer instead of an odor source was used for the experiments

Table 14.1 Specifications of the silkmoth robots

Report	Sensor	Algorithm	Robot size	Reports related to the algorithm in biology
Kuwana et al. (1996)	Dissected antenna	Recurrent neural network with reciprocal inhibition, which generates the flip-flop signal for zigzagging	150 mm (L) × 150 mm (W) × 100 mm (H)	Behavior: Kanzaki et al. (1992)
				Physiology: Olberg (1983) and Kanzaki et al. (1994)
Kanzaki et al. (2004) and Kanzaki (2007)	Dissected antenna	Anatomically constrained network model with reciprocal inhibition based on the bilateral LAL-VPC system, which generates surge-zigzagging-loop	31 mm (L) × 18 mm (W) × 30 mm (H)	Behavior: Kanzaki et al. (1992)
				Physiology: Kanzaki et al. (1994), Mishima and Kanzaki (1999), and Iwano et al. (2010)
Katsumata et al. (2009)	Ion detector	Programmed behavioral sequence of surge-zigzagging-loop with the directional control during surge using bilateral olfaction	55 mm (L) × 45 mm (W) × 40 mm (H)	Behavior: Kanzaki et al. (1992) and Takasaki et al. (2012)

Although the odor-searching behavior of the silkmoth is stereotypical and simple, our recent behavioral and neuroanatomical studies revealed that the behavior is flexible and can be modulated by multiple sensory inputs. Such modulation may increase the efficiency and adaptability of the odor-searching task in the environment, and its implementation is thus also required for autonomous robots. In this chapter, we focus on how moths utilize multiple information for efficient odor-searching and its potential applicability to robotics. We first explain the recent progress in our understanding of the modulation of the odor-searching behavior by visual information in silkmoths. We then introduce our recent study on the insect-driven hybrid robot to investigate whether these modulations enhance the ability to localize an odor source in challenging situations. Lastly, we refer to future perspectives for the application of insect pheromone-searching strategy to robots.

14.2 Visual and Olfactory Integration During Odor-Tracking

The integration of visual information into odor-tracking has been well documented particularly for flying insects (Frye 2003, 2004, 2010; Kaissling 1997; Kennedy 1983; Kennedy and Marsh 1974; Mafra-Neto and Carde 1994; van Breugel and Dickinson 2014), but not for walking insects (Willis et al. 2011). Previous studies on flying moths suggested that moths sense the wind-induced drifting of their own position by optic flow during upwind surge, and correct the drifting by optomotor response, a visually guided compensatory movement against involuntary displacement from an intended course (Götz 1968). This phenomenon, upwind flight with adjustment of direction by wind-sensing and vision, is termed optomotor anemotaxis (Mafra-Neto and Carde 1994; Vickers 2000). Furthermore, visual input is necessary for successful orientation for flying moths even though they are forced to walk by clipping their wings (Willis et al. 2011).

On the other hand, in the silkmoth, it has been difficult to investigate the role of vision during pheromone-searching. Visual stimuli alone do not trigger any locomotor responses in the silkmoth; however, the adult silkmoth is considered to utilize visual stimuli in pheromone source-searching because some descending interneurons that respond to the sex pheromone also respond to visual stimuli such as optic flow and light (Olberg 1983; Kanzaki et al. 1994; Wada and Kanzaki 2005).

We previously demonstrated the use of optic flow during the pheromone-searching behavior of the silkmoth by behavioral experiments (Pansopha et al. 2014). To quantify the effects of optic flow during the pheromone-searching behavior, a vertical grating pattern was presented in front of a tethered moth walking on an air-supported ball (Fig. 14.3a). The ball rotation by locomotion of the moth was measured by an optical mouse. The horizontal motion of the grating was controlled at arbitrary speeds. The speed was not compensated by the moth's own movement under open-loop conditions (moths only receive 'biased' optic flow), whereas it was compensated under closed-loop conditions (the moth received the biased optic flow subtracted from the self-induced optic flow).

To clarify the effects of vision, we separately analyzed locomotion of moths during surge or zigzagging (the loop is regarded as an extension of the zigzagging), based on the assumption that the optic flow is used differently for each behavioral state. During surge, silkmoths exhibit optomotor responses to the biased optic flow in both open- and closed-loop conditions (Fig. 14.3b). In response to the horizontal optic flow, moths tried to compensate for the movement by turning toward the same

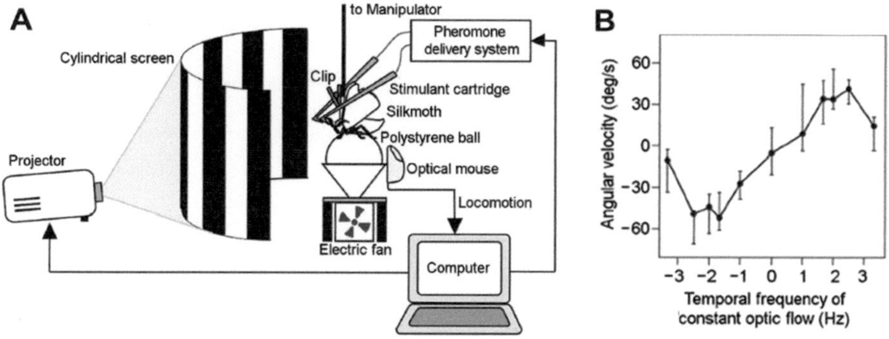

Fig. 14.3 Silkmoths follow optic flow during surge. (**a**) Experimental setup. (**b**) Temporal-frequency tuning curve of the angular velocity during surge under open-loop conditions (N = 7–10 for experimental trials, N = 85 for 0 Hz). The temporal frequency is the rate of flicking by passing the black and white gratings, and no movement at 0 Hz. Dots and error bars represent the medians and 25th–75th percentiles, respectively. (From Pansopha et al. 2014)

direction as the optic flow. The angular turn velocity increased as the optic flow speed increased (shown as temporal frequency, Fig. 14.3b). In contrast, the moth did not exhibit optomotor responses during the zigzagging under open-loop conditions, and there were no significant changes in kinematic parameters with and without optic flow. This suggests that silkmoths alter the use of vision depending on the behavioral states within the pheromone-triggered sequential behavior, in which the states (surge or zigzagging) can be quickly switched by pheromone reception. This is in contrast to previous reports in which the responsiveness to optic flow changes based on the behavioral state with a longer time scale such as quiescent or flying, presence of an attractive odor, or the magnitude of locomotor activity (Maimon et al. 2010; Rosner et al. 2010; Chow and Frye 2008; Chow et al. 2011). Therefore, silkmoths may use the optic flow dynamically during pheromone-searching, and the site of visual integration in the brain may be closely related to activated identified brain areas that are responsible for generating the pheromone-searching behavior.

Although silkmoths do not exhibit optomotor responses during zigzagging, visual input is unlikely completely suppressed during such behavior. Under the closed-loop conditions, the turn angle during zigzagging was prolonged due to a prolonged turn duration when the moth turned to the opposite direction of the biased optic flow (Fig. 14.4). As the optic flow is markedly accelerated when the moth

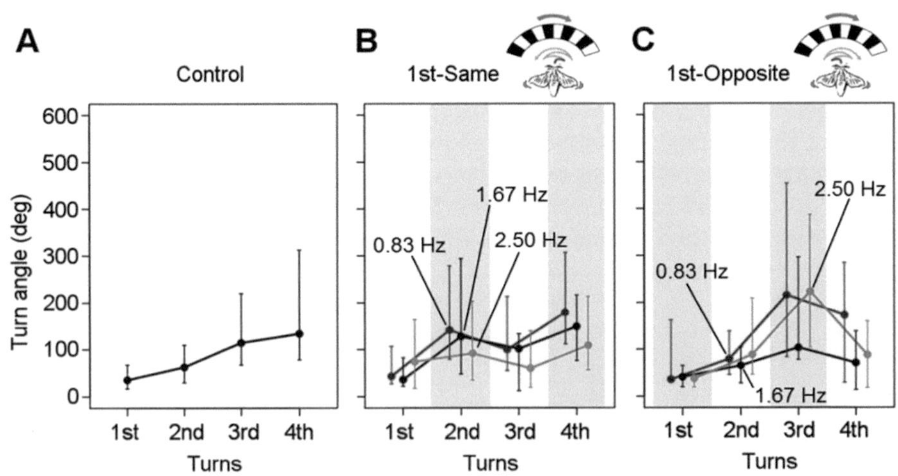

Fig. 14.4 Behavioral responses during zigzagging to simultaneously presented biased optic flow and a single puff of the pheromone under closed-loop conditions. The angle of the first four turns by moths in (**a**) control trials (self-induced optic flow only, no biased flow, $N = 123$), (**b**) experimental trials of the moths that made the 1st turn in the same direction as the biased optic flow (1st-Same; $N = 21$, 21, and 30 for different temporal frequencies of the biased optic flow of 0.83, 1.67, and 2.50 Hz, respectively), and (**c**) those that made the first turn in the opposite direction (first opposite; $N = 13$, 18, and 20 for 0.83, 1.67, and 2.50 Hz, respectively). The turn angle monotonically increased as the number of turns increased in the control, whereas it increased when moths received accelerated optic flow (opposite direction of the biased optic flow, indicated by shaded areas). Dots and error bars represent the medians and 25th–75th percentiles, respectively. (From Pansopha et al. 2014)

switches direction and turns opposite to the biased optic flow (self-induced optic flow is added to the biased flow under the closed-loop conditions), the accelerated optic flow may be involved in the observed modulation during the zigzagging. Although the function of the observed prolongation of the turn by optic flow stimuli under closed-loop conditions is unclear, it is evident that the use of optic flow is dynamically changed by the behavioral state. Based on these results, we hypothesized that there are at least two visual pathways that are responsible for the modulations of surge and zigzagging, respectively (Fig. 14.5a), and the modulations are closely related to the activity state of the neural circuits responsible for each behavioral state. This hypothesis was further investigated by the neuroanatomical study in the next section.

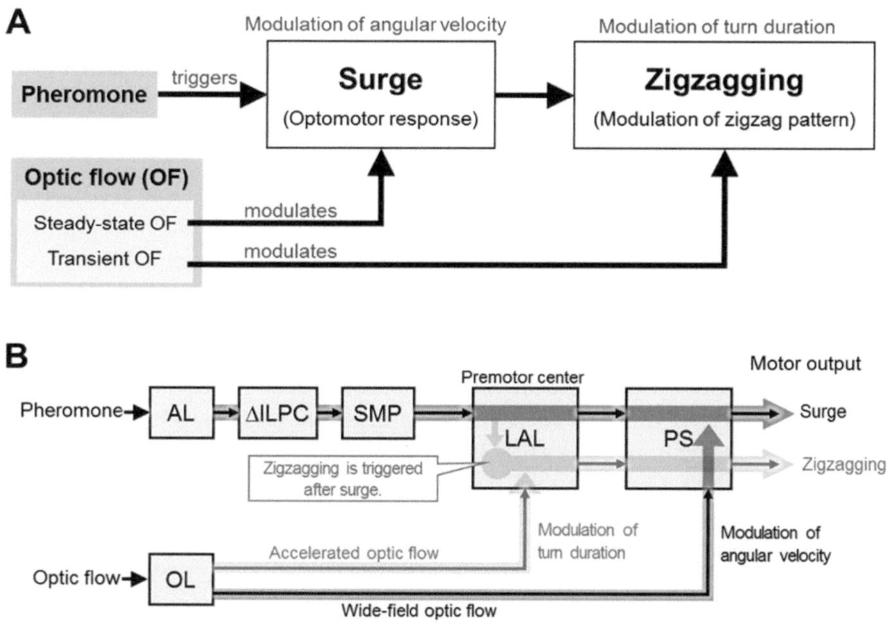

Fig. 14.5 Hypothesis of the interaction between visual and olfactory processing pathways during the odor-searching behaviors of the silkmoth. (**a**) Summary of the behavioral study. Constant-speed biased optic flow (steady-state optic flow) elicits optomotor responses, whereas accelerated optic flow at the beginning of turns against the biased optic flow direction (transient optic flow) modulates the zigzagging pattern. (**b**) Hypothesized information flow of the visual and olfactory integration in the brain. *AL* antennal lobe, *ΔILPC* inferolateral protocerebrum, *SMP* superior medial protocerebrum, *LAL* lateral accessory lobe, *PS* posterior slope, *OL* optic lobe. (From Pansopha et al. 2014; Pansopha Kono 2018)

14.3 Neuroanatomical Basis of Visual and Olfactory Integration

The neural pathway related to olfactory processing has been extensively studied in the silkmoth. It was previously demonstrated that the antennal lobe (AL), a primary olfactory processing center, receives pheromone input from the antenna, and it conveys the information to the mushroom body (MB) and the delta area of the inferolateral protocerebrum (ΔILPC). Then, the pheromone information is relayed to the superior medial protocerebrum (SMP) before being conveyed to the lateral accessory lobe (LAL). The LAL is a premotor center that generates locomotion control signals and sends them via descending neurons to neck muscles and motor centers to control locomotion in the thoracic ganglia (Haupt et al. 2010; Kanzaki et al. 2003; Namiki et al. 2014; Sakurai et al. 2014; Seki et al. 2005). Based on reports on the neural pathway underlying visuomotor control in flies and hawkmoths (Gronenberg and Strausfeld 1990; Milde et al. 1992) and the current knowledge on the pheromone-processing pathway in silkmoths described above, visual input was hypothesized to modulate the turning angular velocity during surge through direct connections between motion-sensitive interneurons from the optic lobe and descending neurons that carry motor control signals (Fig. 14.5b). The potential site for these direct connections is the posterior slope (PS) in the ventroposterior protocerebrum. On the other hand, the modulation in turn duration during zigzagging by accelerated optic flow suggests that the LAL, the area for generation of programmed zigzag walking (Kanzaki and Shibuya 1992), receives visual information from the optic lobe and modulate the temporal characteristics of zigzagging.

The overall pathways from the visual neuropils to the candidate areas, where visual inputs are integrated, were identified by the mass staining technique (Pansopha Kono 2018). The visual pathways from the optic lobe to the central brain are shown in Fig. 14.6. The neurons originating from the optic lobe projected multiple neuropils, including the LAL and the PS, on both sides. For further identification of possible visual pathways responsible for optic flow-sensing, we restricted the injection site to the lobula complex (LoC) where high integration of spatial information occurs. The LoC contains the lobula (Lo) and the lobula plate (LP). The Lo supports detection of object motion (Wu et al. 2016) and the LP contains tangential cells, which process wide-field motion vision (Borst and Haag 2002; Borst et al. 2010). Distinct pathways from the LoC to the central brain, including the LAL and the PS on both sides, were also observed (Fig. 14.7). These results support our hypothesis about the neural connection between them and the optic lobes, and their possible roles in the state-dependent visual modulation of pheromone-searching behavior. More recently, the morphology of visual projection neurons in silkmoths was investigated, employing the single-cell labeling technique. Identification of the projection neurons from the Lo to the LAL and the PS supports the presence of a pathway for olfactory and visual integration (Namiki and Kanzaki 2018).

The proposed neural pathways underlying the multisensory integration in *B. mori* are presented in Fig. 14.8. The direct pathway to the PS is likely responsible for a

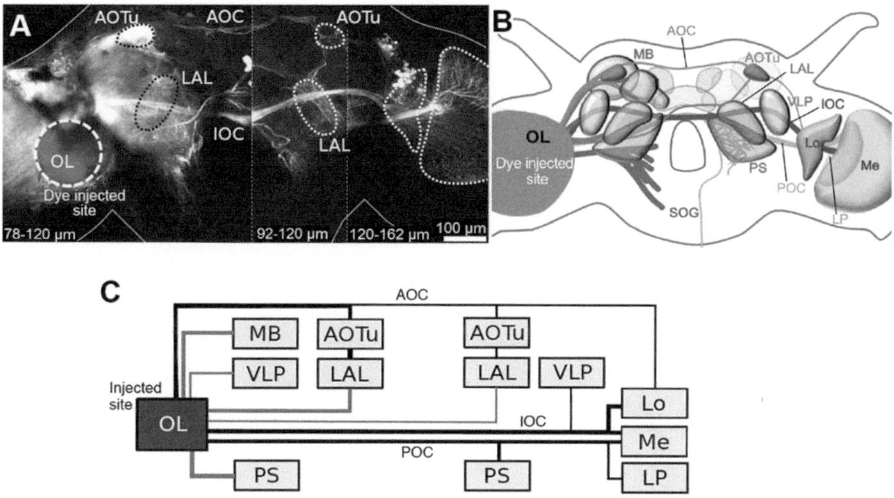

Fig. 14.6 Visual pathways in the silkmoth brain. (**a**) Projection image of mass staining of the visual pathway. Texas Red dextran crystals were injected into the optic lobe (OL). The image was reconstructed from confocal scanning of the left and right hemispheres separately. The border of each image is indicated by a vertical dashed line, and the depths along z-axis used to reconstruct these projection images are labeled. (**b**) A scheme summarizing visual pathway from the optic lobe to different neuropils in the central brain based on the mass staining results. (**c**) A block diagram of visual pathways in the central brain of the silkmoth. *MB* mushroom body, *VLP* ventrolateral protocerebrum, *AOTu* anterior optic tubercle, *Lo* lobula, *Me* medulla, *LP* lobula plate. In bilateral commissures, *AOC* anterior optic commissure, *IOC* inferior optic commissure, *POC* posterior optic commissure. (From Pansopha Kono 2018)

fast and reflex-like pathway for optomotor responses. This pathway may be activated by wide-field optic flow (corresponding to the steady-state biased optic flow in the behavioral experiments) irrespective of pheromone-triggered walking because a quiescent silkmoth moves its head toward the optic flow direction. On the other hand, other visual pathways conveying wide-field optic flow converge into the LAL. During surge, visual information is directly or indirectly input via the anterior optic tubercle (AOTu) to the LAL. As the LAL is the center for generating the walking pattern, visual integration may occur in the LAL and the output is conveyed by descending neurons, which are phasically active in response to the pheromone (Mishima and Kanzaki 1999; Wada and Kanzaki 2005). In addition to direct pathways to the LAL, the MB and the central complex (CC) where multiple types of sensory information are input and regulate locomotion also function in the integration via indirect pathways. These pathways may enhance the optomotor responses and generate effective motor commands for course control during surge. On the other hand, the zigzagging pattern, which occurs after the loss of the pheromone, is generated by reciprocal activity between the bilateral LAL and ventral protocerebrum, and the motor command is sent to the thoracic motor center via tonically

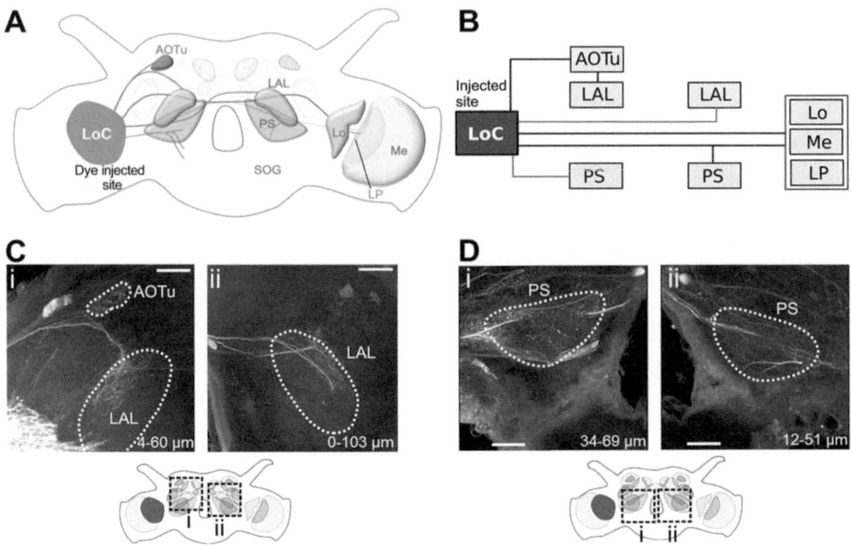

Fig. 14.7 Visual pathways from the lobula complex (LoC) project to neuropils, including the bilateral LAL and PS. (**a**) A Scheme summarizing the visual pathway from the LoC to different neuropils in the central brain depicted based on the mass staining results. SOG, subesophageal ganglion. (**b**) A block diagram of visual pathways related to the LoC in the central brain of the silkmoth. (**c**) A confocal image of the anterior protocerebrum in the ipsilateral (i) and contralateral (ii) hemispheres, in which the AOTus and the LALs on both sides were stained. (**d**) Confocal images of the ventral protocerebrum in the ipsilateral (i) and contralateral (ii) hemispheres taken from the posterior to anterior at different depths show neural connections to the bilateral PS. Scale bars = 50 μm. (From Pansopha Kono 2018)

activated descending neurons (Iwano et al. 2010; Kanzaki et al. 1994; Kanzaki and Shibuya 1992; Mishima and Kanzaki 1999; Wada and Kanzaki 2005). The visual input, the accelerated optic flow caused by the self-generated zigzagging in particular, should be integrated into the LAL and alter the duration of the tonic activity to determine the turn duration.

Taken together, the anatomical observations support the neural pathways presumed based on the behavioral experiments. Although physiological data are lacking, the visual and olfactory integration of the silkmoths is suggested to be closely related to the activity state of the neural circuits responsible for the switch between surge and zigzagging. These results and the presumed circuits also suggest that the surge is well controlled by vision. In the next section, we will explain the importance of directional control during surge.

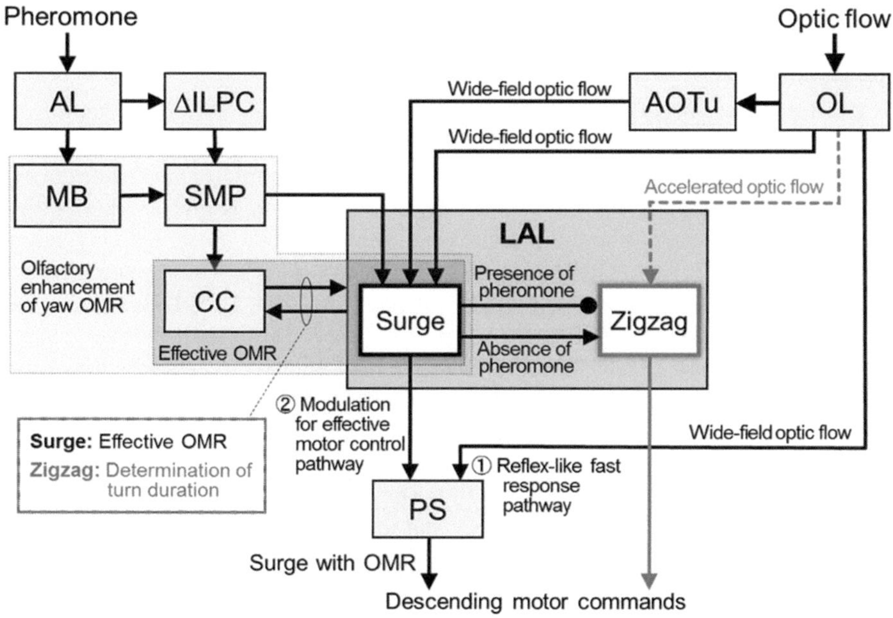

Fig. 14.8 Speculated neural circuitry underlying multisensory integration for tracking pheromone plumes. *CC* central complex, *OMR* optomotor response. (From Pansopha Kono 2018)

14.4 Do Silkmoths Make Decisions?

Surge is represented as straight-line walking. A silkmoth missing one antenna continues to circle in the direction of the remaining antenna (Kellog 1907), indicating that the bilateral olfactory input determines the laterality of the walking pattern. Further behavioral studies revealed that silkmoths select the surge direction according to bilateral olfaction. They choose the side of higher odor concentration during surge, then start the first turn of zigzagging to the same side (Takasaki et al. 2012), and finally continue to circle in the same direction (Fig. 14.10a) (Kanzaki and Shibuya 1992). These observations demonstrate that once the silkmoth detects a difference in the pheromone concentration by bilateral olfaction, they decide on the direction to go and retain that decision in their short-term memory throughout a single behavioral sequence. This is also supported by the electrophysiological 'flip-flop' response in descending neurons originating from the LAL (Kanzaki et al. 1994; Mishima and Kanzaki 1999; Wada and Kanzaki 2005; Olberg 1983).

The directional control by positive chemotaxis (turn toward a higher concentration) is a ubiquitous strategy among organisms to localize an odor source if they can detect a concentration gradient of the odor (Gomez-Marin et al. 2010). In particular, selecting a preferred direction based on simultaneous comparison between bilateral odor concentrations using multiple olfactory organs is termed osmotropotaxis, and is observed in several animal species such as insects (Borst and Heisenberg 1982;

Martin 1965; Hangartner 1967; Duistermars et al. 2009; Gaudry et al. 2013) and mammals (Catania 2013; Rajan et al. 2006; Khan et al. 2012), including humans (Porter et al. 2007).

During locomotion, the optomotor response acts as an automatic stabilizer, which makes the moths go straight once the visual system detects bilaterally asymmetrical optic flow. This may be why silkmoths exhibit optomotor responses during surge but not during zigzagging. The absence of optomotor responses during zigzagging prevents conflict between voluntary turning and optomotor responses during this behavior. However, as previously mentioned, surge is not simple straight-line walking and the direction is modulated by bilateral olfaction. Therefore, surge is the specific phase in which moths actively change their direction at their own discretion. As such, silkmoths have to segregate optic flow induced by perturbations from that induced by their own voluntary movement or they cannot turn voluntarily to their intended direction during surge.

This segregation problem was elegantly solved by von Holst and Mittelstaedt (1950). They rotated the head of a fly 180° and fixed it (i.e., the fly received optic flow in the same direction as movement), and found that the fly continued to rotate once it started to move. This experiment demonstrated that the optomotor response is not suppressed during voluntary movement, and self-induced optic flow that informs the sensory consequences of the insect's own movement refers to a copy of the motor command (corollary discharge or efference copy; here, we use 'corollary discharge' according to Crapse and Sommer (2008)), and the difference between them is then compensated by motor control in the next step. Corollary discharge is important for the segregation of self-induced sensory input from others, and it enables animals to execute precise voluntary motor control as intended. Silkmoths may also have the same motor control scheme during surge because they can voluntarily select the direction using bilateral olfaction and perform optomotor responses. This assumption was experimentally examined using an insect-driven mobile robot, as introduced in the following sections.

14.5 The Use of Mobile Robots to Make Models Interact with Environments

In the previous sections, we provided an overview of the recent advances from studies on pheromone-searching behavior of the silkmoth. The odor-searching behavior is not stereotyped but is modulated by sensory information, which may enhance the performance of odor source localization. However, the observed behavior was a response to an artificially defined sensory stimulus, which is spatially and temporally discontinuous compared with those received in the environment, that is, data are collected under conditions with no environmental interaction. Therefore, whether these sensory–motor controls during surge are effective for successful orientation in the environment should be examined.

To investigate the function of a sensory–motor response for a certain task in the environment, behavioral experiments with manipulation (enable or disable function, or alter the gain) of the specified sensory input or neural circuit are preferable. Two approaches can be used: one is the manipulation of a target sensory system in vivo, and the other is that in silico. Manipulation in vivo can be carried out by physical (covering or ablation of sensory organs, cutting nerves, etc.), pharmacological (injecting an agonist or antagonist of a target neurotransmitter), and genetic methods. In studies on pheromone-searching behavior of the silkmoth, cutting or covering the antennae (Kellog 1907; Kondoh and Obara 1984; Kanzaki et al. 1992), microsurgery of the bilateral commissure of the LAL (Kanzaki et al. 1994), and microinjection of serotonin and its antagonists into the brain (Gatellier et al. 2004) have been performed as physical and pharmacological manipulations. These manipulations are useful to clarify the overall function of the targets, but the disadvantages of these techniques include their invasiveness and the difficulty in pin-pointing the target. The recent development of genetic tools has enabled non-invasive manipulation of specified neural circuits at the cellular level. In the silkmoth, selective activation of the pheromone receptor on the antennae by photo or heat stimuli (Tabuchi et al. 2013; Kiya et al. 2014), and addition or ablation of specified olfactory receptors (Sakurai et al. 2011, 2015) have been reported. However, these techniques were developed for the model insect *Drosophila melanogaster* (Olsen and Wilson 2008), and the application of these techniques to other insect species, including the silkmoth, is still underway.

Mobile robots have been used for in silico experiments, in which behavioral or neural circuit models built based on experimental results were implemented. The advantage of using a mobile robot is that the model can interact with an external environment via the robot as the body even if the model is based on an open-loop experiment (Fig. 14.9). Therefore, sensory–motor models of various insect species have been implemented and tested (Floreano et al. 2014; Webb 2002, 1998; Webb et al. 2004). As robots are artificial, we can quantitatively manipulate the sensory and/or motor properties of the robot. However, 'artificial' may also be the disadvantage of using robots. The major problem is that the response properties of sensors

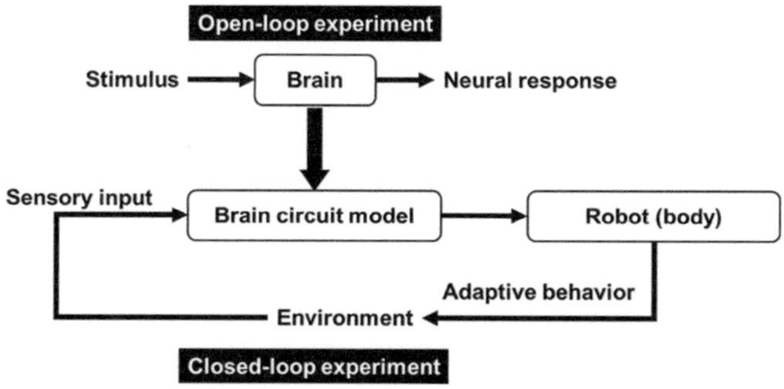

Fig. 14.9 The brain model is embodied by a mobile robot

are different between insects and robots. Even if a robot and an insect are placed in the same environment, internal representations of the world are different between them, that is, different 'Umwelt' (von Uexküll and Kriszat 1956). In odor-searching, in particular, insect antennae have a superior ability to acquire high-speed temporal dynamics of odor concentrations (Szyszka et al. 2012, 2014; Bau et al. 2005), whereas the response and recovery times of conventional gas sensors, such as semiconductor sensors, are slow (Harvey et al. 2006; Martinez et al. 2014). Such differences in temporal responsiveness directly affect the resultant behaviors. To avoid this problem, insect antennae have been used as odor sensors and their electrical responses to odor stimuli (electroantennogram, EAG) were acquired to control the mobile robots (Kuwana et al. 1995, 1996, 1999; Nagasawa et al. 1999; Kanzaki et al. 2004; Martinez et al. 2014). More recently, the use of EAG was attempted for odor sensors of flying robots (Lan et al. 2017). Although the stability of electrical signals and longevity of the antenna should be improved, the use of EAG is advantageous for mobile robots to capture fast odor fluctuations similar to insects, which is an important step for robots to share the same 'Umwelt' as insects.

14.6 Insect-Controlled Robot: An Experimental Tool for Insect Pheromone-Searching Behavior

Expanding the idea of implementing biological tissues for robots, we developed a silkmoth-driven hybrid robot (insect-controlled robot) (Emoto et al. 2007; Ando et al. 2016). The basic design of the insect-controlled robot was the combination of conventional techniques used in neuroethology: a spherical treadmill and mobile robot (Fig. 14.10a). The male silkmoth is dorsally tethered and placed on an air-supported polystyrene ball. The moth performs the odor-searching behavior and rotates the ball once it receives the pheromone. The ball rotation is then measured by an optical mouse sensor, and transformed into the translational and rotational movement of the robot by a microcontroller (Fig. 14.10b). Two flexible tubes with fans and a flow channel inside a canopy function for 'sniffing' to collect and deliver

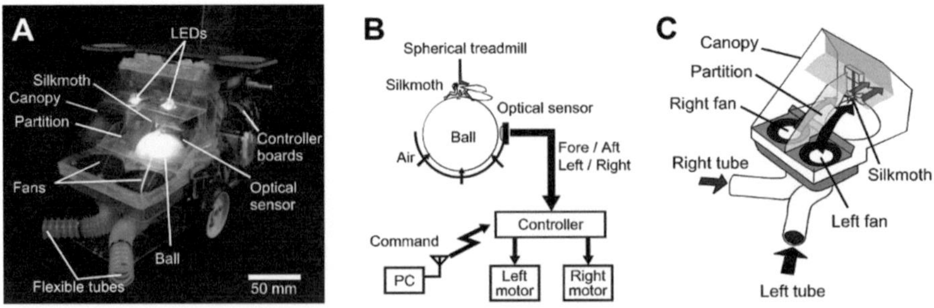

Fig. 14.10 Insect-controlled robot. (**a**) The appearance of the robot. (**b**) Hardware diagram of the robot. (**c**) Odor delivery system. Airflow is indicated by arrows. (From Ando and Kanzaki 2015)

the pheromone to the antennae of the onboard moth (Fig. 14.10c), which produces a similar effect as the self-generated air flow by wing vibration of the silkmoths (Obara 1979; Loudon and Koehl 2000).

The concept of 'insect drives a mobile robot' is to create a robot that has real insect sensory systems and a real brain. As the brain on board interacts with the external environment via the robotic body, the closed-loop between this 'hybrid' robot and the external environment can be quantitatively altered by manipulation of the robotic part. Regarding the conflict between the silkmoth decision by bilateral olfaction and optomotor response, the insect-controlled robot fulfills the requirements for addressing this problem: bilateral olfaction and visual input can be noninvasively manipulated (Ando and Kanzaki 2015; Ando et al. 2016). The flexible suction tube samples an odor on each side and the intake can be placed at an arbitrary position to manipulate the bilateral olfactory input. Inverting the position of the suction tube on each side effectively switches the robot's response to the side with the higher odor concentration from attraction to avoidance. Inverting motor rotation on each side also switches the response to the higher concentration side. It also inverts the direction of self-induced optic flow. Therefore, if silkmoths use corollary discharge for voluntary directional control, behaviors of the insect-controlled robot will differ between the inverted olfactory input and the inverted motor output. Thus, the robot with the inverted motor output will continue circling like the head-inverted flies (von Holst and Mittelstaedt 1950).

14.7 Sensory–Motor Control for Efficient Localization to the Pheromone Source

Odor-tracking experiments were conducted using the robot to investigate the sensory–motor control during surge. We released bombykol at 2 Hz (200 ms of bombykol puffing alternating between 300 ms pauses) in a wind tunnel, which is high enough to elicit repetitive surges of moths in a pheromone plume (Kanzaki et al. 1992; Pansopha et al. 2014). The robot under the control conditions (normal olfactory input and motor output) demonstrated straight trajectories (the straightness of movement is indicated by the mean length of the synthetic vectors of robot movement (R) ranging from 0 to 1; in control R = 0.81) with a success rate of 100% (Figs. 14.11b and 14.12a). We then compared the odor-searching behavior of the robot with four different combinations of olfactory input and motor output (Figs. 14.11b–e and 14.12a): control (normal input and normal output), inverted olfactory input (inverted input and normal output), inverted motor output (normal input and inverted output), and both inverted (inverted input and inverted output). Compared with the control conditions, the inverted olfactory input or inverted motor output led to complicated trajectories. The success rate and R were reduced to 61.5%–58.3% and 0.29–0.32, respectively. The success rate and R improved slightly to 73.3% and 0.65, respectively, under the conditions of both inverted input and

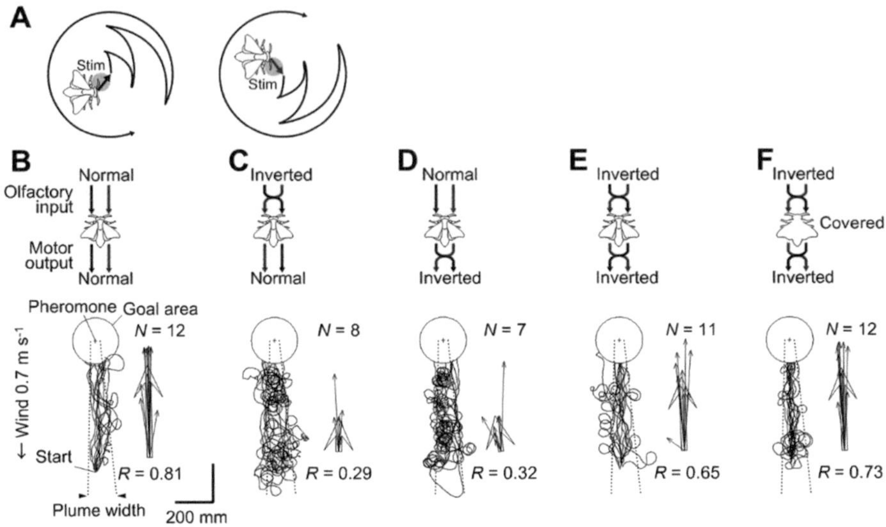

Fig. 14.11 Male silkmoths use bilateral olfaction and optomotor responses for efficient localization to the pheromone source. (**a**) Laterality of the pheromone-searching behavior depends on which of the two antennae is stimulated (pheromone stimulus on the left or right antenna). (**b–e**) Odor-tracking trajectories of the robot with control (**b**), inverted olfactory input (**c**), inverted motor output (**d**), or inverted olfactory input and motor output (**e**). Thin solid arrows indicate synthetic vectors of robot movement normalized by time to localization in trials. Large open arrowheads indicate the mean vector of trials, and its length R is shown for each condition. The goal area was defined to judge successful localization to the pheromone source. (**f**) Trajectories with inverted input and output, and the visual field covered. (From Ando and Kanzaki 2015)

output. These results revealed that directional control of surge by bilateral olfaction is necessary for efficient odor source localization.

The difference between the effects of inverted olfactory input and inverted motor output was not clear in terms of the success rate or straightness of movement. However, when we investigated the tendency of circling by calculating the mean turn angle (mean angle of consecutive turns in a given direction by each trial; a large angle indicates continuous circling in the same direction; for details, see Ando and Kanzaki (2015)), the angle under the inverted motor conditions was significantly larger than that under the control conditions. On the other hand, no significant difference was noted between the angle in the control and the inverted olfactory input (Fig. 14.12b). These results indicate that the inverted self-induced optic flow under the inverted motor output additively affected the resultant motor control (i.e., positive feedback) and enhanced the optomotor response of the moths. Furthermore, the difference in the mean turn angle between the control and the inverted motor output conditions was reduced after covering the visual field, and the success rate and the straightness under inverted olfactory input and motor output conditions were restored to 100% and $R = 0.73$, respectively (Figs. 14.11f and 14.12a). These results are consistent with the assumption that the moth compares the corollary discharge

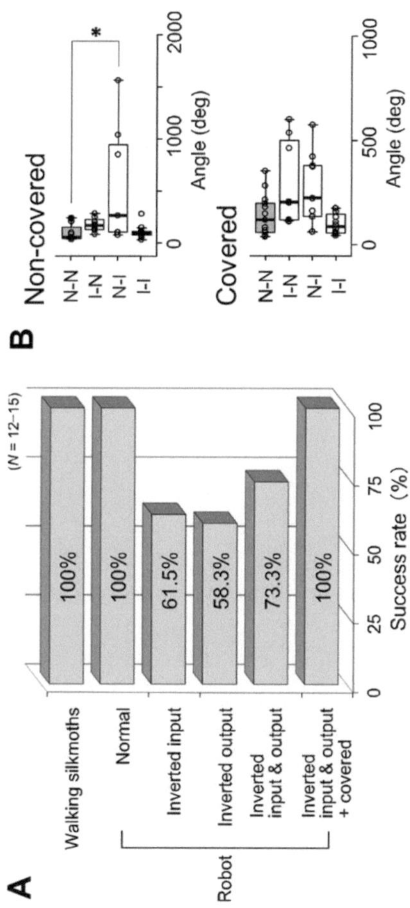

Fig. 14.12 Experimental results of manipulation of sensory–motor relationships. (**a**) Success rate of localization. The success rate of walking silkmoths is shown in comparison with those of the robot. (**b**) The mean turn angle (index of continuous circling to either side) during orientations without covering of the visual field (upper) and with covering (lower). A significant difference is marked by an asterisk (Steel test, $P < 0.05$). N–N, normal olfactory input and motor output; I–N, inverted input and normal output; N–I, normal input and inverted output; I–I, inverted input and output. (From Ando and Kanzaki 2015)

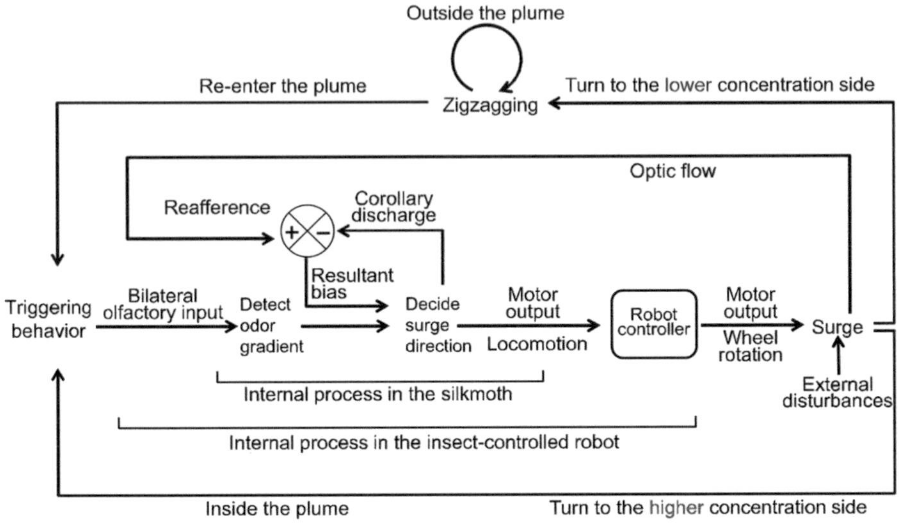

Fig. 14.13 Schematic diagram of the sensory–motor control of the odor-searching behavior of the silkmoth. (From Ando and Kanzaki 2015)

and reafference of self-induced optic flow. Taken together, the experiments using the insect-controlled robot suggest that the ubiquitous strategy for motor control across the animal kingdom, which is based on corollary discharge and reafference signals (Webb 2004; Crapse and Sommer 2008), is also implemented in the directional control during silkmoth surge.

A schematic diagram of the information flow in the sensory–motor system in the insect-controlled robot is shown in Fig. 14.13. When the robot enters the pheromone plume, an onboard moth decides the direction of surge based on the concentration gradient detected by bilateral olfaction. The motor output from the moth is then translated into robot movement. The corollary discharge of the motor command comprising surge direction cancels the reafference of the self-induced sensory signals, and the resultant bias is added to the motor command for course control. At the same time, the robot with the inverted olfactory input or motor output performs zigzagging and looping until it re-enters the plume. This strategy effectively helps the robot re-enter an odor plume even under the perturbed sensory–motor conditions, and the robot successfully localized the odor source in more than half of the trials (success rates, 58.3–75.0%; Fig. 14.11a). This result is expected because odorants are intermittently distributed in the natural environment and insects cannot always detect reliable odor gradients (Murlis and Jones 1981). Thus, silkmoths use 'surge' for an accurate orientation toward the odor source based on corollary discharge and reafference, and 'zigzagging' for enhancing the possibility of re-entering the odor plume even when it cannot acquire the concentration gradient. The use of multiple searching strategies, such as switching between serial and bilateral odor sampling in different odor contexts, has been reported in insects (Hangartner 1967;

Martin 1965), crustaceans (Reeder and Ache 1980), and mammals (Catania 2013). In contrast to these animals, the two strategies of silkmoths are implemented as a sequential behavior triggered by a single puff of the pheromone. Consequently, the frequency of resetting the behavior, which was determined by external (spatiotemporal odor distribution around an insect) and internal states (behavioral threshold of the insect), is an important factor that differentiates the use of the two strategies from others.

14.8 What Can the Future Insect-Mimetic Robots Do?

From a biological point of view, the mobile robots implemented with behavioral or neural circuit models are useful to understand the behavioral mechanisms behind pheromone-searching behaviors. In robotics, such robots can be regarded as prototypes of biomimetic robots. An insect-mimetic robot equipped with a neural circuit model of the brain may track and localize a target odor like a sniffing dog if the brain circuit is completely understood. Although neurophysiological and computational studies on the silkmoth have been performed to reconstruct the large-scaled neural circuit in the brain (Namiki et al. 2009; Sato et al. 2010; Ikeno et al. 2012; Miyamoto et al. 2012), the research has not yet finished and no complete model of the brain is currently available. The insect-controlled robot can be regarded as a future insect-mimetic robot because it acquires external information with real insect sensory systems and is driven by the real insect brain. Therefore, we can analyze how the future insect-mimetic robot functions by evaluating the insect-controlled robot.

If the optomotor response of the silkmoth is implemented to an odor-tracking robot, it will make the robot behave more adaptively. Such a hypothesis can be investigated by the insect-controlled robot with perturbations of the motor system. In our experiments, an asymmetrical turning bias (either side of turning was amplified by fourfold) was applied to examine whether an onboard moth was able to compensate for the bias (Ando et al. 2013). The trajectories of the manipulated robot, which had either a clockwise or counterclockwise bias as compared with those under control conditions, are presented in Fig. 14.14a and b. Even with the turning bias, the manipulated robot had a success rate of 80.8%. The contribution of optomotor response to this ability can be easily clarified by covering the visual field of the onboard moth. The success rate of the manipulated robot was reduced to 53.8% after covering the visual field (Fig. 14.14c). We also confirmed that this compensatory turning against the turning bias was reflexive and not learned because it was observed within 1 s of switching to the biased conditions during locomotion (Ando et al. 2013).

It is worth noting that localization was successful in more than half of the trials (53.8%) with the turning bias and covered visual field. This suggests a non-visual mechanism for the compensation because the robot with the bias tracked the boundary of the pheromone plume on the ipsilateral side of the imposed bias even when

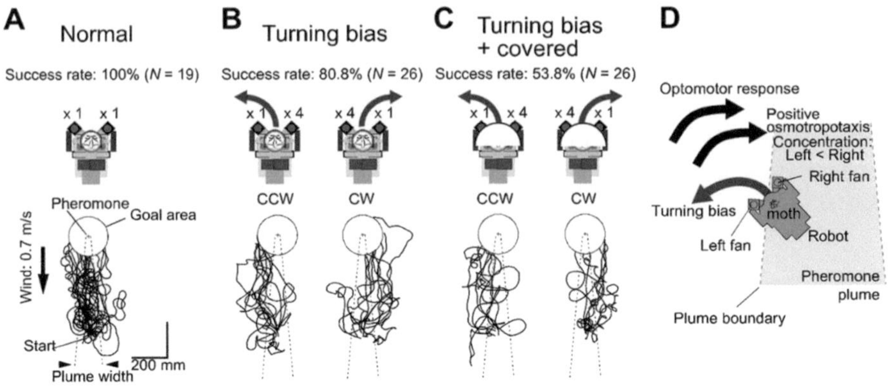

Fig. 14.14 Silkmoths can overcome the imposed turning bias using optomotor responses and bilateral olfaction. (**a**) Trajectories of the insect-controlled robot without manipulation. (**b**) Trajectories after manipulating the turning bias. *CCW* turning bias in counterclockwise rotation, *CW* turning bias in clockwise rotation. (**c**) Trajectories of the robot with turning bias and covering of the visual field. Note that the robot with the bias followed either side of the plume boundary under both non-covered and covered conditions. (**d**) Mechanisms of the compensatory turning by optomotor responses and osmotropotaxis using bilateral olfaction against the turning bias at the plume boundary. (From Ando et al. 2013)

the visual field was covered. This behavior can be explained by the osmotropotaxis during surge. At the plume boundary, steering toward a higher odor concentration (i.e., inside plume) is a driving force that can cancel out the imposed steering due to the turning bias (Fig. 14.14d). A similar behavior was reported by the flying moth with a single antenna (Vickers and Baker 1991).

From a biomimetic perspective, the insect-controlled robot is a useful step that demonstrates the goal performance of the future insect-mimetic robot. The evaluations of the insect-controlled robot suggest that the future insect-mimetic robots will be able to localize an odor source even with internal or external perturbations. Further manipulations of the robot will enable us to evaluate its performance from different aspects. For example, the time delay of the robot movement in response to insect locomotion reveals the computation time needed for sensory–motor processing (Ando et al. 2013) or the fine-tuning of motor gains may increase the performance beyond that of intact insects.

14.9 Conclusion and Perspectives

The pheromone-searching behavior of male silkmoths is a useful model for the study of the behavioral and neuronal basis of animal odor-searching, and robotic odor-searching. The relevant behavioral response and neural pathways responsible for the behavior have been well studied. The accumulation of biological knowledge is also helpful for robotics, not only for the evaluation of models in the environment, but also for application of the insect's ability to autonomous robots. The pheromone-triggered behavior of the silkmoth is typified as surge, zigzagging, and looping, which can be implemented as behavior-based or abstracted neural circuit models simply according to input–output relationships. The mobile robots implemented with such models can perform the fundamental strategy of odor tracking and localize the odor source in some defined environments; however, it is unclear whether such robots have the same odor-tracking capability as silkmoths because modulation of the behavior is still poorly understood.

The recent behavioral and anatomical studies introduced in this chapter suggest that the pheromone-searching behavior is not fixed, but flexible. The moth can voluntarily control locomotion during surge by utilizing multiple sensory stimuli. This control is suspected to increase the efficiency of tracking the pheromone plume and its robustness against perturbations. This hypothesis will be examined using a mobile robot if this alteration is correctly modeled. However, as the neural mechanism of alteration is still unknown except for some anatomical information regarding visual pathways, many suppositions are needed to fill in the missing information for modeling.

Although the internal processes in the brain of an onboard moth remain as a 'black box,' the insect-controlled robot enables us to directly manipulate multiple sensory inputs and demonstrates the importance of behavioral modulation during the odor-searching task. We demonstrated that the decision of surge direction (using bilateral olfaction) and precise course control (using corollary discharge and visual feedback) effectively enhance the moth's capability to localize the pheromone source. Furthermore, the robot experiment also revealed that the programmed pattern of zigzagging enables the moth to effectively recontact the pheromone plume even with perturbation, during which the moth cannot compensate by sensory feedback.

The insect-controlled robot further demonstrates the ability of the future robot implemented with a complete brain model, which is the goal for neuroscience aiming at comprehensively understanding the brain and for robotics trying to capture any technological advantage from biological systems. We strongly believe that biomimetic robots should not be limited to simple mimicry of an input–output relationship within a sensory–motor system. Our hybrid robot reminds us of the importance of internal computation in the brain toward the application of the biological system to robotics, and will facilitate the bottom-up approach toward a comprehensive understanding of the brain.

Acknowledgments We are grateful to Professor Ryohei Kanzaki for supervising the studies introduced in this chapter. We also thank Dr. Michiyo Kinoshita for technical guidance during the neuroanatomical study on the visual pathways in the brain.

References

Ando N, Kanzaki R (2015) A simple behaviour provides accuracy and flexibility in odour plume tracking – the robotic control of sensory-motor coupling in silkmoths. J Exp Biol 218(23):3845–3854. https://doi.org/10.1242/jeb.124834

Ando N, Emoto S, Kanzaki R (2013) Odour-tracking capability of a silkmoth driving a mobile robot with turning bias and time delay. Bioinspir Biomim 8(1):016008. https://doi.org/10.1088/1748-3182/8/1/016008

Ando N, Emoto S, Kanzaki R (2016) Insect-controlled robot: a Mobile robot platform to evaluate the odor-tracking capability of an insect. J Vis Exp 118:e54802. https://doi.org/10.3791/54802

Baker TC, Willis MA, Haynes KF, Phelan PL (1985) A pulsed cloud of sex-pheromone elicits upwind flight in male moths. Physiol Entomol 10(3):257–265. https://doi.org/10.1111/j.1365-3032.1985.tb00045.x

Bau J, Cardé RT (2015) Modeling optimal strategies for finding a resource-linked, windborne odor plume: theories, robotics, and biomimetic lessons from flying insects. Integr Comp Biol 55(3):461–477. https://doi.org/10.1093/icb/icv036

Bau J, Justus KA, Loudon C, Carde RT (2005) Electroantennographic resolution of pulsed pheromone plumes in two species of moths with bipectinate antennae. Chem Senses 30(9):771–780. https://doi.org/10.1093/chemse/bji069

Belanger JH, Arbas EA (1998) Behavioral strategies underlying pheromone-modulated flight in moths: lessons from simulation studies. J Comp Physiol A 183(3):345–360. https://doi.org/10.1007/s003590050261

Borst A, Haag J (2002) Neural networks in the cockpit of the fly. J Comp Physiol A Neuroethol Sens Neural Behav Physiol 188(6):419–437. https://doi.org/10.1007/s00359-002-0316-8

Borst A, Heisenberg M (1982) Osmotropotaxis in *Drosophila melanogaster*. J Comp Physiol 147(4):479–484

Borst A, Haag J, Reiff DF (2010) Fly Motion Vision. Annu Rev Neurosci 33:49–70. https://doi.org/10.1146/annurev-neuro-060909-153155

Cardé RT (2016) Moth navigation along pheromone plumes. In: Allison JD, Carde RT (eds) Pheromone communication in moths evolution, behavior, and application. University of California Press, Oakland, pp 173–189

Catania KC (2013) Stereo and serial sniffing guide navigation to an odour source in a mammal. Nat Commun 4:1441. https://doi.org/10.1038/ncomms2444

Chow DM, Frye MA (2008) Context-dependent olfactory enhancement of optomotor flight control in *Drosophila*. J Exp Biol 211(Pt 15):2478–2485. https://doi.org/10.1242/jeb.018879

Chow DM, Theobald JC, Frye MA (2011) An olfactory circuit increases the fidelity of visual behavior. J Neurosci 31(42):15035–15047. https://doi.org/10.1523/JNEUROSCI.1736-11.2011

Crapse TB, Sommer MA (2008) Corollary discharge across the animal kingdom. Nat Rev Neurosci 9(8):587–600. https://doi.org/10.1038/nrn2457

Duistermars BJ, Chow DM, Frye MA (2009) Flies require bilateral sensory input to track odor gradients in flight. Curr Biol 19(15):1301–1307. https://doi.org/10.1016/j.cub.2009.06.022

Edwards S, Rutkowski AJ, Quinn RD (2005) Willis MA moth-inspired plume tracking strategies in three-dimensions. In: Proceedings of the 2005 IEEE international conference on robotics and automation, 18–22 April 2005. pp 1669–1674. https://doi.org/10.1109/ROBOT.2005.1570353

Elkinton JS, Schal C, Onot T, CardÉ RT (1987) Pheromone puff trajectory and upwind flight of male gypsy moths in a forest. Physiol Entomol 12(4):399–406. https://doi.org/10.1111/j.1365-3032.1987.tb00766.x

Emoto S, Ando N, Takahashi H, Kanzaki R (2007) Insect-controlled robot—evaluation of adaptation ability. J Robot Mechatron 19(4):436–443

Fabre J-H, Miall B (1912) Social life in the insect world, vol viii, 327 p. Century, New York

Floreano D, Ijspeert AJ, Schaal S (2014) Robotics and neuroscience. Curr Biol 24(18):R910–R920. https://doi.org/10.1016/j.cub.2014.07.058

Frye MA (2003) Odor localization requires visual feedback during free flight in *Drosophila melanogaster*. J Exp Biol 206(5):843–855. https://doi.org/10.1242/jeb.00175
Frye MA (2004) Motor output reflects the linear superposition of visual and olfactory inputs in *Drosophila*. J Exp Biol 207(1):123–131. https://doi.org/10.1242/jeb.00725
Frye MA (2010) Multisensory systems integration for high-performance motor control in flies. Curr Opin Neurobiol 20(3):347–352. https://doi.org/10.1016/j.conb.2010.02.002
Gatellier L, Nagao T, Kanzaki R (2004) Serotonin modifies the sensitivity of the male silkmoth to pheromone. J Exp Biol 207(Pt 14):2487–2496. https://doi.org/10.1242/jeb.01035
Gaudry Q, Hong EJ, Kain J, de Bivort BL, Wilson RI (2013) Asymmetric neurotransmitter release enables rapid odour lateralization in *Drosophila*. Nature 493(7432):424–428. https://doi.org/10.1038/Nature11747
Gomez-Marin A, Duistermars BJ, Frye MA, Louis M (2010) Mechanisms of odor-tracking: multiple sensors for enhanced perception and behavior. Front Cell Neurosci 4:6. https://doi.org/10.3389/fncel.2010.00006
Götz KG (1968) Flight control in *Drosophila* by visual perception of motion. Kybernetik 4(6):199–208. https://doi.org/10.1007/Bf00272517
Gronenberg W, Strausfeld NJ (1990) Descending neurons supplying the neck and flight motor of diptera: physiological and anatomical characteristics. J Comp Neurol 302(4):973–991. https://doi.org/10.1002/cne.903020420
Hangartner W (1967) Spezifität und Inaktivierung des Spurpheromons von Lasius fuliginosus Latr. und Orientierung der Arbeiterinnen im Duftfeld. Z Vergl Physiol 57(2):103–136. https://doi.org/10.1007/BF00303068
Harvey D, Lu TF, Keller M (2006) Odor sensor requirements for an insect inspired plume tracking mobile robot. In: The 2006 IEEE international conference on robotics and biomimetics, 17–20 Dec. 2006, pp 130–135. https://doi.org/10.1109/ROBIO.2006.340328
Haupt SS, Sakurai T, Namiki S, Kazawa T, Kanzaki R (2010) Olfactory information processing in Moths. In: Menini A (ed) The neurobiology of olfaction. Frontiers in neuroscience. CRC Press, Boca Raton
Ikeno H, Kazawa T, Namiki S, Miyamoto D, Sato Y, Haupt SS, Nishikawa I, Kanzaki R (2012) Development of a scheme and tools to construct a standard moth brain for neural network simulations. Comput Intell Neurosci 2012:795291. https://doi.org/10.1155/2012/795291
Ishida H, Wada Y, Matsukura H (2012) Chemical sensing in robotic applications: a review. IEEE Sensors J 12(11):3163–3173. https://doi.org/10.1109/jsen.2012.2208740
Iwano M, Hill ES, Mori A, Mishima T, Mishima T, Ito K, Kanzaki R (2010) Neurons associated with the flip-flop activity in the lateral accessory lobe and ventral protocerebrum of the silkworm moth brain. J Comp Neurol 518(3):366–388. https://doi.org/10.1002/cne.22224
Kaissling KE (1997) Pheromone-controlled anemotaxis in moths. In: Lehrer M (ed) Orientation and communication in arthropods, vol 84. EXS. Birkhäuser, Basel, pp 343–374. https://doi.org/10.1007/978-3-0348-8878-3_12
Kanzaki R (2007) How does a microbrain generate adaptive behavior? Int Congr Ser 1301:7–14. https://doi.org/10.1016/j.ics.2006.12.037
Kanzaki R, Shibuya T (1992) Long-lasting excitation of protocerebral bilateral neurons in the pheromone-processing pathways of the male moth *Bombyx mori*. Brain Res 587(2):211–215. https://doi.org/10.1016/0006-8993(92)90999-P
Kanzaki R, Sugi N, Shibuya T (1992) Self-generated zigzag turning of *Bombyx mori* males during pheromone-mediated upwind walking. Zool Sci 9(3):515–527
Kanzaki R, Ikeda A, Shibuya T (1994) Morphological and physiological properties of pheromone-triggered flipflopping descending interneurons of the male silkworm moth, *Bombyx mori*. J Comp Physiol A 175(1):1–14. https://doi.org/10.1007/BF00217431
Kanzaki R, Soo K, Seki Y, Wada S (2003) Projections to higher olfactory centers from subdivisions of the antennal lobe macroglomerular complex of the male silkmoth. Chem Senses 28(2):113–130

Kanzaki R, Nagasawa S, Shimoyama I (2004) Neural basis of odor-source searching behavior in insect Microbrain system evaluated with a mobile robot. In: Kato N, Ayers J, Morikawa H (eds) Bio-mechanisms of swimming and flying, 1st edn. Springer, Tokyo, pp 155–170

Katsumata S, Ando N, Kanzaki R (2009) An insect-sized atmospheric ion source localization robot for the evaluation of odor source localization algorithms of insects. J Robot Soc Jpn 27(7):711–717. https://doi.org/10.7210/jrsj.27.711

Kellog VL (1907) Some silkworm moth reflexes. Biol Bull 12(3):152–154. https://doi.org/10.2307/1535862

Kennedy JS (1983) Zigzagging and casting as a programmed response to wind-borne odor – a review. Physiol Entomol 8(2):109–120. https://doi.org/10.1111/j.1365-3032.1983.tb00340.x

Kennedy JS, Marsh D (1974) Pheromone-regulated anemotaxis in flying moths. Science 184(4140):999–1001. https://doi.org/10.1126/science.184.4140.999

Khan AG, Sarangi M, Bhalla US (2012) Rats track odour trails accurately using a multi-layered strategy with near-optimal sampling. Nat Commun 3:703. https://doi.org/10.1038/ncomms1712

Kiya T, Morishita K, Uchino K, Iwami M, Sezutsu H (2014) Establishment of tools for neurogenetic analysis of sexual behavior in the silkmoth, *Bombyx mori*. PLoS One 9(11):e113156. https://doi.org/10.1371/journal.pone.0113156

Kondoh Y, Obara Y (1984) Lesion studies of neural mechanisms underlying the mating dance of the silkmoth, *Bombyx mori*. J Ethol 2(1):11–16. https://doi.org/10.1007/bf02348202

Kowadlo G, Russell RA (2008) Robot odor localization: a taxonomy and survey. Int J Robot Res 27(8):869–894. https://doi.org/10.1177/0278364908095118

Kramer E (1997) A tentative intercausal nexus and its computer model on insect orientation in windborne pheromone plumes. In: Cardé R, Minks A (eds) Insect pheromone research. Springer, Boston, pp 232–247. https://doi.org/10.1007/978-1-4615-6371-6_22

Kuwana Y, Shimoyama I, Miura H (1995) Steering control of a mobile robot using insect antennae. In: Proceedings 1995 IEEE/RSJ international conference on intelligent robots and systems. Human Robot Interaction and Cooperative Robots, 5–9 Aug. 1995. pp 530–535 vol.532. https://doi.org/10.1109/IROS.1995.526267

Kuwana Y, Shimoyama I, Sayama Y, Miura H (1996) Synthesis of pheromone-oriented emergent behavior of a silkworm moth. In: IEEE/RSJ international conference on intelligent robots and systems, 4–8 Nov 1996. pp 1722–1729. https://doi.org/10.1109/IROS.1996.569043

Kuwana Y, Nagasawa S, Shimoyama I, Kanzaki R (1999) Synthesis of the pheromone-oriented behaviour of silkworm moths by a mobile robot with moth antennae as pheromone sensors. Biosens Bioelectron 14(2):195–202. https://doi.org/10.1016/S0956-5663(98)00106-7

Lan B, Ando N, Kanzaki R (2017) Flying odour tracking robot with insect antennae. In: The 8th international symposium on adaptive motion of animals and machines, Sapporo, Japan, June 27–30 2017. pp 88–89

Loudon C, Koehl MAR (2000) Sniffing by a silkworm moth: wing fanning enhances air penetration through and pheromone interception by antennae. J Exp Biol 203(19):2977–2990

Mafra-Neto A, Carde RT (1994) Fine-scale structure of pheromone plumes modulates upwind orientation of flying moths. Nature 369(6476):142–144. https://doi.org/10.1038/369142a0

Maimon G, Straw AD, Dickinson MH (2010) Active flight increases the gain of visual motion processing in *Drosophila*. Nat Neurosci 13(3):393–399. https://doi.org/10.1038/nn.2492

Martin H (1965) Osmotropotaxis in honey-bee. Nature 208(5005):59–63. https://doi.org/10.1038/208059a0

Martinez D, Arhidi L, Demondion E, Masson JB, Lucas P (2014) Using insect electroantennogram sensors on autonomous robots for olfactory searches. J Vis Exp 90:e51704. https://doi.org/10.3791/51704

Milde JJ, Gronenberg W, Strausfeld NJ (1992) The head-neck system of the blowfly calliphora: 2. Functional organization and comparisons with the sphinx moth manduca sexta. In: Alain B, Werner G, Vidal PP (eds) The head-neck sensory motor system. Oxford University Press, New York, pp 64–70. https://doi.org/10.1093/acprof:oso/9780195068207.003.0007

Mishima T, Kanzaki R (1999) Physiological and morphological characterization of olfactory descending interneurons of the male silkworm moth, *Bombyx mori*. J Comp Physiol A 184(2):143–160. https://doi.org/10.1007/s003590050314

Miyamoto D, Kazawa T, Kanzaki R (2012) Neural circuit simulation of Hodgkin-Huxley type neurons toward Peta scale computers. In: 2012 SC companion: high performance computing, networking storage and analysis, 10–16 Nov. 2012. pp 1541–1541. doi:https://doi.org/10.1109/SC.Companion.2012.314

Murlis J, Jones CD (1981) Fine-scale structure of odor plumes in relation to insect orientation to distant pheromone and other attractant sources. Physiol Entomol 6(1):71–86. https://doi.org/10.1111/j.1365-3032.1981.tb00262.x

Murlis J, Elkinton JS, Cardé RT (1992) Odor plumes and how insects use them. Annu Rev Entomol 37(1):505–532. https://doi.org/10.1146/annurev.en.37.010192.002445

Naeem W, Sutton R, Chudley J (2007) Chemical plume tracing and odour source localisation by autonomous vehicles. J Navig 60(2):173–190. https://doi.org/10.1017/S0373463307004183

Nagasawa S, Kanzaki R, Shimoyama I (1999) Study of a small mobile robot that uses living insect antennae as pheromone sensors. In: IEEE/RSJ international conference on intelligent robots and systems, 1999. pp 555–560. https://doi.org/10.1109/IROS.1999.813062

Namiki S, Kanzaki R (2018) Morphology of visual projection neurons supplying premotor area in the brain of the silkmoth *Bombyx mori*. Cell Tissue Res 374:497. https://doi.org/10.1007/s00441-018-2892-0

Namiki S, Haupt SS, Kazawa T, Takashima A, Ikeno H, Kanzaki R (2009) Reconstruction of virtual neural circuits in an insect brain. Front Neurosci 3(2):206–213. https://doi.org/10.3389/neuro.01.028.2009

Namiki S, Iwabuchi S, Pansopha Kono P, Kanzaki R (2014) Information flow through neural circuits for pheromone orientation. Nat Commun 5:5919. https://doi.org/10.1038/ncomms6919

Neumann PP, Hernandez Bennetts V, Lilienthal AJ, Bartholmai M, Schiller JH (2013) Gas source localization with a micro-drone using bio-inspired and particle filter-based algorithms. Adv Robot 27(9):725–738. https://doi.org/10.1080/01691864.2013.779052

Obara Y (1979) *Bombyx-mori* mating dance - essential in locating the female. Appl Entomol Zool 14(1):130–132

Olberg RM (1983) Pheromone-triggered flip-flopping interneurons in the ventral nerve cord of the silkworm moth, *Bombyx mori*. J Comp Physiol 152(3):297–307

Olsen SR, Wilson RI (2008) Cracking neural circuits in a tiny brain: new approaches for understanding the neural circuitry of *Drosophila*. Trends Neurosci 31(10):512–520. https://doi.org/10.1016/j.tins.2008.07.006

Pansopha Kono P (2018) A study on state-dependent multisensory integration for odor-source localization. Dissertation, The University of Tokyo, Toyko

Pansopha P, Ando N, Kanzaki R (2014) Dynamic use of optic flow during pheromone tracking by the male silkmoth, *Bombyx mori*. J Exp Biol 217(Pt 10):1811–1820. https://doi.org/10.1242/jeb.090266

Porter J, Craven B, Khan RM, Chang SJ, Kang I, Judkewitz B, Volpe J, Settles G, Sobel N (2007) Mechanisms of scent-tracking in humans. Nat Neurosci 10(1):27–29. https://doi.org/10.1038/nn1819

Pyk P, Bermúdez i Badia S, Bernardet U, Knüsel P, Carlsson M, Gu J, Chanie E, Hansson BS, Pearce TC, PFMJ V (2006) An artificial moth: chemical source localization using a robot based neuronal model of moth optomotor anemotactic search. Auton Robot 20(3):197–213. https://doi.org/10.1007/s10514-006-7101-4

Rajan R, Clement JP, Bhalla US (2006) Rats smell in stereo. Science 311(5761):666–670. https://doi.org/10.1126/science.1122096

Reeder PB, Ache BW (1980) Chemotaxis in the Florida spiny lobster, *Panulirus argus*. Anim Behav 28(Aug):831–839. https://doi.org/10.1016/S0003-3472(80)80143-6

Rosner R, Egelhaaf M, Warzecha AK (2010) Behavioural state affects motion-sensitive neurones in the fly visual system. J Exp Biol 213(2):331–338. https://doi.org/10.1242/Jeb.035386

Russell RA, Bab-Hadiashar A, Shepherd RL, Wallace GG (2003) A comparison of reactive robot chemotaxis algorithms. Robot Auton Syst 45(2):83–97. https://doi.org/10.1016/s0921-8890(03)00120-9

Sakurai T, Mitsuno H, Haupt SS, Uchino K, Yokohari F, Nishioka T, Kobayashi I, Sezutsu H, Tamura T, Kanzaki R (2011) A single sex pheromone receptor determines chemical response specificity of sexual behavior in the silkmoth *Bombyx mori*. PLoS Genet 7(6). https://doi.org/10.1371/journal.pgen.1002115

Sakurai T, Namiki S, Kanzaki R (2014) Molecular and neural mechanisms of sex pheromone reception and processing in the silkmoth *Bombyx mori*. Front Physiol 5. https://doi.org/10.3389/fphys.2014.00125

Sakurai T, Mitsuno H, Mikami A, Uchino K, Tabuchi M, Zhang F, Sezutsu H, Kanzaki R (2015) Targeted disruption of a single sex pheromone receptor gene completely abolishes in vivo pheromone response in the silkmoth. Sci Rep 5:11001. https://doi.org/10.1038/srep11001

Sato Y, Haupt SS, Kazawa T, Namiki S, Takashima A, Ikeno H, Nishikawa I, Kanzaki R (2010) Large-scale realistic network simulation of pheromone-processing circuits in the silkmoth brain. Front Neurosci Conf Abstr: Neuroin 4. https://doi.org/10.3389/conf.fnins.2010.13.00050

Seki Y, Aonuma H, Kanzaki R (2005) Pheromone processing center in the protocerebrum of *Bombyx mori* revealed by nitric oxide-induced anti-cGMP immunocytochemistry. J Comp Neurol 481(4):340–351. https://doi.org/10.1002/cne.20392

Szyszka P, Stierle JS, Biergans S, Galizia CG (2012) The speed of smell: odor-object segregation within milliseconds. PLoS One 7(4):e36096. https://doi.org/10.1371/journal.pone.0036096

Szyszka P, Gerkin RC, Galizia CG, Smith BH (2014) High-speed odor transduction and pulse tracking by insect olfactory receptor neurons. Proc Natl Acad Sci U S A 111(47):16925–16930. https://doi.org/10.1073/pnas.1412051111

Tabuchi M, Sakurai T, Mitsuno H, Namiki S, Minegishi R, Shiotsuki T, Uchino K, Sezutsu H, Tamura T, Haupt SS, Nakatani K, Kanzaki R (2013) Pheromone responsiveness threshold depends on temporal integration by antennal lobe projection neurons. Proc Natl Acad Sci U S A 110(38):15455–15460. https://doi.org/10.1073/pnas.1313707110

Takasaki T, Namiki S, Kanzaki R (2012) Use of bilateral information to determine the walking direction during orientation to a pheromone source in the silkmoth *Bombyx mori*. J Comp Physiol A 198(4):295–307. https://doi.org/10.1007/s00359-011-0708-8

van Breugel F, Dickinson MH (2014) Plume-tracking behavior of flying *Drosophila* emerges from a set of distinct sensory-motor reflexes. Curr Biol 24(3):274–286. https://doi.org/10.1016/j.cub.2013.12.023

Vickers NJ (2000) Mechanisms of animal navigation in odor plumes. Biol Bull 198(2):203–212

Vickers NJ, Baker TC (1991) The effects of unilateral antennectomy on the flight behavior of Male Heliothis-Virescens in a Pheromone Plume. Physiol Entomol 16(4):497–506. https://doi.org/10.1111/j.1365-3032.1991.tb00589.x

Vickers NJ, Baker TC (1996) Latencies of behavioral response to interception of filaments of sex pheromone and clean air influence flight track shape in Heliothis virescens (F.) males. J Comp Physiol A 178(6):831–847. https://doi.org/10.1007/BF00225831

von Holst E, Mittelstaedt H (1950) Das Reafferenzprinzip – (Wechselwirkungen Zwischen Zentralnervensystem Und Peripherie). Naturwissenschaften 37(20):464–476

von Uexküll J, Kriszat G (1956) Streifzüge durch die Umwelt von Tieren und Menschen/Bedeutungslehre. Rowohlt, Hamburg

Wada S, Kanzaki R (2005) Neural control mechanisms of the pheromone-triggered programmed behavior in male silkmoths revealed by double-labeling of descending interneurons and a motor neuron. J Comp Neurol 484(2):168–182. https://doi.org/10.1002/cne.20452

Wall C, Perry JN (1987) Range of action of moth sex-attractant sources. Entomol Exp Appl 44(1):5–14. https://doi.org/10.1111/j.1570-7458.1987.tb02232.x

Webb B (1998) Robots, crickets and ants: models of neural control of chemotaxis and phonotaxis. Neural Netw 11(7–8):1479–1496. https://doi.org/10.1016/S0893-6080(98)00063-X

Webb B (2002) Robots in invertebrate neuroscience. Nature 417(6886):359–363. https://doi.org/10.1038/417359a

Webb B (2004) Neural mechanisms for prediction: do insects have forward models? Trends Neurosci 27(5):278–282. https://doi.org/10.1016/j.tins.2004.03.004

Webb B, Harrison RR, Willis MA (2004) Sensorimotor control of navigation in arthropod and artificial systems. Arthropod Struct Dev 33(3):301–329. https://doi.org/10.1016/j.asd.2004.05.009

Wei L, Farrell JA, Shuo P, Arrieta RM (2006) Moth-inspired chemical plume tracing on an autonomous underwater vehicle. Ieee T Robot 22(2):292–307. https://doi.org/10.1109/TRO.2006.870627

Willis MA (2008) Chemical plume tracking behavior in animals and mobile robots. Navigation 55(2):127–135

Willis MA, Avondet JL, Zheng E (2011) The role of vision in odor-plume tracking by walking and flying insects. J Exp Biol 214(Pt 24):4121–4132. https://doi.org/10.1242/jeb.036954

Wu M, Nern A, Williamson WR, Morimoto MM, Reiser MB, Card GM, Rubin GM (2016) Visual projection neurons in the *Drosophila lobula* link feature detection to distinct behavioral programs. elife 5. https://doi.org/10.7554/eLife.21022

Index

Y. Ishikawa (ed.), *Insect Sex Pheromone Research and Beyond*, Entomology Monographs, https://doi.org/10.1007/978-981-15-3082-1

Taxonomic Index

Y. Ishikawa (ed.), *Insect Sex Pheromone Research and Beyond*, Entomology Monographs, https://doi.org/10.1007/978-981-15-3082-1

MIX
Papier aus verantwortungsvollen Quellen
Paper from responsible sources
FSC® C105338

If you have any concerns about our products,
you can contact us on
ProductSafety@springernature.com

In case Publisher is established outside the EU,
the EU authorized representative is:
Springer Nature Customer Service Center GmbH
Europaplatz 3, 69115 Heidelberg, Germany

Printed by Libri Plureos GmbH
in Hamburg, Germany